The Geology of
Southern New Mexico's
Parks, Monuments, and Public Lands

The Geology of Southern New Mexico's Parks, Monuments, and Public Lands

Edited by Peter A. Scholle, Dana S. Ulmer-Scholle, Steven M. Cather, and Shari A. Kelley

New Mexico Bureau of Geology and Mineral Resources

A Research Division of New Mexico Institute of Mining and Technology

Socorro, New Mexico

2020

The Geology of Southern New Mexico's Parks, Monuments, and Public Lands

Edited by Peter A. Scholle, Dana S. Ulmer-Scholle, Steven M. Cather, and Shari A. Kelley

A Research Division of New Mexico Institute of Mining and Technology
Stephen G. Wells, *President*
801 Leroy Place
Socorro, NM 87801-4796
(575) 835-5490
geoinfo.nmt.edu

ISBN 978-1-883905-48-4

MANAGER PUBLICATIONS PROGRAM: Brigitte Felix

DESIGN: Christina Watkins

LAYOUT: Stephanie Chavez and Dana Ulmer-Scholle

CARTOGRAPHY & GRAPHICS: Stephanie Chavez and Dana Ulmer-Scholle

GIS & CARTOGRAPHIC SUPPORT: Mark Mansell and Phil Miller

EDITING: Peter A. Scholle, Dana S. Ulmer-Scholle, Steven M. Cather, Shari A. Kelley,

Belinda Harrison, and Jennifer Eoff

PRODUCTION SUPPORT: Gina D'Ambrosio, Elena Taylor, Dante Padilla Romero,

Adam Read, and Cynthia Connolly

COVER PHOTOGRAPHS: Front cover: Organ Mountains–Desert Peaks National Monument © Wayne Suggs.

Back cover: Carlsbad Caverns National Park and Fra Cristobal Range by Elephant Butte Reservoir

© New Mexico Bureau of Geology and Mineral Resources, Peter A. Scholle.

First Edition 2020

Preface

The label "Land of Enchantment" is often used to summarize the touristic appeal of New Mexico. That term is an apt, if somewhat simplified encapsulation of the many wonders of this state—the rich mix of cultural history, diverse climatic zones, and beautiful mountains, valleys and even, in good years, rivers. Although all these components of the state's enchantment can be taken at face value, one can also look "behind the landscapes" and "beyond the cultural aspects" to see how geology has molded the geography and evolution of this state over a much longer period than that in which humans have been influential. We have rocks as old as 1.7 billion years exposed in New Mexico, as compared to a roughly 12,000-year history of human settlement. Geologic studies have revealed a long and complex record of mountain building and erosion, of continental collisions and rifting events, entry and departure of oceans, and episodes of spectacular volcanism (with future such eruptions promised as well). Such geologic events not only underlie our scenery, they created many of the resources we exploit today, from oil and gas to industrial and metallic minerals and even to "fossil groundwater" that we extract to sustain life in our arid climate. It is these behind-the-scenery stories that we want to tell in this book. The book is not a dry and systematic recitation of geologic studies, but rather is a fascinating tale of specifically what created the scenery visible at some of the most beautiful and best-protected places in our state.

The book is the second of two volumes on the geology of New Mexico's public lands—the first, published in 2010, dealt with the northern half of the state and, correspondingly, this volume deals exclusively with the southern regions. Nearly 35% of New Mexico's land is Federally owned (a relatively low percentage when compared with the 47% Federal ownership of the 11 coterminous western states including New Mexico). In that context, however, it should also be noted that the percentage of Federal lands in southern New Mexico is far higher than in northern New Mexico. An additional 11% of New Mexico's area is State owned and administered. Although not all State or Federal lands are available for public access, we nonetheless had to make difficult choices on which of the enormous reserved land areas to include in this book. All national parks and national monuments (under either National Park Service or Bureau of Land Management administration) are included in this book. All state parks, and many state monuments and historical sites also are covered. Geologically important national forest areas, wilderness and wilderness study areas, and scenic byways are included as well and in this volume, unlike the prior one, we also feature a few additional areas of special geological interest that have public access. Those include areas owned or operated by the Nature Conservancy, the National Radio Astronomy Observatory (Very Large Array) and Freeport-McMoRan. On the other hand, we excluded many geologically interesting areas that have little or no public access, especially the extensive military reservations.

Because this is a book primarily about the geology (and hydrology) of public lands, the organization of the volume is based on geologic provinces. This is a necessity for minimizing repetition in geologic descriptions, but perhaps an unfamiliar way to organize things from a touristic perspective. We have divided the southern part of the state into six areas that each have distinctive and common geologic histories. They are:

1. The Mogollon Slope/Colorado Plateau (the latter part well represented in northern New Mexico but less extensive in the southern part of the state);

2. The largely inaccessible Mogollon-Datil Volcanic Field/ Mogollon Plateau;

3. The Basin and Range (a province that stretches from southwestern New Mexico to California, Nevada and other states);

4. The Rio Grande rift and adjacent mountains that occupy most of the central part of New Mexico;

5. The Permian Basin/eastern Basin and Range transition in the southeastern part of the state; and finally,

6. The Great Plains region, a topographically- and hydrologically-challenged area that forms a transition from the generally arid lands of southern New Mexico to the rich farm and ranchlands of the U.S. midcontinent.

Thus, the chapters in the book roughly form a counterclockwise tour around southern New Mexico. They also emphasize very different aspects of the state's diverse geologic history. Sections 1 to 4 deal with the spectacular and relatively young volcanic and tectonic history of southwestern and central New Mexico; sections 5 to 6 mainly deal with the older Paleozoic history of the state and with the gradual erosion and submergence of those older landscapes under younger strata.

Geology is a complex science with many specialized terms (a polite way to say that it is filled with specialized technical vocabulary), but we have endeavored to minimize that jargon here (or, where technical terms had to be used, we included definitions in a glossary located near the back of the volume). The book is aimed at a general audience

with an interest in, and perhaps a bit of experience with, geology but
with little or no formal training in the subject. Nonetheless, the book
includes geologic maps drawn on a shaded relief base for each of the
selected areas and its surrounding lands. Those are best understood by
also looking at stratigraphic columns showing the names of mapped
formations and the relative ages of exposed rocks that are generally
located in the overview articles for each of the six geological provinces
around which the book is organized.

Although written for the general public, the articles in this book
all are authored by well-qualified geoscientists from the New Mexico
Bureau of Geology and Mineral Resources (NMBGMR)—the State's
Geologic Survey—or from geology departments at universities or
science museums both within and outside the state. Brief biographies
of those authors are included near the back of the book. The articles
have all been peer reviewed for accuracy, but unlike standard technical
papers, we have not included formal citations to sources of information. We have included, however, some mention of other less technical
information sources that may be useful for follow-up reading. The
NMBGMR offers many publications that give more formal and technical information (papers, books and maps) about the geology covered
in this book. Those can be accessed at our offices, bookstore, and
mineral museum on the campus of New Mexico Tech (more formally
known as the New Mexico Institute of Mining and Technology). They
are also available at our website: geoinfo.nmt.edu.

We hope that this book is both educational and enjoyable. Almost
all geologists love what they do for a living and, if they had to do it all
over again, would choose the same career. We hope that enthusiasm
and love shines through in this book and that it inspires the young to
choose similar careers to ours.

Contents

PART 5   The Permian Basin...285

PART 6   The Southern Great Plains...347

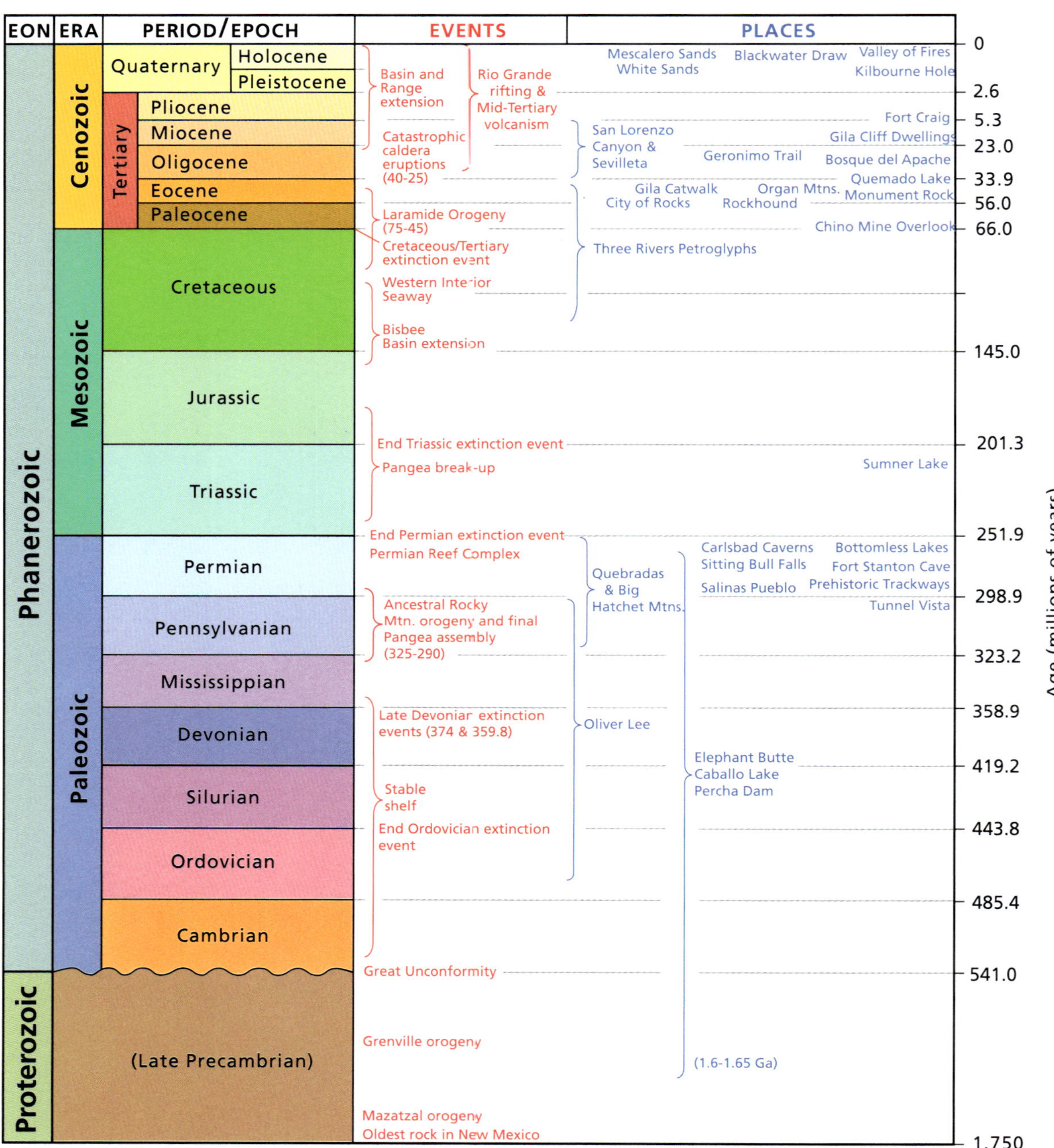

EON | ERA | PERIOD/EPOCH | EVENTS | PLACES
Quaternary | Holocene
Pleistocene
Pliocene
Miocene
Oligocene
Eocene
Paleocene
Tertiary
Cenozoic
Cretaceous
Jurassic
Triassic
Mesozoic
Permian
Pennsylvanian
Mississippian
Devonian
Silurian
Ordovician
Cambrian
Paleozoic
Phanerozoic
(Late Precambrian)
Proterozoic
Basin and Range extension
Rio Grande rifting & Mid-Tertiary volcanism
Catastrophic caldera eruptions (40-25)
Laramide Orogeny (75-45)
Cretaceous/Tertiary extinction event
Western Interior Seaway
Bisbee Basin extension
End Triassic extinction event
Pangea break-up
End Permian extinction event
Permian Reef Complex
Ancestral Rocky Mtn. orogeny and final Pangea assembly (325-290)
Late Devonian extinction events (374 & 359.8)
Stable shelf
End Ordovician extinction event
Great Unconformity
Grenville orogeny
Mazatzal orogeny
Oldest rock in New Mexico
Mescalero Sands
White Sands
Blackwater Draw
Valley of Fires
Kilbourne Hole
Fort Craig
Gila Cliff Dwellings
San Lorenzo Canyon & Sevilleta
Geronimo Trail
Bosque del Apache
Quemado Lake
Gila Catwalk
City of Rocks
Organ Mtns.
Rockhound
Monument Rock
Chino Mine Overlook
Three Rivers Petroglyphs
Sumner Lake
Carlsbad Caverns
Sitting Bull Falls
Salinas Pueblo
Bottomless Lakes
Fort Stanton Cave
Prehistoric Trackways
Tunnel Vista
Quebradas & Big Hatchet Mtns.
Oliver Lee
Elephant Butte
Caballo Lake
Percha Dam
(1.6-1.65 Ga)
0
2.6
5.3
23.0
33.9
56.0
66.0
145.0
201.3
251.9
298.9
323.2
358.9
419.2
443.8
485.4
541.0
1,750
Age (millions of years)

A Brief Introduction to the Geology of Southern New Mexico

Southern New Mexico encompasses portions of five broad, overlapping geologic provinces: the Mogollon Slope/Colorado Plateau, the Basin and Range, the Rio Grande rift, the Mogollon-Datil volcanic field, and the Great Plains. This book is organized around these provinces, each with a different landscape and geologic history. The Permian Basin, which is commonly grouped with the southern Great Plains, has such economic importance to New Mexico that its story is told in a separate section.

The enormous diversity of landscapes in New Mexico is the result of the complex geologic history of the state. In order to decipher that history, geologists interpret the rocks themselves, their relationship to one another, and their structural features. Radiometric dating techniques allow us to measure numerical ages that quantify the rates and timings of geologic processes recorded by the rocks. Although the scenery of New Mexico was largely shaped by events that occurred in the past 75 million years, the geologic evolution of this part of North America encompasses a much longer story.

Much of the history of the North American continent (and the surface of the Earth in general) can be understood in the light of plate tectonics, a unifying geologic theory that was developed in the last half of the twentieth century thanks to our increased understanding of the geology and geophysics of both the continents and the oceans. We now know that the surface of our planet is divided into a series of rigid but mobile plates made of the Earth's crust and uppermost mantle that float on a semi-molten layer of the Earth's interior known as the asthenosphere. The interaction of these plates as they move across the surface of our globe is responsible for most geologic activities, including uplifts, earthquakes, and volcanoes. Early continents bore little resemblance to today's; they have changed size and shape, and have grown through both the accretion of crustal material and the volcanic eruption of new crust. As the continents have moved, their climate has changed in response to their position on the Earth's surface relative to the equator. Plate margins and their interactions have been responsible for shaping much of our geologic history.

The Precambrian Era

Our planet is 4.5 billion years old, but the oldest rocks we can find anywhere on the planet are about 4.3 billion years old. One of the interesting features of plate tectonics is the continual recycling of older crust, as dense oceanic crust is subducted beneath continental crust, and older

OPPOSITE: Generalized geologic time scale for New Mexico, showing major divisions and timing of major events (in red). Ages (in millions of years) are shown at right. Important tectonic events that shaped the landscapes of New Mexico and major episodes of extinction that influenced the history of life on our planet are shown in red. Shown in blue are selected places where strata recording these events can be seen; the list is by no means comprehensive. The Phanerozoic, which is shown in the most detail here, occupies but a small fraction of the geologic history of our 4.5 billion-year-old planet. Ages are from the Geological Society of America's *2018 Geologic Time Scale*.

continental crust is subjected to uplift and erosion. This destruction of evidence makes the early history of our planet difficult to decipher. Older (Precambrian) rocks constitute only a small fraction of the exposed rocks in New Mexico. Where we do find them, often the details of their history are obscure. However, the broad outlines of the story have emerged intact.

Late Precambrian (Proterozoic) rocks in New Mexico record at least two major events or orogenies, the term geologists use to describe these major episodes of deformation, faulting, and uplift that occur on a grand scale. The oldest event (about 1.7 to 1.6 billion years ago) is termed the Mazatzal orogeny. It records the collision of ancient volcanic islands with the southern edge of the continent of Laurentia, an early piece of what would become the North American continent. This process of accretion is how continents grow, as pieces of crust rafted by plate tectonics across the face of the globe become welded onto the margins of other continental masses. Evidence for accretion is particularly well preserved in the Manzano Mountains and traces are also found in the Burro and San Andres Mountains. A younger tectonic event (1.48 to 1.35 billion years ago) is characterized by widespread volcanism. Most of the plutons associated with this significant magmatic episode are granitic. Granitic plutons emplaced during this time frame are well exposed in the Burro, Caballo, and San Andres Mountains and in several Basin and Range mountain blocks in southwestern New Mexico. Proterozoic tectonism in New Mexico culminated in the Grenville orogeny about 1.1 billion years ago. Deposition of limestones, sandstones, and shales and volcanic eruptions accompanied Grenville deformation.

Toward the end of Precambrian time (from about 1,100 million to 750 million years ago) the Earth's continents were united in a single supercontinent that we refer to as Rodinia. The exact configuration of this landmass is open to interpretation, but the coming together of these older continental landmasses into a single supercontinent, and the subsequent breakup of that supercontinent, explains a number of features we see in the Proterozoic rocks of New Mexico. The breakup of Rodinia occurred about 750 million years ago. A localized pluton in the Florida Mountains was emplaced at 510 million years ago as the result of crustal stretching across a broad north-northeast trending zone in New Mexico and Colorado as Rodinia broke apart.

The Paleozoic Era

By the beginning of the Paleozoic, some pieces of fragmented continental crust were moving toward one another once again. This would ultimately result in a series of continental collisions that shaped much of North America, culminating in the Late Paleozoic. However, the first few hundred million years of the Paleozoic in this part of the world were relatively quiet. The Cambrian was marked by global transgressions, incursions of marine waters that accompanied a worldwide sea-level rise. The explosion of invertebrate life that began in the late Precambrian continued, and these early Paleozoic seas were home to a rich variety of invertebrate faunas.

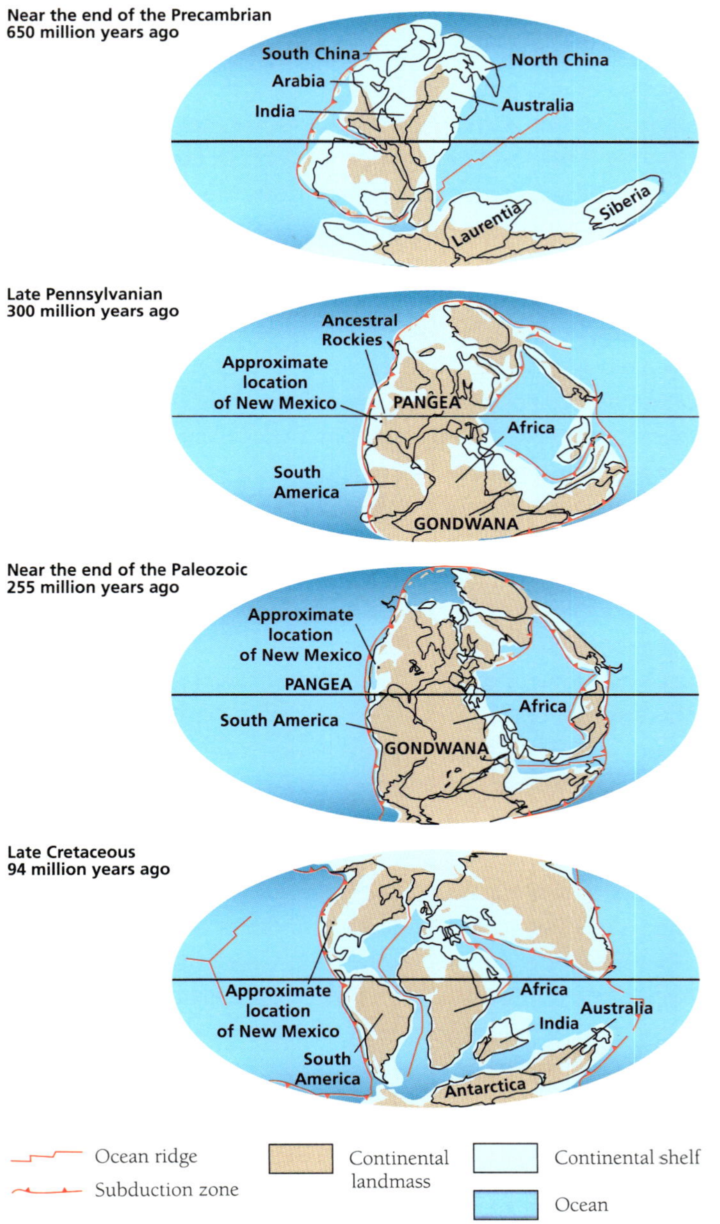

Starting about 500 million years ago, lower Paleozoic sedimentary rocks in southern New Mexico were deposited on or near a shallow marine shelf southwest of the Transcontinental Arch, a broad, northeast-trending uplifted region that stretched across the midcontinent of North America. The gap in the geologic record between the Precambrian igneous and metamorphic rocks and the Paleozoic strata directly above them—between about 1,400 million years ago and 500 million years ago in southern New Mexico—is known as the Great Unconformity. This feature is perhaps best known from the Grand Canyon in Arizona, where it was first recognized and named, but the Great Unconformity is also beautifully exposed on the west side of the Caballo Mountains in south-central New Mexico. Early Paleozoic rocks were never deposited on, or were eroded from, the Transcontinental Arch in northern New Mexico. The Early Paleozoic rocks of southern New Mexico pinch out to the north and are absent north of the approximate latitude of Socorro.

In southern New Mexico, thick, widespread Mississippian limestones, most notably the Lake Valley Limestone, formed in shallow seas and preserve a wide variety of marine fossils. The Mississippian seas lapped northward onto developing highlands in northern and central New Mexico, leaving thin beds of limestone, sandstone, shale, and locally, even gypsum. However, much of the Mississippian rock record in northern and central New Mexico has been lost to subsequent erosion.

By Pennsylvanian time. as much as 80 percent of New Mexico was submerged beneath warm, shallow, tropical seas (New Mexico lay close to the equator at that time). Pennsylvanian strata are well exposed in many ranges in southern New Mexico, including the Sacramento, San Andres, Fra Cristobal, Caballo, and Manzano Mountains. The Pennsylvanian was a time of cyclical fluctuations in sea level (related to southern hemisphere episodes of glaciation and deglaciation) and increasing aridity; we see evidence of both of these processes in the Pennsylvanian sediments that remain.

During Late Pennsylvanian–Early Permian time, the major continents formed a single landmass we call Pangea. The final assembly of Pangea involved the collision of Laurentia (North America and Europe) and Gondwanaland (Africa, Australia, Antarctica, and South America). This event was responsible for enormous upheavals in what is now the eastern part of North America that gave rise to the Appalachian and Ouachita Mountains. Here in western North America we see evidence of mountains and basins that developed during the Ancestral Rocky Mountains orogeny. That orogeny began in the Late Mississippian and extended into the Early Permian (320–290 million years ago), with the most intense uplift occurring in the Middle Pennsylvanian. Ancestral Rockies deformation extended from what is now Utah and Colorado into southern New Mexico, west Texas, and Oklahoma. Despite their name and location, the Ancestral Rockies had little to do with the Rocky Mountains of today. Their erosion provided the raw material for Pennsylvanian and Permian sediments in this part of the continent. The Pedernal (south-central New Mexico), Central Basin Platform (southeast New Mexico and west Texas), and Florida (southwest New Mexico) uplifts were the most significant highlands affecting sedimentation in this region. By the end of the Permian, the Ancestral Rockies were largely eroded away and buried by younger sediments.

The Permian record in New Mexico is rich. In southern New Mexico and Texas, world-class exposures of the Permian reef complex tell a story of marine and marginal marine sedimentation on the flank of the Delaware Basin. Permian (and older) strata in the subsurface of southeast New Mexico and Texas have provided both source rocks and reservoir rocks for one of the most important oil and gas provinces in the world. Trackways in Permian rocks, especially well exposed near Las Cruces, preserve important clues about the reptiles that roamed the landscape during this time. The end of the Permian was a time of aridity, as the ocean retreated from New Mexico. By the beginning of the Mesozoic, New Mexico was above sea level, and the earliest Mesozoic sediments are largely terrestrial.

The Mesozoic Era

Shortly after the beginning of the Mesozoic, the continent of Pangea began to break up. Development of a northwest-trending Jurassic rift, known as the Bisbee Basin, heralded the commencement of the breakup of Pangea in southwestern New Mexico. The rift was the dominant tectonic feature controlling erosion and deposition in southern and central New Mexico during middle Mesozoic time. Triassic to Jurassic strata were either never deposited or were eroded from the Silver City–Las Cruces region, because this area was a rift-flank highland. The Bisbee Basin to the south, however, preserves both Jurassic and Early Cretaceous sedimentary rocks. By Early Cretaceous, as Pangea continued to break apart, marine waters once again invaded the continental interior. As Pangea broke up, North America moved away from Europe as the Atlantic Ocean began to open, and the continents began to move toward the positions we see them in today. A subduction zone off the west coast of North America, which would later play such an important role in the evolution of western North America, was well established by this time.

The Late Cretaceous, at the end of the Mesozoic, was a time of rising sea levels throughout the world. In North America, this resulted in the growth of one of the greatest inland seas of all time, as marine waters moved in across the interior of the continent. Over a thousand feet deep in places, the Western Interior Seaway extended across the continent from the Gulf of Mexico to the Arctic Ocean. Stretching across the midcontinent region, the Seaway prevailed from about 100 million years ago until 70 million years ago, reaching its greatest extent about 90 million years ago. The Dakota Sandstone represents the initial transgression of marine waters across the continent. Marginal marine sediments that accumulated along the edge of the Western Interior Seaway are well exposed at the Three Rivers Petroglyph site in the Sierra Blanca Basin and near the southernmost part of the Quebradas Back Country Byway (along US 380). Uplift associated with the Laramide orogeny caused the Western Interior Seaway to retreat to the east and eventually vanish from North America.

The Laramide Orogeny

Much of the geologic structure of the American West is related to a regionally-extensive mountain-building event known as the Laramide orogeny, which occurred at the end of the Mesozoic and continued into the early Cenozoic. Laramide deformation was felt broadly throughout the American West, from Alaska to Mexico, and from the west coast to the Great Plains. Compressive deformation associated with the Laramide orogeny was responsible for much of the development of the Rocky Mountains. Many of the structural features in southern New Mexico are Laramide in origin, including the Baca, Love Ranch, and Sierra Blanca basins, and the folds and thrust faults in the Burro and Caballo mountains. Beginning about 75 million years ago, the Laramide orogeny lasted for about 30 million years. Although not everyone agrees on the details, many geologists believe that Laramide deformation was related to the subduction of a shallow-dipping oceanic plate (known as the Farallon plate) beneath the west coast of North America, which caused compression within the overriding North American plate.

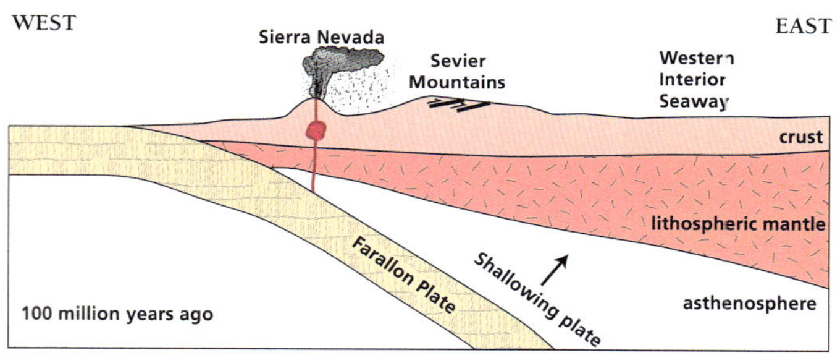

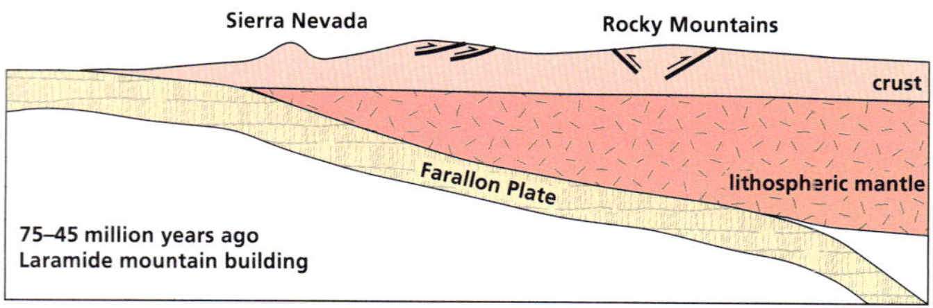

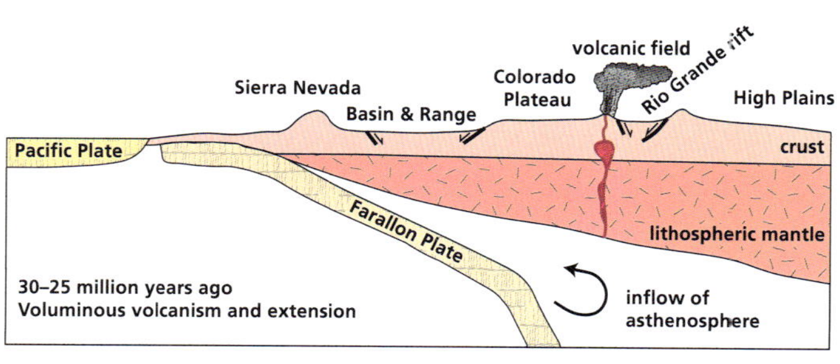

Cenozoic Volcanism and the Rio Grande Rift

At the start of the Cenozoic, New Mexico lay above sea level. This was a time of widespread erosion and uplift associated with the waning phases of the Laramide orogeny. Following the end of the Laramide orogeny, two major periods of volcanic activity occurred in New Mexico: one in the mid-Cenozoic lasted from about 40 to 25 million years ago, and a younger period from about 15 million years ago to the present. Remnants of mid-Cenozoic volcanic activity are best seen in the southern part of the state, particularly in the Mogollon–Datil and Sierra Blanca volcanic fields. The angle of the subducting oceanic plate began to steepen as the Laramide orogeny ended, and hot upper mantle flowed into the space between the descending plate and the North American plate, triggering melting and the development of enormous volcanic fields. Large andesitic stratovolcanoes erupted lavas starting at 40 million years ago, shedding cold and hot debris aprons that covered the older Laramide highlands and basins. Starting at 36 million years ago, caldera eruptions from supervolcanoes in the Mogollon-Datil field blanketed the region with rhyolitic tuff and tuffaceous volcaniclastic sediments, while basaltic andesite erupted from shield volcanoes and fissures and flowed across the landscape. Erosion of one of these fractured tuffs created the fantastic natural rock sculptures at City of Rocks State Park. Much later, cliff dwellers in the Gila region created homes carved into younger tuffs formed by these cataclysmic eruptions. Perhaps the most significant mid-Cenozoic event to affect New Mexico was the development of the Rio Grande rift, the defining topographic feature of New Mexico and the major control on the course of the Rio Grande. This deep-seated crustal feature, which stretches from central Colorado to Texas and Mexico, is

FEATURE	WHERE BEST TO SEE	AGE (YEARS)
Carrizozo flow	Carrizozo flow / Valley of Fires Recreational Area	5,200
Jornada volcanic field	Ft. Craig	820,000–78,000
Potrillo volcanic field	Kilbourne Hole	1.2 million–20,000
Elephant Butte	Elephant Butte State Park	3–2 million
Capitan Pluton	Capitan Mountain Wilderness	28 million
Sierra Blanca volcanic field	Three Rivers Petroglyphs / White Mountain Wilderness	38–28 million
Bootheel volcanic field	Geronimo Trail	33–27 million
Mogollon–Datil volcanic field	Quemado Lake / Monument Rock/Sawtooth / Datil Well Recreation Area / Gila Cliff Dwellings National Monument / Catwalk/Gila Wilderness / Gila Lower Box / Fort Cummings / City of Rocks State Park / Rockhound State Park / Magdalena/San Mateo Mountains / Organ Mountains–Desert Peaks National Monument	40–25 million
Laramide plutons	Chino Mine Overlook	60–58 million

A summary of some of the significant volcanic features and events in southern New Mexico.

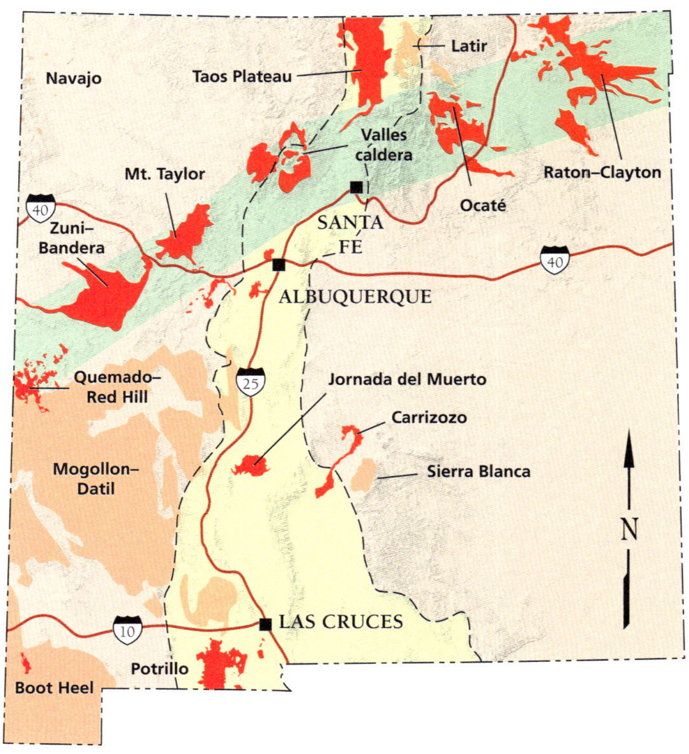

The major volcanic fields in New Mexico. In northern New Mexico late Cenozoic volcanism tends to follow the Jemez lineament, a zone of crustal weakness considered by some to be the remnant of a Precambrian suture zone, where two pieces of the Earth's crust were welded together.

considerably wider toward the south. The rift represents an ongoing episode of east-west crustal stretching that began about 25 million years ago, eventually causing major basins to develop during the Miocene. The very fabric of the continent has been ripped open along the Rio Grande rift, resulting in a thinning of the Earth's crust. The volcanism that is associated with the Rio Grande rift occurred as the upper mantle moved closer to the Earth's surface during crustal thinning. The rift is considered an extension of the Basin and Range province, a zone of lateral stretching that covers broad portions of the western U.S. The basins of the Rio Grande rift are bounded by rift-flank uplifts (the Sacramento and Magdalena mountains are good examples), which today are partially buried by sediment eroded from the mountains and carried into the rift.

Late Cenozoic Volcanism and Faulting

The most recent episodes of volcanism and tectonism occurred toward the end of the Cenozoic. During the Quaternary, young basalt flows erupted from scattered vents across southern New Mexico, including a small flow on the east side of the Peloncillo Mountains (in southwestern New Mexico in the Basin and Range province) and three volcanic fields in south-central New Mexico in the Rio Grande rift—the Jornada del Muerto, West Potrillo, and Carrizozo sites). The Carrizozo flow is only 5200 years old! In addition, southern New Mexico hosts one of the longest continuous Quaternary normal fault scarps on the planet, the fault scarp on the west side of the Tularosa Basin at the foot of the San Andres Mountains. Hot springs and hot wells across southwestern and south-central New Mexico are used to heat greenhouses, raise fish (tilapia), and for stress-reducing soaking. All of these features—youthful volcanoes, recent faulting, and bubbling hot springs—remind us that we live in a geologically active area.

The geologic processes responsible for sculpting the remarkable scenery of New Mexico are ongoing. Some landscape-changing events. like erosion as the result of an intense rainstorm, are easy to observe. Other processes, such as the slow movement of plates across the globe, are harder to perceive, except with the help of sensitive scientific instruments. Although we live in a landscape that looks timeless and eternal to us, it is constantly being shaped by both rapid, punctuated and slow, relentless forces that have acted on the Earth through its long history.

—Shari A. Kelley and L. Greer Price

Suggested Reading

For more detailed information on the geologic evolution of New Mexico, we recommend two very fine volumes, both of which are up-to-date and provide a great deal more detail than is possible here:

Geology of the American Southwest, by W. Scott Baldridge. Cambridge University Press, 2004. An excellent comprehensive overview of the geologic history of the Southwest.

The Geology of New Mexico: A Geologic History, edited by Greg H. Mack and Katherine A. Giles. New Mexico Geological Society Special Publication 11, 2004. A more technical and detailed look at selected topics in the geologic history of New Mexico, but an invaluable source of data.

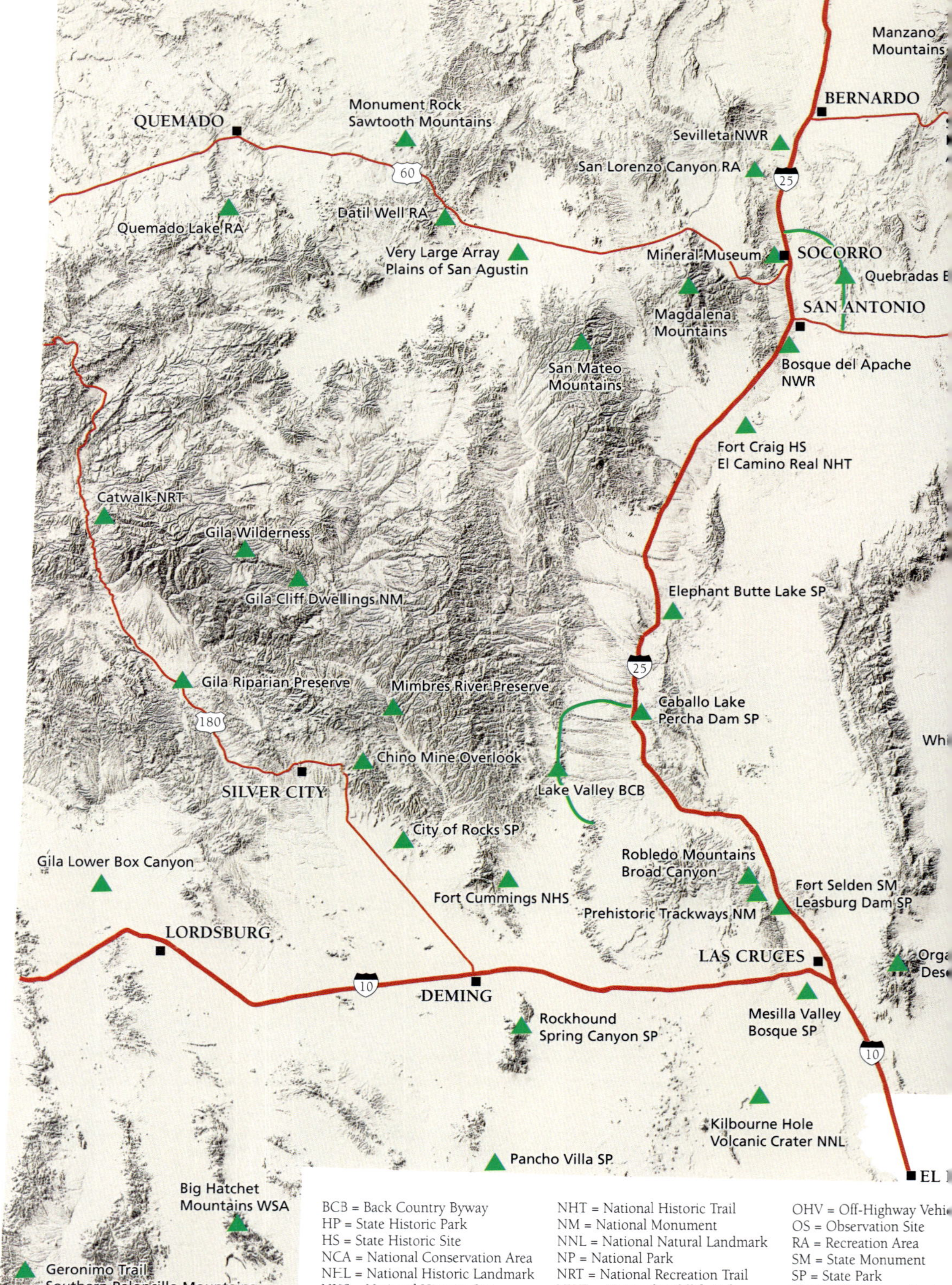

Manzano Mountains
BERNARDO
QUEMADO
Monument Rock
Sawtooth Mountains
Sevilleta NWR
San Lorenzo Canyon RA
25
60
Datil Well RA
Quemado Lake RA
Very Large Array
Plains of San Agustin
Mineral Museum
SOCORRO
Quebradas B
SAN ANTONIO
Magdalena Mountains
San Mateo Mountains
Bosque del Apache NWR
Fort Craig HS
El Camino Real NHT
Catwalk NRT
Gila Wilderness
Gila Cliff Dwellings NM
Elephant Butte Lake SP
Gila Riparian Preserve
Mimbres River Preserve
180
25
Caballo Lake
Percha Dam SP
Whi
Chino Mine Overlook
SILVER CITY
Lake Valley BCB
City of Rocks SP
Robledo Mountains
Broad Canyon
Gila Lower Box Canyon
Fort Selden SM
Leasburg Dam SP
Fort Cummings NHS
Prehistoric Trackways NM
LORDSBURG
LAS CRUCES
Orga
Des
10
DEMING
Rockhound
Spring Canyon SP
Mesilla Valley
Bosque SP
10
Kilbourne Hole
Volcanic Crater NNL
Pancho Villa SP
EL
Big Hatchet
Mountains WSA
BCB = Back Country Byway
HP = State Historic Park
HS = State Historic Site
NCA = National Conservation Area
NHL = National Historic Landmark
NHS = National Historic Site
NHT = National Historic Trail
NM = National Monument
NNL = National Natural Landmark
NP = National Park
NRT = National Recreation Trail
NWR = National Wildlife Refuge
OHV = Off-Highway Vehic
OS = Observation Site
RA = Recreation Area
SM = State Monument
SP = State Park
WSA = Wilderness Study
Geronimo Trail
Southern Peloncillo Mountains

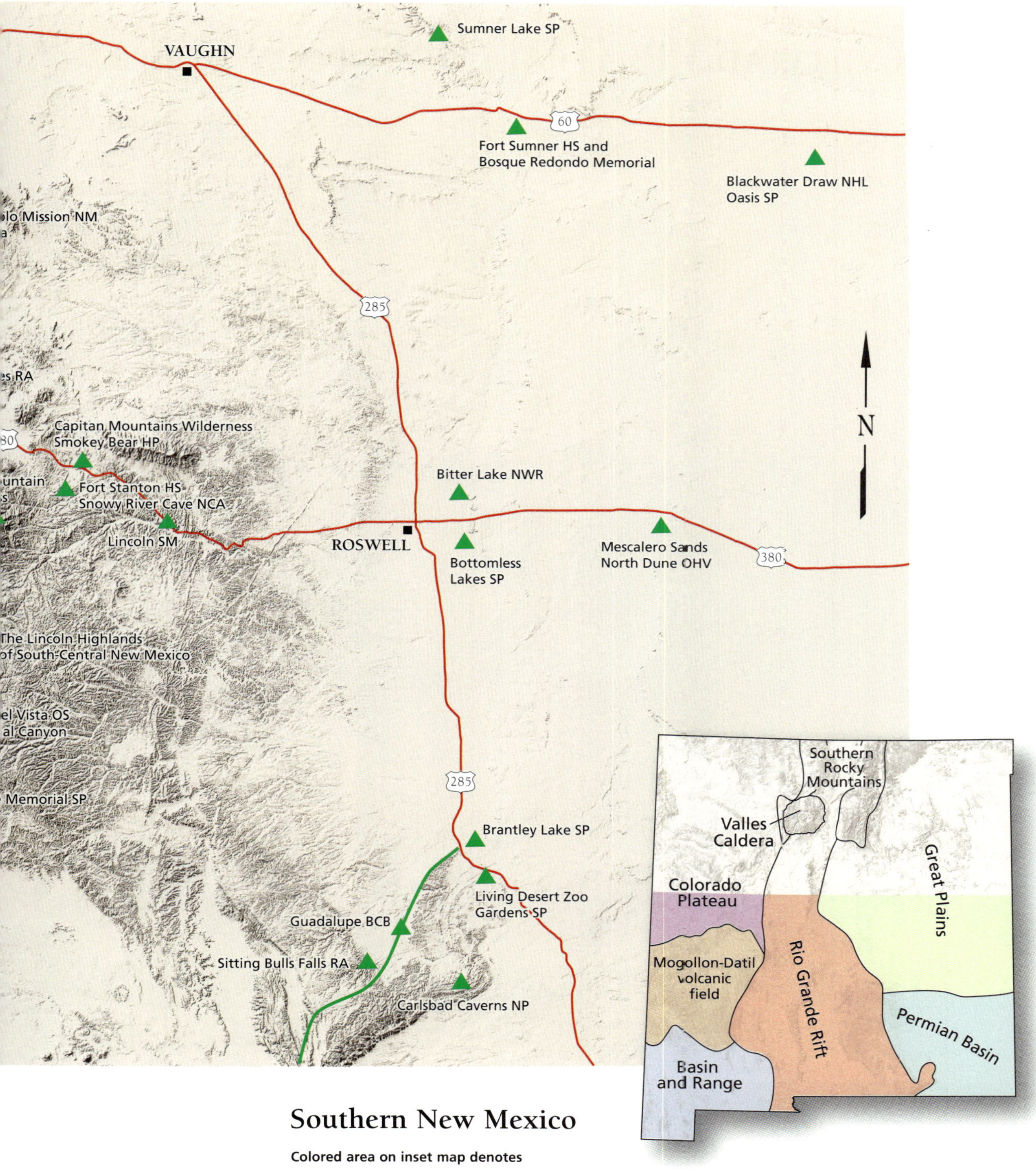

Southern New Mexico

Colored area on inset map denotes portion of New Mexico covered in this volume. Individual chapters are keyed by color to the geologic provinces.

COLORADO PLATEAU/MOGOLLON SLOPE

The Colorado Plateau is a comparatively stable and only slightly deformed crustal block located in northwestern New Mexico, northeastern Arizona, southeastern Utah, and southwestern Colorado. The Plains of San Agustin, which are sometimes considered to be an arm of the Rio Grande rift, mark the southeastern edge of the Plateau. The parks and the recreational facilities discussed in this chapter lie along the downwarped southeastern margin of the Colorado Plateau in an area that forms a transition zone to the Mogollon–Datil volcanic field and the Basin and Range province to the south and east. This part of the Colorado Plateau is variously called the Mogollon Slope or the Mogollon–Datil section of the Plateau. The thickness of the continental crust decreases across this area, from 25 miles on the north to 22 miles on the south, and elevations are relatively high (7,000–8,000 feet) compared to other parts of the Colorado Plateau. The Continental Divide crosses this elevated region about 1.5 miles east of Pie Town, separating drainages that flow to the Rio Grande from those that flow into the Little Colorado River in Arizona.

Generally, the more central parts of the Colorado Plateau, in Arizona and Utah, expose flat-lying and very colorful Paleozoic to Mesozoic sedimentary rocks. In the Mogollon Slope area, these older and more scenic rocks are broken by north to northeast-striking normal faults and buried by younger volcanic rocks and drab-colored sediments eroded from volcanic highlands to the south. These volcanic and sedimentary rocks record the effects of dramatic and catastrophic volcanic and tectonic events.

Geologic History

The Colorado Plateau in west-central New Mexico has a fascinating geologic history shaped by four major events: 1) Jurassic to Early Cretaceous rifting in southwestern New Mexico; 2) Late Cretaceous to early Cenozoic Laramide compressional deformation within the area; 3) eruptions of numerous stratovolcanoes and calderas just to the south in the Mogollon–Datil volcanic field during middle Cenozoic time; and 4) crustal extension in the Rio Grande rift and the Basin and Range province to the east and south during middle to late Cenozoic time. The result of those diverse and partially overlapping processes created an equally diverse assemblage of Mesozoic and Cenozoic rocks.

The oldest rocks exposed in this area belong to the Triassic Moenkopi Formation, which is mainly composed of red sandstone and

OPPOSITE: **A much-needed rainstorm over the Plains of San Agustin, south of the Very Large Array.**

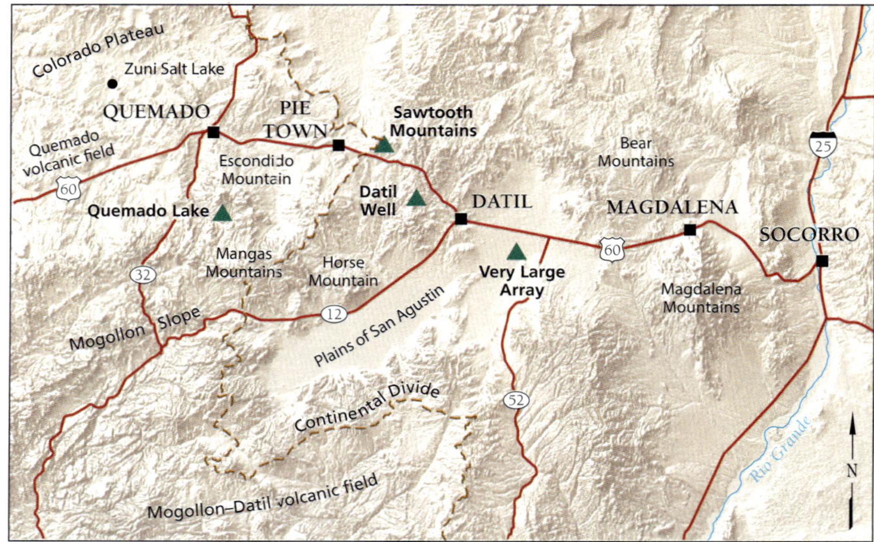

Map showing the recreational facilities and areas of interest on the Mogollon Slope in the vicinity of US 60.

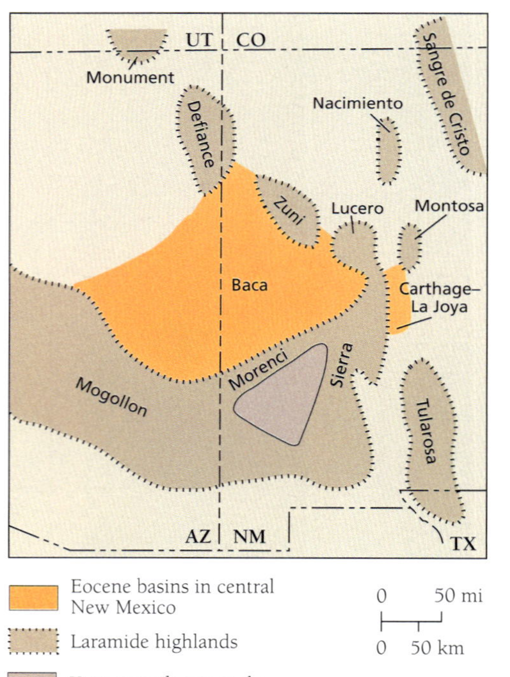

Map of the Laramide highlands surrounding the Baca Basin.

mudstone deposited in streams flowing to the northwest. The red mudstones, sandstones, and conglomerates in the Chinle Formation were also carried by river systems that originated from highlands in Texas flowing toward the northwest. Cretaceous rocks rest directly on Triassic rocks in this area. The Jurassic section was not deposited, or was eroded, because this part of New Mexico was on the high-standing northern flank of the northwest-trending Bisbee Basin rift that crossed the southwestern corner of the state. The sandstones, shales, and coalbeds preserved in the Late Cretaceous section record oscillations in the position of the western shoreline of the Western Interior Seaway, an inland sea that covered the central North American continent from Canada to the Gulf of Mexico.

The Late Cretaceous seaway retreated during the onset of the Laramide deformation. As the dip angle of the subducting Farrallon plate decreased, compressional forces in the crust created the Baca Basin and reverse faults like the Red Lake and the Hickman fault zones. The Zuni, Lucero, and Defiance uplifts to the north and northeast, the Mogollon Rim in Arizona to the southwest, and the now-buried Sierra and Morenci uplifts to the south and east were raised as a result of this deformation. These highlands shed debris into the Laramide Baca Basin,

AGE	UNIT	ROCK TYPE	DEPOSITIONAL SETTING	TECTONIC EVENT
Quaternary	Zuni–Salt Lake lava flow	basalt flows	maar	Minor extension
Pliocene	Red Hill–Quemado lava flows	basalt flows	volcanoes	
Miocene	Horse Mountain volcano	felsic flows	lava dome	
	Santa Fe Group	conglomerate, sandstone	alluvial fill	
	Fence Lake Formation	conglomerate, sandstone	ancestral Little Colorado River	
Oligocene	Mogollon Group	ash-flow tuffs basaltic-andesite lavas	volcanoes and calderas	Ignimbrite flareup
	Datil and Spears groups	ash-flow tuffs debris flows andesite lavas		Slab rollback begins
Eocene	Baca Formation	sandstone, conglomerate, mudstone	streams and lakes	Laramide
Paleocene	Unnamed paleosol	oxidized (reddened) Cretaceous strata	tropical weathering	
Cretaceous	Crevasse Canyon Formation/ Moreno Hill Formation	sandstone, shale, coal	swamp near shoreline	Western Interior Seaway
	Atarque Sandstone	sandstone, with marine fossils	marginal marine	
	Mancos Shale	shale with thin sandstone beds	marine	
	Dakota Sandstone	sandstone, shale, cong.	marginal marine	Bisbee Basin ← rifting
Triassic	Chinle Formation	sandstone, siltstone, mudstone	rivers flowing NW	
	Moenkopi Formation	sandstone, siltstone	rivers flowing NW	

the fill of which is exposed at many sites near the US 60 roadway, especially in the vicinity of the town of Quemado. The redbeds of the Eocene Baca Formation include conglomerates, sandstones, and mudstones deposited in sediment-choked streams and in lakes.

As plate velocities slowed, the shallowly subducted Farallon plate began to sink, causing hot mantle to come into contact with the base of the North American plate and triggering volcanism in the Mogollon–Datil volcanic field to the south of US 60. First, debris and rare lava flows from andesitic stratovolcanoes erupting 40 to 36 million years ago lapped onto the Mogollon Slope. These units are overlain by ash-flow tuffs that erupted 36 to 25 million years ago from numerous calderas located to the south of the plateau. Later, basaltic andesite lavas flowed across this area from volcanoes in the Mogollon–Datil field to the south and from a 26 million-year-old volcano in the Mangas Mountains, located about 20 miles southwest of Pie Town on the Continental Divide. At about the same time, large dikes of basaltic andesite radiated outward from some volcanic centers, heating and altering the sedimentary rocks they intruded. Among the most spectacular of these is the Pie Town dike, which is located on the west

Stratigraphic units exposed on the Mogollon Slope section of the Colorado Plateau.

21

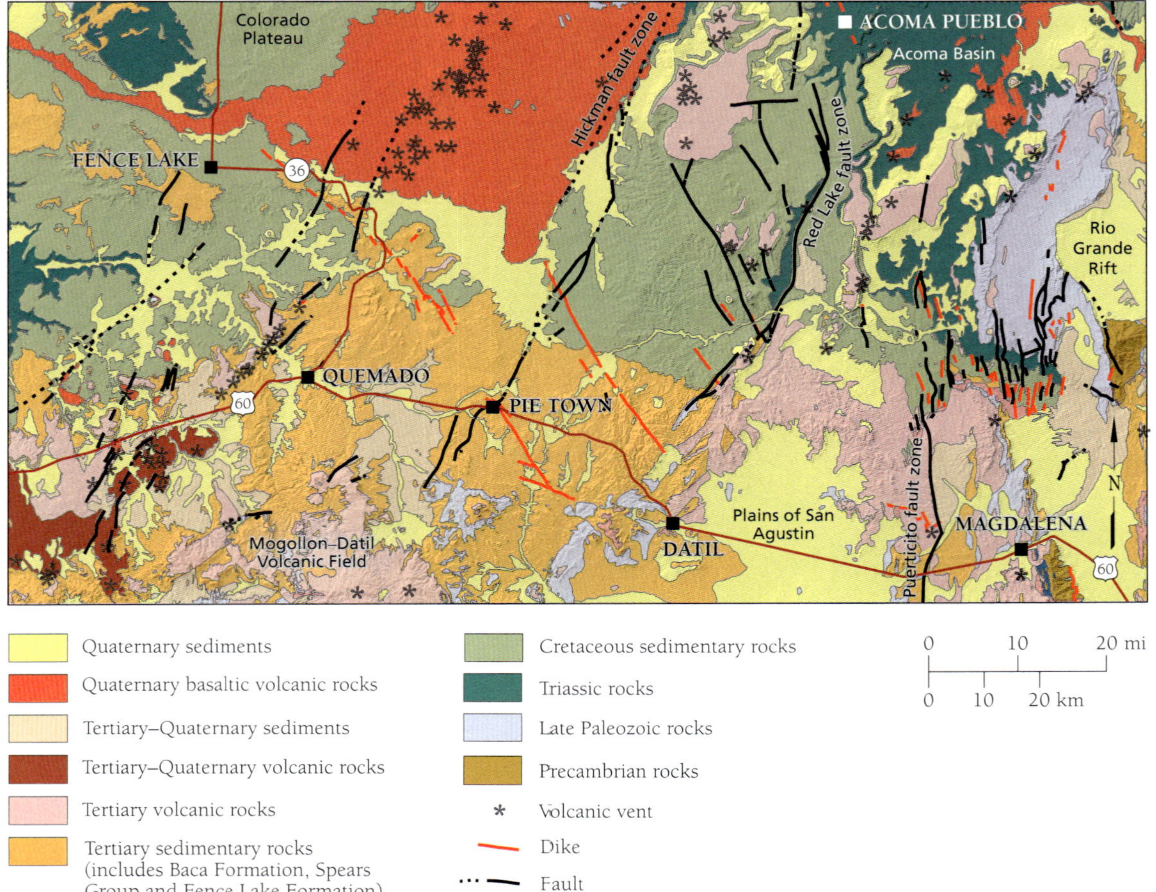

Simplified geologic map of the southeastern Colorado Plateau and Mogollon Slope showing the location of the Acoma Basin, Hickman fault zone, and Red Lake fault zone.

side of that village both north and south of US 60, stretching across the landscape as an imposing wall of rock. That 29 million-year-old dike has been traced eastward for more than 40 miles toward its source in a caldera complex near Magdalena.

During that general time interval, a large Oligocene eolian dune field occupied significant portions of western New Mexico and eastern Arizona and covered the Mogollon Slope with dune sands. This dune field is called the Chuska erg and dates to about 33.5 and 27 million years ago. Those deposits interfinger with the volcaniclastic debris eroded from the Mogollon–Datil volcanic field and with regional tuff deposits.

Crustal extension related to the Rio Grande rift and Basin and Range roughly began during the formation of the Mogollon–Datil volcanic field, about 36 million years ago. Many of the earlier Laramide reverse faults in this area were reactivated as Neogene normal faults, and down-dropped areas were filled with alluvial deposits of the

Santa Fe Group and Fence Lake Formation. Some localized volcanism is associated with this extension. For example, 12.6 million-year-old felsic lava flows were derived from the Horse Mountain volcano. The basalts of the Red Hill–Quemado volcanic field erupted during two distinct intervals from 8 to 5 million years ago, and 2.5 million years to 26,000 years ago; these basalts are considered to be part of the Jemez lineament, a NE-trending zone of less than 10 million-year-old mafic to silicic volcanism that transects the southeastern margin of the Colorado Plateau, the north-central Rio Grande rift, and the southern High Plains. Younger lavas, just 26,000–200,000 years old, are associated with a maar eruption at Zuni Salt Lake. A maar develops when ascending magmas interact with groundwater, creating an explosive eruption, a crater, and finely-layered hydromagmatic deposits. The Red Hill–Quemado lava flows have been cut by normal faults, but the younger Zuni Salt Lake flows have not been faulted; these flows thus bracket the time of most recent faulting in this area.

Economic Resources

Coal resources in the Cretaceous Crevasse Canyon Formation in the mountains north of Datil and in the Cretaceous Moreno Hill Formation north of Quemado were examined in detail during the 1980s and 1990s, but the deposits were never developed because the coalbeds are generally thin and discontinuous. Similarly, the oil and gas potential of this region was evaluated during the same time frame, but economic oil or gas reservoirs were not found. The small uranium deposits that formed at the eroded and weathered top of the Cretaceous section just below the Baca Formation are also currently uneconomic. The main material mined in the area is scoria from the cinder cones in the Red Hills. In addition, many small sand and gravel deposits were exploited, mainly for local use.

—Shari A. Kelley

Quemado Lake Recreation Area

The 800-acre Quemado Lake Recreation Area is located approximately
20 miles south of Quemado, New Mexico, within the northern reaches
of the Gila National Forest. The man-made lake is impounded behind
an earthen embankment dam constructed across Largo Creek. The
dam was built in 1971 to promote recreational activities for the area.
The lake is stocked with a variety of fish and is a popular spot for
New Mexico anglers. The lake and the surrounding campgrounds are
located at an elevation of about 7,650 feet, providing a reprieve from
the summer heat. Rock outcrops in the adjacent mountains and
prominent bluffs near the lake contain evidence for outpourings of
lava from the volcanoes of the Mogollon–Datil volcanic field, as well as
sedimentary deposits from millions of years of erosion.

OPPOSITE: **View of the eastern half of the Quemado Lake Recreational Area where Largo Creek enters the lake.**

Regional Setting

Quemado Lake is located near the boundary of the Mogollon–Datil
volcanic field and the Colorado Plateau. The Mogollon–Datil field,
located largely to the south, erupted a wide variety of volcanic rocks
between approximately 40 and 25 million years ago. These volcanic
rocks lap onto the Mesozoic and Cenozoic sedimentary rocks of the
southern Colorado Plateau. In turn, the weight of the volcanic rock pile
has warped the southern edge of the plateau downward to produce a
feature called the Mogollon Slope. Quemado Lake sits within a small
valley on the Mogollon Slope, bounded by the Agua Fria Mountains and
Escondido Mountain.

The Rock Record

The oldest rocks in the area are the Eocene–Oligocene Spears Group
consisting of three distinct rock types deposited about 40–25 million
years ago. Gray andesitic sandstones that comprise the lower Spears
Group are well exposed along the road south of Quemado. Light-gray,
ash-rich sandstones of the middle Spears Group are well exposed at
Castle Rock. Tan cross-bedded sandstones, blown into a large dune
field (Chuska erg) by westerly winds about 30 million years ago, form
the upper Spears Group.

Sedimentary rocks of the lower and middle Spears Group contain
much volcanic material, such as pumice and lava fragments, and are
commonly referred to as volcaniclastic (clastic refers to rocks composed
of fragments or broken pieces of older rocks). Light-brown sandstone

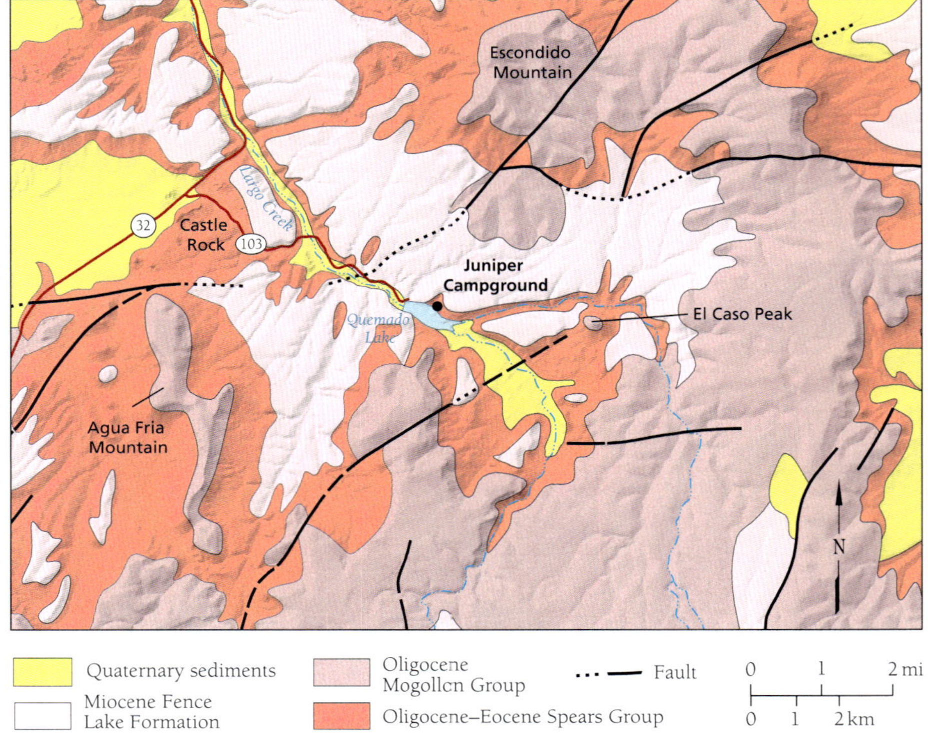

Generalized geologic map of Quemado Lake and the surrounding area. The Mogollon Group includes lava flows and tuffs of the Mogollon–Datil volcanic field.

outcrops of the upper Spears Group visible along the switchback north of Juniper campground contain high-angle crossbeds distinctive of wind-blown sands, here part of the Chuska erg. Other sedimentary rocks in this formation are also crossbedded but were instead deposited in river systems. Coarser sandstones and conglomerates were deposited in the fast flowing waters of stream channels, whereas finer mudstones were deposited in the adjacent floodplains.

Volcanic rocks of the area can be found in the cliffs south of the lake and on the nearby peaks. Several stacked lava flows define El Caso Peak to the east of the lake. Most of these lava flows are composed of andesite that was erupted between 29 and 26 million years ago. At some places, rhyolite tuffs erupted from supervolcanoes south of the lake within the Gila Wilderness are found interbedded with the lava flow sequence. These tuffs represent the distal reaches of pyroclastic flows that would have covered much of the area following an eruption.

The youngest rocks in the area are the Fence Lake Formation of Miocene age. The large basaltic andesite cobbles and boulders in the conglomerate indicate that the sediments have not traveled very far from their sources. The most likely sources are the adjacent high terrains such as the Gallo and Escondido Mountains to the south and north.

Geologic History

The Mogollon–Datil volcanic field is one of the largest mid-Tertiary eruptive centers in the southwestern U.S. Approximately 20 catastrophic caldera (supervolcano) eruptions occurred about 36–25 million years ago. However, prior to and during these caldera eruptions, large outpourings of andesitic lava flows also were erupted from smaller volcanoes. The Spears Formation mostly represents the deposition of sediments eroded from these volcanoes.

The andesitic lava flows found in the vicinity of Quemado Lake document the pre- and post-caldera volcanic activity. Although individual lava flows are volumetrically small compared to the caldera-related pyroclastic tuffs, together the lava flows represent a significant portion of the Mogollon–Datil field. In some places the stacks of lava flows are greater than 1,000 feet thick. The lava flows were erupted from volcanoes that probably looked very similar to those found in the

Outcrop of the Fence Lake Formation just north of the Quemado Lake parking lot. Most of the cobbles and boulders in the unit are composed of andesite and basaltic andesite lava, eroded from adjacent volcanic highlands.

Multi-colored sediments of the Spears Group exposed at Castle Rock near the intersection of NM 32 and NM 103. These ash-rich sediments were deposited 34 million years ago.

Outcrop showing crossbeds in wind-blown deposits in the upper Spears Group.

Cascades of the Pacific Northwest. Following the end of volcanism, large valleys were eroded into the northern part of the Mogollon–Datil field. Backfilling of these valleys during the Miocene produced the Fence Lake Formation, which represent the highland tributaries to the ancestral Little Colorado River deposited about 10 million years ago.

The youngest geologic activity is the transport and deposition of Quaternary alluvium by Largo Creek. Sediment is mostly transported during the spring snowmelt runoff and the summer monsoon. This sediment will eventually fill the reservoir.

Geologic Features

OUTCROPS AT DAM SITE—Directly north of the dam parking lot are spectacular exposures of Fence Lake Formation conglomerate. Imbricated boulders indicate that stream flow was to the northwest. Boulders are composed of lavas with different mineralogy, textures, and vesicularity (gas-bubble content).

QUEMADO LAKE OVERLOOK TRAIL—Take the trail heading south over the dam and continue on uphill until it reaches the cliffs of andesite lava. Enjoy the distant views of the Spears Group at Castle Rock and the numerous lava flows exposed near the summit of El Caso Peak to the east.

—Matthew J. Zimmerer

If You Plan to Visit

From the town of Quemado take US 60 west for 0.5 miles. Turn left (south) on NM 32 and travel 14.2 miles the Quemado Lake/NM 103 sign. Turn left (east) on NM 103 and go 4 miles to where FR 13 (gravel) begins. Continue straight on FR 13 for 1 mile. For more information:

Gila National Forest
Quemado Ranger Station
1 Forest Service Rd 13
Quemado, NM 87829
(575) 773-4678
www.fs.usda.gov/recarea/gila/recarea/?recid=79490

View looking north toward Quemado Lake, including the dam, and large cliffs of andesite above the lake that erupted between the caldera-forming eruptions.

Monument Rock and the Sawtooth Mountains
CIBOLA NATIONAL FOREST

Monument Rock and the Sawtooth Mountains are in a scenic and geologically unique part of the Cibola National Forest on the southeastern margin of the Colorado Plateau. Monument Rock is a 100-foot-tall pillar of sandstone at the east end of a series of peaks called the Sawtooth Mountains. These features are the eroded remnants of a broad sedimentary apron deposited about 37 million years ago on the northwest flank of a large Mount St. Helens-type volcano. Volcanic rocks and sediments derived from this volcano are called the Dog Springs Formation, and the volcano itself is here informally referred to as the "Dog Springs volcano." Dog Springs is the oldest formation in the Spears Group.

The spectacular exposures of intensely deformed sandstone beds exposed in the cliff faces of the Sawtooth Mountains and the overlying slabs of tilted volcanic conglomerate most likely represent a major earthquake-induced liquefaction event that occurred some 37 million years ago. The deformed beds now exposed in these cliffs have been "frozen" into the Dog Springs Formation since shortly after the earthquake. Liquefaction events triggered by large earthquakes are well documented from modern events, including the 1964 "Good Friday" earthquake in Anchorage, Alaska and the February 2011 earthquake in Christchurch, New Zealand.

Regional Setting of the Dog Springs Volcano

The Dog Springs volcano formed near the northern edge of the Mogollon–Datil volcanic field. This field was active from about 40 to 25 million years ago. In western New Mexico, the volcanic field laps onto the southern margin of the relatively stable Colorado Plateau. In the Sawtooth Mountains only the northern sedimentary apron of the Dog Springs volcano is exposed.

OPPOSITE: **The east side of Monument Rock exposing horizontally bedded sandstone near the base of the lower Dog Springs Formation.**

Generalized geologic map of the Dog Springs Formation showing the inferred location of the Eocene Dog Springs volcano. Megabreccia zone is interpreted as a Mount St. Helens-type sector collapse and landslide deposit.

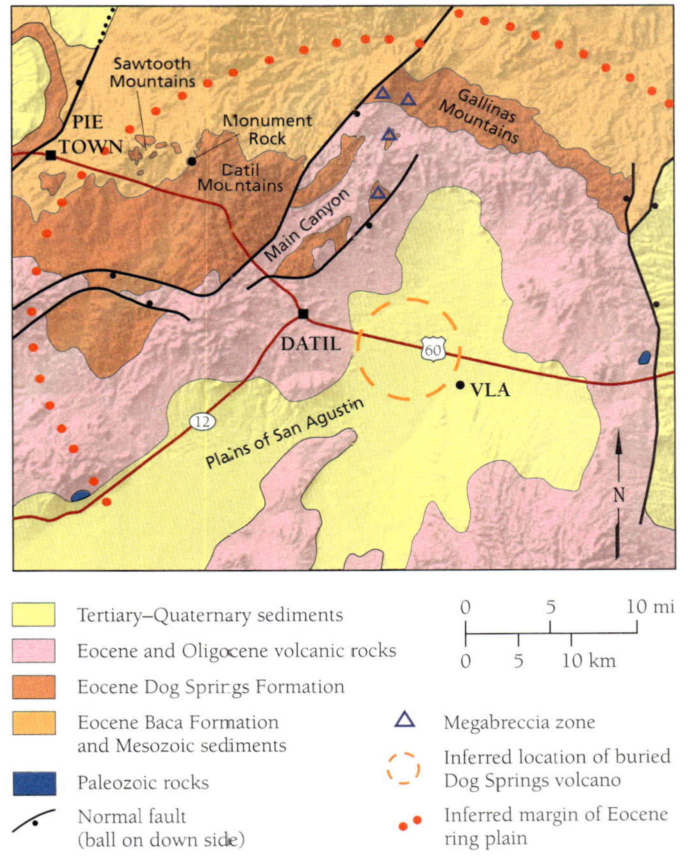

Looking northeast at cliff face 0.5 miles northeast of Monument Rock. Intricately folded (liquified) sandstone beds are at the top of the lower Dog Springs Formation. Massive rock capping the hill is a volcanic conglomerate bed in the upper Dog Springs Formation that was faulted and tilted to near vertical, but not liquified.

The distribution of rock types in the Dog Springs Formation suggests that the central cone of the Dog Springs volcano is now buried beneath the Plains of San Agustin, about seven miles east of Datil.

The original geometry of the Dog Springs volcano was probably similar to present-day stratovolcanoes such as Mount Fuji in Japan that have a large central cone of andesitic lava and pyroclastic flows that tower as much as 1.5 miles above the surrounding landscape. The steep slopes of stratovolcanoes rise from a circular base as much as six miles in diameter. Surrounding the central cone is a "ring plain" of ash-rich sand and mud deposited by streams, plus coarse debris flows emanating from the rising volcano. Sedimentary ring plains around mature stratovolcanoes are as much as 40 miles in diameter. Studies of Cascade volcanoes in Oregon indicate such features have an active lifetime of about a half million years.

Cenozoic Geologic History

About 50 million years ago southwestern New Mexico and southeastern Arizona were in a zone of Laramide crustal shortening along the rigid southern margin of the Colorado Plateau. Streams that drained Laramide highlands carried mud, sand, and gravel eastward across the southern margin of the Colorado Plateau into the Baca Basin. These stream deposits of red floodplain mudstones and coarse-grained channel sandstones are now preserved in the Baca Formation of Middle Eocene age.

The Dog Springs volcano first erupted near Datil about 37 million years ago. As the Dog Springs volcano grew, ash-rich sands were transported radially away from the volcano to form the lower sandy portion of the Dog Springs Formation, about 300 to 400 feet thick.

The growing volcanic apron may have blocked streams of the older Baca Basin to create small lakes. If so, then basal sandstones of the Dog Springs Formation were almost certainly saturated with groundwater early in the history of the growing volcano. Soon after deposition, the lower Dog Springs sandstone beds probably consisted of a loose mix of feldspar crystals and fine clay (altered ash) with small pore spaces filled with water.

The volcano became more explosive as is grew in height, so the upper Dog Springs Formation consists of a stack of coarse-grained, debris-flow conglomerates and thin pyroclastic flow breccias. Near Monument Rock the upper conglomeratic beds are at least 300 feet thick.

Thus, about 37 million years ago, the stage was set for a major liquefaction event. All that was needed was a large earthquake to trigger widespread liquefaction, folding, and faulting of the Dog Springs Formation. Liquefaction of the lower Dog Springs Formation was unusually deep-seated, extending at least 400 feet below the Eocene land surface. Soon after a liquefaction event, excess water escapes from the sandy sediment and it solidifies to form a compact mass that cannot be easily liquified again, thus "freezing in" the deformation.

The Eocene rock record near Datil, New Mexico. The tightly folded and faulted beds of the Dog Springs Formation are sandwiched between nearly horizontal beds of the Baca Formation and the Chavez Canyon Formation. Diagram is not drawn to scale.

AGE (million years)	LITHOLOGY		UNIT NAME	ROCK TYPE	ENVIRONMENT OF DEPOSITION
35.3			Datil Well Tuff	ash-flow tuff	caldera eruption
~36			Chavez Canyon Formation	andesitic sandstone and conglomarates	stream deposits eroded from volcano
~37.5–37		Spears Group	Dog Springs Formation (volcanic complex)	upper beds of mudflow conglomerates and pyroclastic flows	sedimentary apron of growing andesitic stratovolcano
				lower beds of andesitic sandstones	lower sandstones liquefied by large earthquake
~50–38			Baca Formation	lake deposits	minor lake deposits near top
				sandstone beds red mudstone and conglomeratic beds	stream deposits eroded from Laramide highlands

Following the liquefaction event, the Dog Springs Formation was buried by about 200 feet of andesitic sandstones of the Chavez Canyon Formation and then by the 35 million-year-old Datil Well Tuff (a volcanic ash-flow tuff). Regional uplift and erosion of Cenozoic volcanic strata along the southeast margin of the Colorado Plateau has occurred in pulses over the last 20 million years. The most recent pulse, during the last few million years, led to rapid erosion and formation of cliffs along the north flank of the Datil Mountains and Sawtooth Mountains.

—*Richard M. Chamberlin*

Sawtooth Mountains from US 60, east of Pie Town.

Additional Reading

Liquefaction: **en.wikipedia.org/wiki/Soil_liquefaction**

If You Plan to Visit

Monument Rock is located 77 miles west of Socorro and 4 miles north of US 60 in the eastern Sawtooth Mountains of Catron County, New Mexico. The area can be reached from Datil by traveling 12.3 miles west on US 60 (between mile markers 64 and 65) to FR 6A. Turn right on FR 6A and drive 4.1 miles north to the prominent peaks of the eastern Sawtooth Mountains and to the junction with FR 325. Turn right on FR 325 and drive 0.8 miles to the center of a grassy meadow. Monument Rock is on the right.

The Very Large Array and Plains of San Agustin

The Plains of San Agustin occupy a topographically closed basin surrounded by forested mountains and hills. The plains are best known for hosting the National Science Foundation-funded Karl G. Jansky Very Large Array (VLA) radio telescope observatory, where twenty-seven large (82-feet-wide), movable, dish-shaped antennas are arranged in a three-armed pattern. This extremely sensitive system detects naturally emitted radio signals from objects in space. Astronomers use the VLA to study a variety of celestial phenomena, including stellar evolution, planet formation, and black holes. Datil, the only town in the area, is nestled in a small valley northwest of the eastern plains. The remoteness of the plains (it is far from cities and their radio interference) provides an ideal location for the VLA.

Regional Setting

Geographically, the Plains of San Agustin lie between highlands of the Mogollon Slope to the north and the Mogollon Plateau (the center of the Mogollon–Datil volcanic field) to the south. A slight topographic high extends south of the town of Datil and west of the VLA, dividing

Satellite image showing the Plains of San Agustin, adjoining mountains, and physiographic regions (e.g., Colorado Plateau and Mogollon Plateau). The various mountains north of the plains occupy the Mogollon Slope. The three arms of the VLA are depicted by white dashed lines.

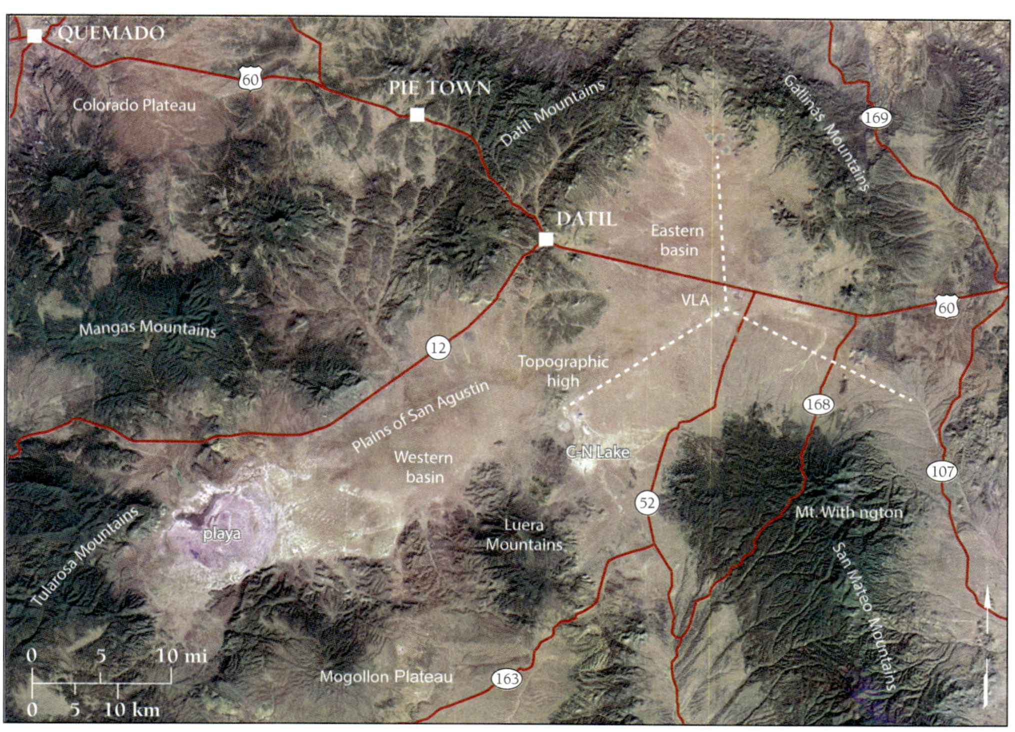

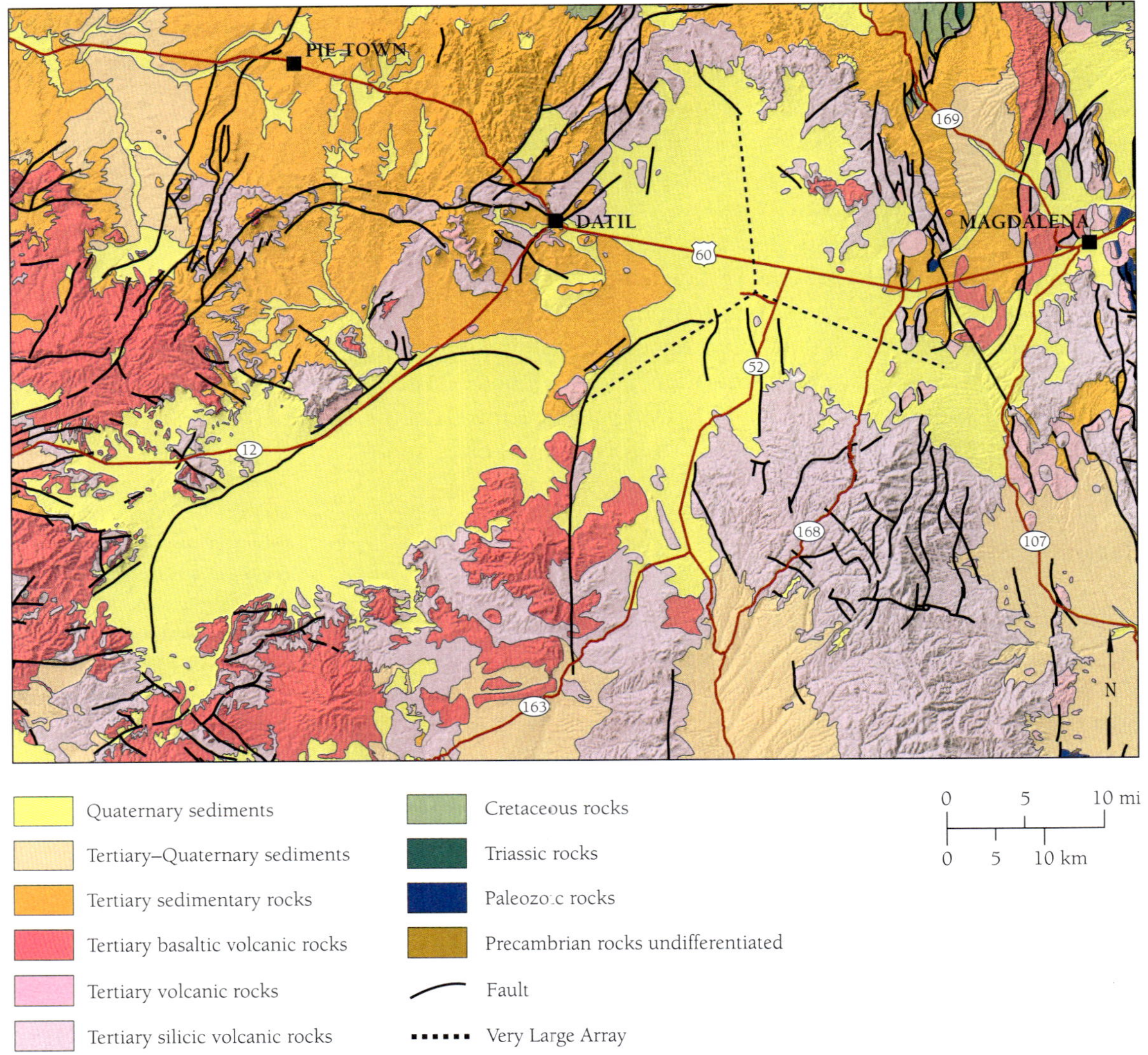

Generalized geologic map of the Plains of San Agustin and surrounding highlands.

the Plains of San Agustin into eastern and western basins. The western basin is elongated northeast-southwest and has a ephemeral lake (playa) in its southwestern part. The eastern basin contains an embayment on its southwest side, where a smaller playa called C–N Lake is located. These playas and a smaller one northeast of the VLA headquarters temporarily hold water up to three feet deep after heavy rains, but are otherwise dry.

The Rock Record

The Plains of San Agustin are underlain by basin-fill sediment hundreds to thousands of feet thick, reaching a maximum of about

4,000 feet at its southwestern end. Over the past 25 million years, this sediment was transported into the basin by streams draining the surrounding mountains. Mostly gravel and sand have been deposited along the edges of the basin and this coarse sediment transitions laterally to finer sand, silt, and clay towards the center of the basins and near modern-day playas.

The surrounding highlands are underlain by bedrock composed of volcanic rocks and sedimentary rocks derived from them. These rocks are about 40–25 million years old and include regionally extensive, thick ash-flow tuffs that erupted from nearby supervolcanoes to the east and south. Data from deep wells indicate that volcanic bedrock lies beneath the younger basin-fill sediments within the eastern and western basins.

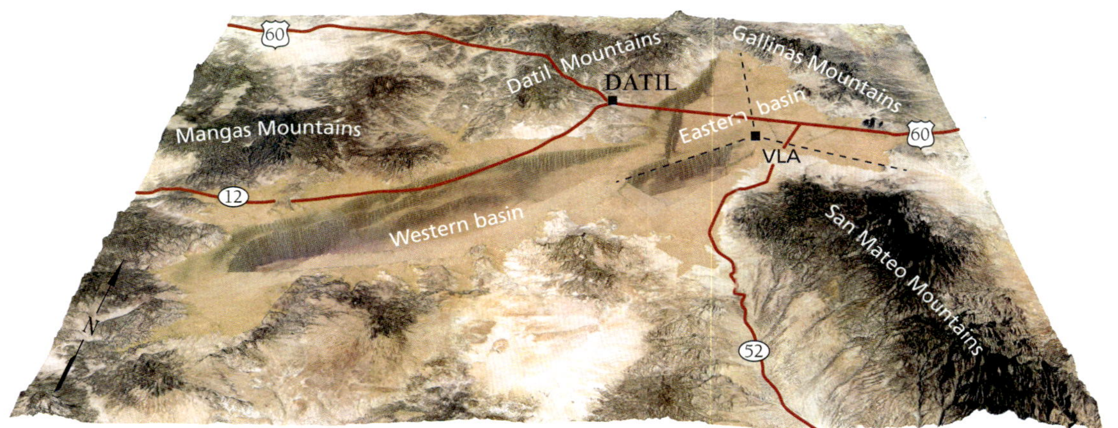

Geologic History

During the past approximately 25 million years, tectonic stretching in the region has down-dropped crustal blocks along faults, creating a complex basin. These faults can be divided into: 1) older, buried faults that generally trend east-northeast, and 2) young faults that have broken the Earth's surface, forming visible scarps (low linear cliffs) that generally trend north-south. The older, buried faults have been inferred using data from wells or geophysical surveys, such as measurements of the Earth's gravity or electrical fields.

Three-dimensional depiction of the geologic structure beneath the Plains of San Agustin. View is to the north. The sand, gravel, clay, and silt filling the eastern and western basins have been removed in this image, showing the elevation of the underlying bedrock surface. The same package of volcanic rocks that compose the surrounding mountains underlies the floor of these basins.

Hydrology

Because the Plains of San Agustin are topographically closed, no surface water flows out of the basin. In the western basin, most runoff drains towards the low-lying, clayey playa at its southeastern end, the surface of which is commonly covered with a white crust of alkaline salts.

Photograph looking south at one of the north-trending fault scarps southwest of the VLA headquarters (photo taken 2.6 miles north-north-west of VLA headquarters). The steep slopes of the scarp lie in the shade from the early morning sun.

In the eastern basin, shallow drainage channels convey storm water from mountain canyons out onto the wide basin floor, where it collects in several small ephemeral lakes such as C–N playa and the unnamed playa near the VLA.

Recent hydrologic studies suggest that the Plains of San Agustin may not be "closed" with respect to groundwater. The general groundwater flow direction is toward the southwest, and some groundwater may leak out of the western basin towards the south. The water table lies about 25–300 feet below the surface, with the shallowest levels beneath the southwestern playa.

Groundwater under the plains is generally too deep to be taken up by plants or to be lost to evaporation. Dating of this groundwater indicates most of it percolated into the aquifer during wetter climatic conditions of the last ice age (26,000–10,000 years ago). Therefore, it appears that the vast majority of recharge comes in very slowly through the groundwater system. Most surface runoff into the basin is used by plants or evaporated from playas. During the mid-2010s, considerable

Satellite image showing past shorelines associated with Lake San Agustin. Waves beating against the lakeshore carved notches into slopes and built-up beach berms alongside these notches. The highest wave-cut notch and berm is labeled by the number six and lies at an elevation of 6,910–6,940 feet. Lower shorelines are consecutively labeled 6 to 1. These lower shorelines formed as the lake progressively shrank after 16,000 years ago. The highest definitive shoreline at 6,940 feet is not visible here but is preserved elsewhere in the basin.

controversy has arisen because of formal requests by Agustin Plains Ranch, LLC, to pump large volumes of this deep groundwater and transport it to the Rio Grande valley.

Geologic Features: Ancient Lakes

Periodically during the past one million years, perennial (year-round) lakes occupied the floor of the Plains of San Agustin. This time period coincides with the Pleistocene epoch, which was characterized by dramatic climate fluctuations between glacial episodes (ice ages) and interglacial episodes (like the modern-day climate). Past lakes are interpreted to have formed and expanded during glacial periods in association with cooler temperatures and increased precipitation. Glacial climatic conditions caused increased recharge to the aquifer, raising groundwater levels and allowing lakes to form and grow.

A particularly large lake existed between 25,000 to 16,000 years ago. That lake, known as Lake San Agustin, locally breached the topographic divide between the western and eastern basins. Shoreline features created by Lake San Agustin are still visible in some places, particularly the southern end of the western basin. A smaller version of the lake existed 8,000 years ago, but since about 5,000 years ago only scattered playas, like those seen today, have been present on the basin floor.

Depiction of the lake that once covered much of the Plains of San Agustin during the last ice age. The light blue shade represents the largest extent of the lake at about 18,000 years ago, based on the elevation of the highest preserved, definitive shoreline (6,940 feet). The darker blue areas show smaller lakes (elevations of 6,810 and 6,790 feet) that were present during the drying of the basin (between 18,000 and 8,000 years ago).

Geologic Features: Faults

In the low hills southwest of the VLA headquarters, one can see low but steep slopes (scarps) created by two north-south faults. The last movement on these (and the accompanying large earthquake) occurred about 100,000 years ago.

—Daniel Koning and Alex Rinehart

If You Plan to Visit

The VLA and its great geeky gift shop are open daily during daylight hours, with special guided tours offered on the first Saturday of the month. To get to the Visitor Center from Socorro drive west on US 60 for about 50 miles; then turn south on NM 52 for 2.5 miles, and then turn right (west) on the road that leads to the Visitor Center.

The largest playa and the best preserved ancient lake shorelines are seen at the southwest end of the Plains of San Agustin. Drive west on US 60 to Datil; then turn left on NM 12 at the crossroad in Datil; drive southwest for 34 miles and turn left (southeast) on an unmarked dirt road. After six miles, the playa is within 0.5 miles on the left, and shorelines can be observed on hillslopes to the right as the road winds around the south end of the large playa. For more information:

NRAO Array Operations Center
P.O. Box O
Socorro, NM 87801-0387
(575) 835-7000
public.nrao.edu/visit/very-large-array

The Very Large Array has 27 active radio antennas with a dish size of 82 feet.

Datil Well Recreation Area Campground

The Datil Well Campground is located at the southern edge of the Datil Mountains at an elevation of about 7,500 feet. The campground includes 1 of 15 wells drilled by the Civilian Conservation Corps in the 1930s to provide water for cattle and sheep herded along the Magdalena Livestock Driveway (also termed the Hoof Highway) and preserves some of the original watering troughs. The driveway operated from 1885 through 1971, before truck transport of livestock was the norm. It extended eastward 125 miles from Springerville, Arizona, to Magdalena, New Mexico, where it terminated at stockyards (still preserved) located near the former Atchison, Topeka, and Santa Fe railhead. In its peak year, 1919, 150,000 sheep and 21,000 cattle were herded along the trail.

Geologic map of Datil Well Recreation area and hiking trails.

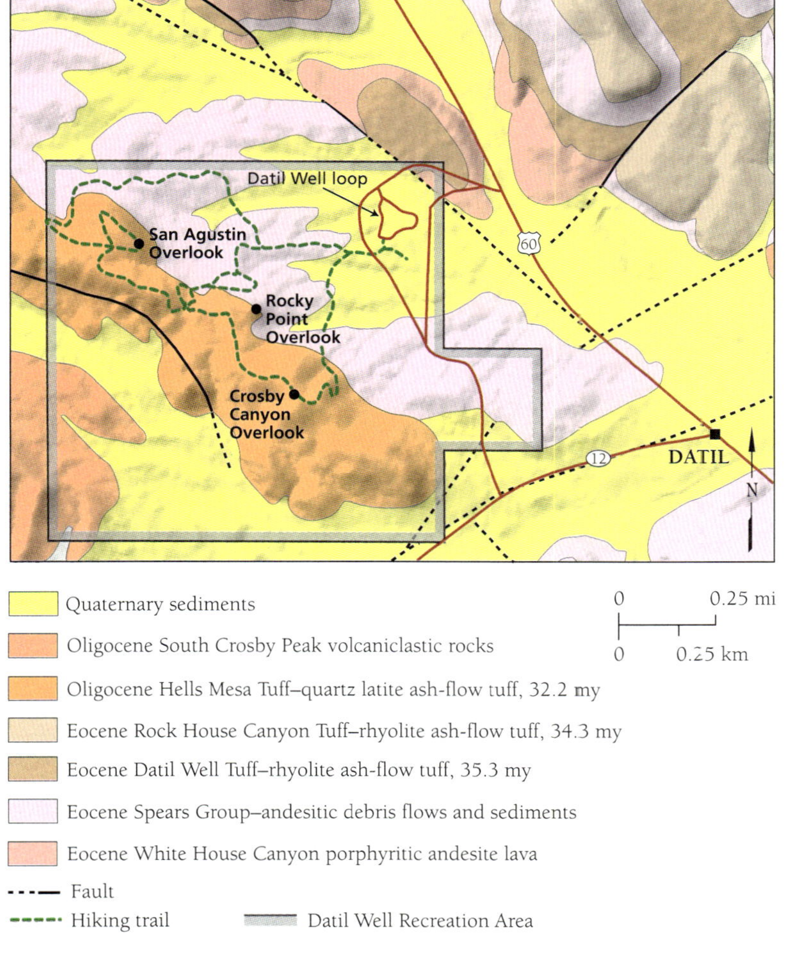

Facilities at Datil Well Campground include campsites, picnic tables, fire pits, drinking water, and restrooms. Three miles of trails through a mixed forest of piñon, juniper, and ponderosa provide numerous vistas and good exposures of the Hells Mesa Tuff, one of New Mexico's larger regional ash-flow tuffs.

Regional Setting

Datil Well Campground is located in the northern Mogollon–Datil volcanic field just northwest of a cluster of caldera supervolcanoes that extended from Socorro to the San Mateo Mountains. Volcanic rocks of the Mogollon–Datil field lap northward over the Mesozoic and Cenozoic sedimentary rocks of the southern Colorado Plateau and are overlain by the younger basalts of the Springerville volcanic field far to the west in Arizona.

The Rock Record

Four main rock units are present in and near Datil Well Campground. The oldest is the White House Canyon Andesite, a crystal-rich lava exposed near the north campground entrance. Overlying this lava is the 35.3 million-year-old Datil Well Tuff, best exposed in low cliffs on the south side of US 60 just west of the western entrance to the Datil Well Campground. The Datil Well Tuff is overlain by deposits of the Spears Group—pink volcanic breccias made of angular fragments of volcanic rock in a finer matrix of sand and mud. These breccias, exposed along trails near the starting point of the park's trail system (at the Trail Gazebo), probably formed as large mudflows derived from stratovolcanoes in the area. Overlying the Spears Group breccias is the

View east of Datil Well across the Plains of San Agustin

32.2 million-year-old Hells Mesa Tuff. Both the Datil Well Tuff and the Hells Mesa Tuff were catastrophically erupted as rhyolitic ash flows from Mogollon–Datil calderas to the east and south. Nearly all of the prominent pink to tan bedrock ridges and cliffs within the Datil Well Campground are formed by the Hells Mesa Tuff.

Geologic History

The rocks exposed within Datil Well Campground record a slice of the early history of the Mogollon–Datil volcanic field, a huge volcanic complex that included 20 supervolcanoes as well as numerous andesite stratovolcanoes, active between about 40 and 25 million years ago. The first recorded event at Datil Well Campground occurred about 36 million years ago, when a nearby stratovolcano erupted a lava now exposed as the Whitehouse Canyon Andesite near the campground entrance. This lava was subsequently buried by a series of far more catastrophic eruptions from supervolcano calderas to the east and south. The largest of these eruptions occurred 32.2 million years ago, producing the Hells Mesa Tuff, which is well exposed along the Datil Well Campground trails. The Hells Mesa Tuff was explosively erupted in a single brief caldera collapse from a 20-mile-wide supervolcano near what is now Socorro, New Mexico. The erupted volume of hot pumice, ash, gas, and rock fragments in the Hells Mesa eruption was more than 1,000 times larger than that of the 1980 eruption of Mount St. Helens. The incandescent ash flow spread rapidly across the land surface. When it came to rest, glassy particles within it welded together, producing the solid rocks

Flattened pumice and angular lithic fragments in densely welded Hells Mesa Tuff are well exposed along hiking trails and in boulders in the campground area (image is approximately 2.5 inches wide).

now exposed at Datil Well Campground. The underlying Datil Well Tuff is a similar large-volume pyroclastic flow that was erupted from an older supervolcanc caldera to the south. The two explosive tuff units are separated by mudflow (lahar) breccias derived from a collapse event on a nearby andesite stratovolcano.

Geologic Features

ROCKY POINT OVERLOOK—Rocky Point Overlook can be accessed from either of two trails that extend west or south from the Trail Gazebo. The walking loop is about two miles long. Lichen-encrusted boulders at and near the overlook provide excellent examples of flattened pumice and layering produced by welding of the hot mixture of pumice, ash, and gas after the Hells Mesa Tuff was emplaced. Some feldspar crystals in the tuff have a distinctive shimmering blue moonstone appearance due to unmixing of potassium and sodium during cooling. Angular fragments in the tuff are pieces of older rock ripped from the caldera wall during the eruption. The overlook also provides an excellent view to the Plains of San Agustin and the Very Large Array discussed in the previous chapter.

CROSBY CANYON OVERLOOK AND TRAIL—Pink mudflow breccias are exposed along the first half of the trail from Trail Gazebo to Crosby Canyon Overlook. The second half of the trail climbs into tan, platy Hells Mesa Tuff. The view to the south across Crosby Canyon includes a prominent white ledge of Hells Mesa Tuff overlain by younger tuffs and lavas.

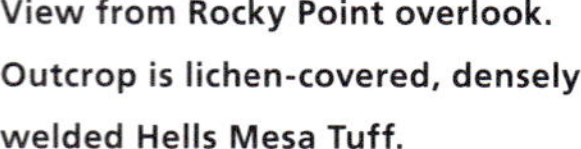

View from Rocky Point overlook. Outcrop is lichen-covered, densely welded Hells Mesa Tuff.

US 60 ROADSIDE OUTCROPS—Just north of the western entrance tc Datil Well Campground, on the south side the highway, are low cliffs of Datil Well Tuff that show several features typical of welded tuffs, similar to those visible in the Hells Mesa Tuff. These include flattened pumice, layered welding textures, rectangular feldspar crystals, and angular rock fragments. The Datil Well Tuff is named for this area, where it was first recognized.

—William C. McIntosh

Additional Reading

The Magdalena Trail (BLM pamphlet): **www.blm.gov/download/file/fid/23638**

If You Plan to Visit

The Datil Well Campground is located near the town of Datil and can be accessed either from US 60 (one mile west of Datil) or from NM 12 (about 0.5 miles south of the junction with US 60). For more information:

Socorro Field Office
Bureau of Land Management
901 S. Highway 85
Socorro, NM 87801
(575) 835-0412
www.blm.gov/office/socorro-field-office

Crystal-rich White House Canyon Andesite lava near northern campground entrance. Crystals are feldspar (white), hornblende (black), and biotite (bronze).

PART 2

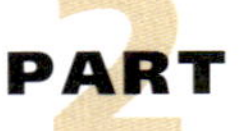

THE MOGOLLON–DATIL VOLCANIC FIELD

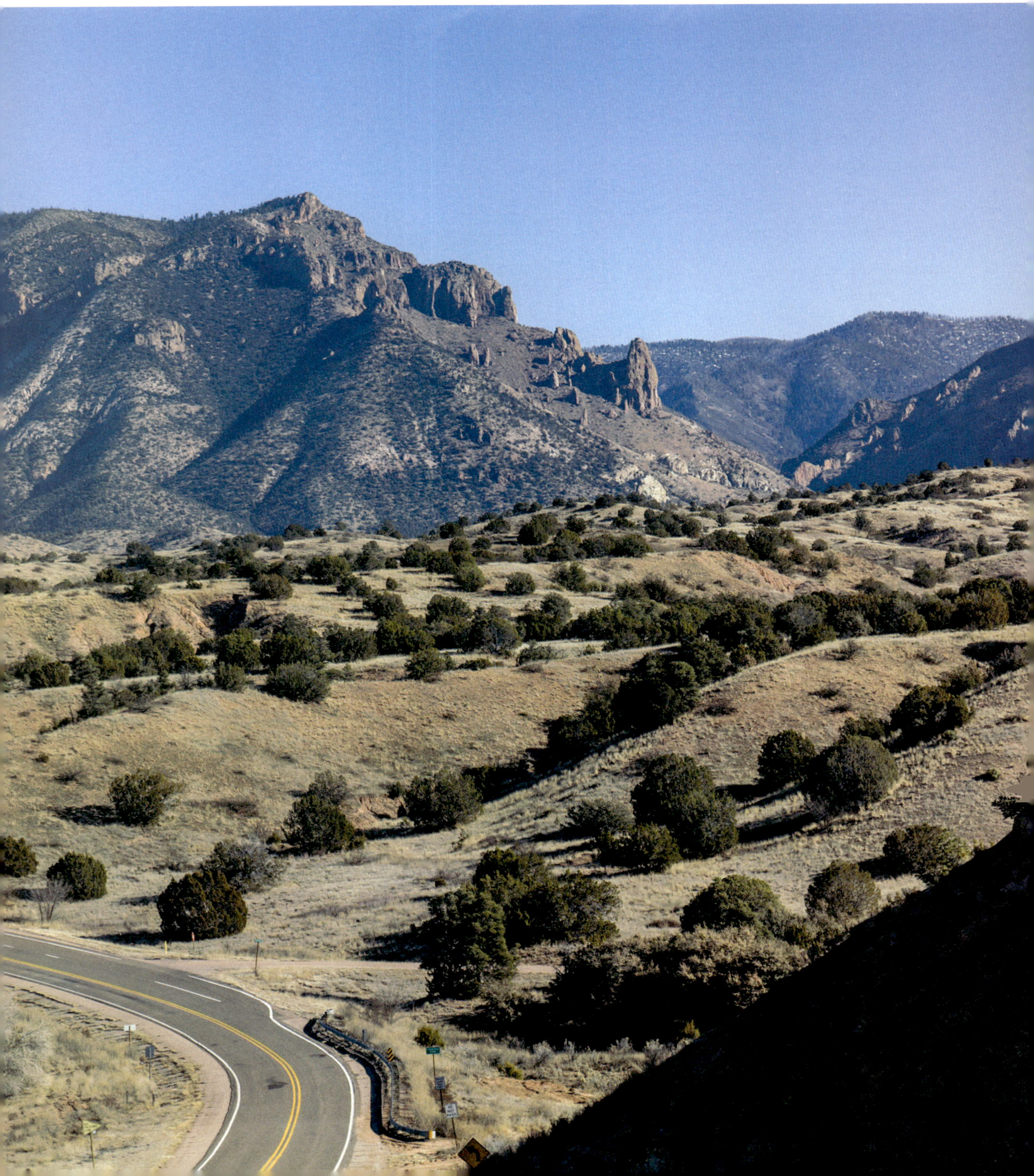

The Mogollon–Datil volcanic field of southwestern New Mexico encompasses some of the most remote and rugged terrain in the Land of Enchantment. The volcanic field covers an area of approximately 15,000 square miles from Socorro in the northeast to Las Cruces in the southeast, and extends into eastern Arizona. This is a landscape filled with wild rivers, scenic valleys, narrow slot canyons, and towering mountains, many of which rival the topographic relief of even the highest peaks of the southern Rocky Mountains in northern New Mexico. The volcanic field contains numerous wilderness areas totaling more than 850,000 acres, including the Blue Range, Mount Withington, Apache Kid, Aldo Leopold, and the Gila wildernesses, the last of which is the state's largest and the country's first protected wilderness area. The volcanic field sits at the conjunction of three major physiographic provinces in New Mexico, the Colorado Plateau to the north, the Rio Grande rift to the east, and the Basin and Range to the south and west. Although different rock types of nearly every age can be found in the area, the majority of the public lands, parks, and monuments showcase the diverse volcanic and tectonic activity that took place here.

The Mogollon–Datil volcanic field is the erosional remnant of a vast volcanic highland built by a series of eruptions between 40 and 25 million

OPPOSITE: **Western edge of the Gila massif from US 180 near Aldo Leopold overlook. Gila Conglomerate deposits underlie the yellowish hills in the foreground. The rugged peaks in the distance are remnants of calderas.**

Map showing the recreational facilities and areas of interest in and near the Gila Wilderness.

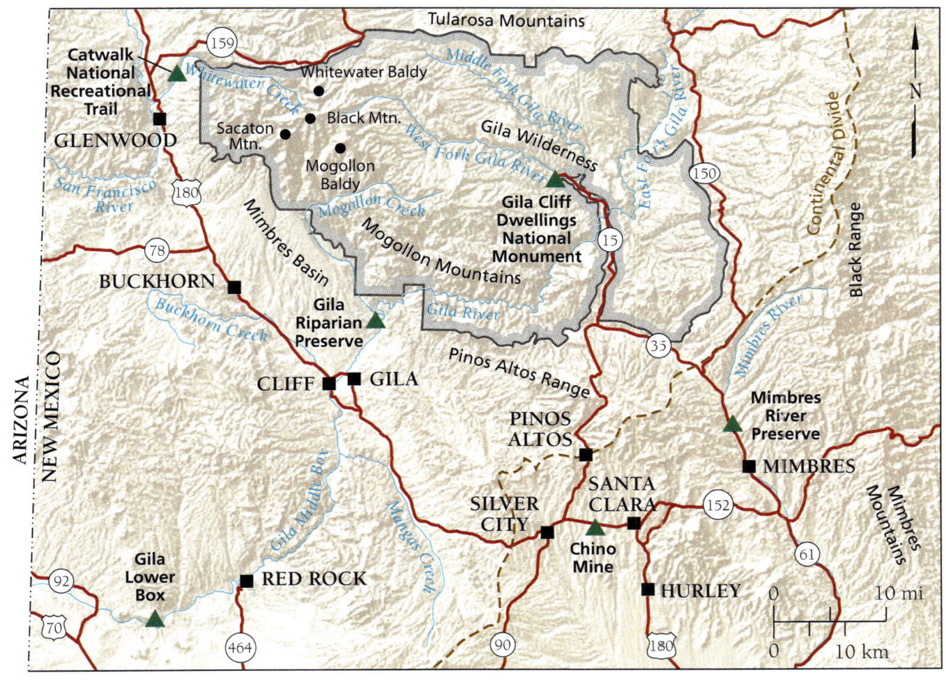

years ago. During this period, hundreds of volcanoes erupted magmas of various compositions to produce almost every type of known volcanic landform. Most notable, however, are the eruptions from approximately 20 calderas, also known as supervolcanoes. The calderas of the Mogollon–Datil field are the extinct equivalents to the dormant Valles caldera of northern New Mexico or the active Yellowstone caldera in Wyoming. Powerful caldera eruptions release several tens to several hundreds of cubic miles of volcanic gas, ash, and pumice. These eruptions produce a variety of voluminous pyroclastic (pyro, meaning fire; clastic, meaning broken) deposits, chiefly pyroclastic flows known as ash-flow tuffs, where pyroclastic material surges along the ground prior to deposition, and pyroclastic falls, where ash is ejected into the atmosphere before settling out to form ash layers.

The Mogollon–Datil volcanic field is one of several major volcanic fields that covered much of western North America during Cenozoic time, extending from the northern Rockies to central Mexico. To geologists, this major pulse of volcanism is commonly referred to as the "mid-Tertiary ignimbrite flare-up" because of the hundreds of ignimbrite (ash-flow) eruptions from nests of calderas that erupted in mid-Cenozoic time. Toward the end of volcanic activity, the Earth's crust began to stretch and pull apart, tearing the volcanic field into a series of large sediment-filled valleys, called grabens, and adjacent dissected mountain ranges that are prominently visible today. This tectonic activity and subsequent erosion has destroyed many of the original volcanic landforms. This is caldera country.

Building the Mogollon–Datil Volcanic Field

Outcrops of Paleozoic and Mesozoic sedimentary rocks dominated the landscape of southwestern New Mexico before the development of the volcanic field. During these times, the coming and going of vast seas and extensive networks of ever-changing rivers deposited thousands of feet of sediments on top of Precambrian metamorphic and igneous rocks. Also prior to volcanism, numerous intrusions of granites and similar magmas were emplaced into the middle and upper crust during the widespread Laramide Orogeny (from approximately 75 to 45 million years ago). Most of these older rocks are now covered by a several mile-thick blanket of volcanic rocks that erupted during the lifespan of the Mogollon–Datil field. However, these pre-existing

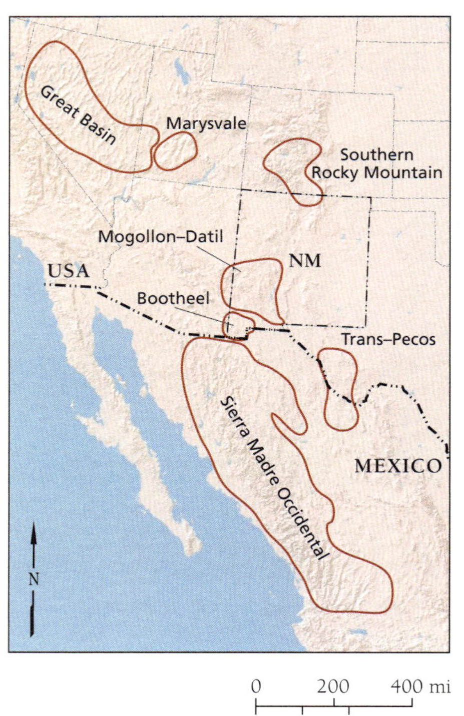

Regional map showing caldera-related centers formed during the mid-Tertiary ignimbrite flare-up.

rocks are exposed in outcrops along the margins of the field, such as the Quebradas region east of Socorro. Fragments of these older rocks were picked up as magmas moved through the crust, or as lavas and ignimbrites flowed across the landscape.

Each caldera in the Mogollon–Datil volcanic field experienced a similar cycle of volcanic activity. Prior to each caldera eruption, there was a stage of activity known as pre-caldera volcanism. This period was characterized by numerous, small-volume eruptions of compositionally diverse magmas, leading to the formation of stratovolcanoes (composite volcanoes) that may have looked much like the present-day Mount St. Helens. These early eruptions did not form calderas because of their relatively small volumes (typically less than a few cubic miles) and the low viscosity of their magmas, which controls the eruptive style. Although the composition of pre-caldera volcanic rocks was diverse, andesite volcanism was fairly common. Thus, many of the earlier eruptions produced lava flows, along with lesser amounts of silicic pyroclastic rocks. Assessment of pre-caldera volcanism from calderas around the world indicates that the duration of this stage can last as little as a few tens of thousands of years to as long as several millions of years.

Although there have not been supervolcano eruptions in historical times (thankfully!), detailed investigations of exposed calderas, such as those in the Mogollon–Datil volcanic field, shed light on the processes related to caldera creation. During caldera formation, several dozen to several hundred of cubic miles of rhyolitic and dacitic magma are erupted from upper crustal magma chambers. The high viscosity of the rhyolite and dacite magmas results in extremely explosive eruptions that send material as high as the stratosphere. Ash deposits from caldera eruptions can be found thousands of miles downwind from the source caldera. For example, ash from the latest Yellowstone caldera eruption 640,000 years ago can be found throughout central and southern New Mexico and into the Gulf of Mexico.

Geologists estimate that most catastrophic caldera eruptions are completed within only a few days to perhaps a few months. Because such a large volume of material is erupted within a very short amount of time, the crust above the evacuating magma chamber becomes unstable and collapses into the magma chamber during the eruption. This collapse

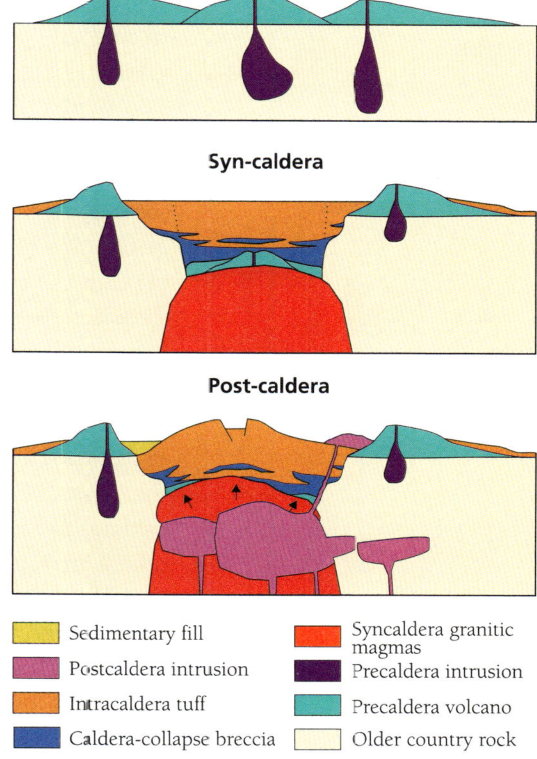

Schematic diagram showing caldera stages.

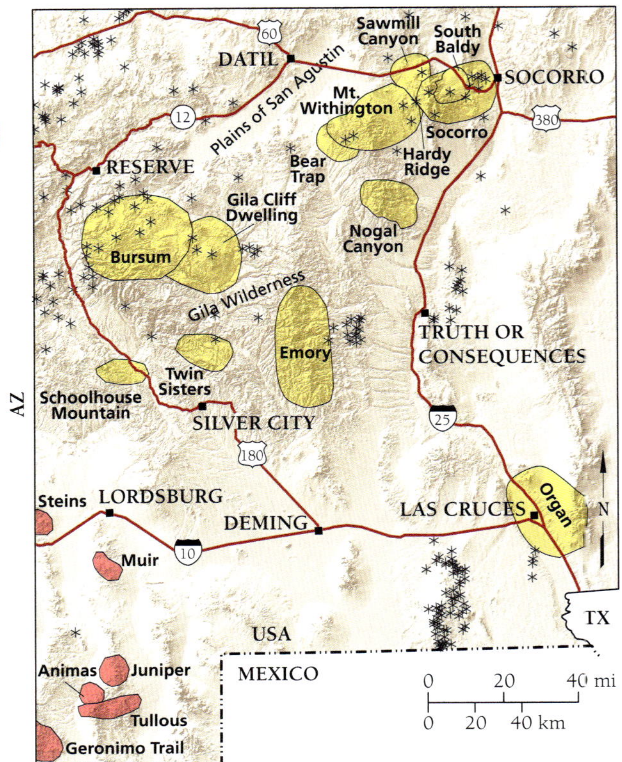

Bootheel calderas

Mogollon–Datil calderas

* Volcanic vent

Map of southwest New Mexico showing caldera locations.

forms a very large depression, or caldera, at the Earth's surface, which is continuously filled by the ongoing eruption of volcanic material. Over-steepened walls of the collapsing caldera catastrophically slide into the depression, forming landslide-like deposits within the intracaldera ignimbrite, known as megabreccias (breccia is rock composed of broken rock fragments). Despite having some common general characteristics, each caldera eruption is different. Individual calderas can be many tens of miles in diameter, and the subsidence of the caldera is typically between 1 and 3 miles. Pyroclastic flows emanating from the caldera send ash, rock fragments, and hot gases across the landscape in all directions. If a caldera were to erupt today, everything within the caldera and in the near vicinity would be totally destroyed and buried in pyroclastic material.

Approximately 20 calderas erupted catastrophically between 36 and 25 million years ago in the Mogollon–Datil volcanic field. Some of the calderas are well exposed and relatively easy to identify in the rock record. Thick accumulation of ignimbrites, deposits of megabreccia, and intense alteration are a few of the diagnostic features that geologists use to identify calderas. However, some source calderas are not well exposed, being either completely or partially buried by younger volcanic deposits or perhaps hidden beneath the even younger rift basins. The only pieces of evidence for these catastrophic eruptions are the regional sheet-like ignimbrite deposits that spilled outside the caldera.

The identified calderas of the Mogollon–Datil field are found in two zones—a northern one between Socorro Peak (just west of the town of Socorro) and the Plains of San Agustin and a southern one between the Organ Mountains (near Las Cruces) and the Gila Wilderness. Some calderas are found in clusters of grouped or nested calderas, whereas others are found as isolated single calderas. In general, the ages of the calderas suggest a migration of volcanism, because the calderas to the east tend to be older than the calderas to the west. However, there are some exceptions to this pattern. An additional and similar nest of calderas is found south of the Mogollon–Datil volcanic field in the Bootheel region of southwestern New Mexico (called the Bootheel volcanic field) and may represent a transition to the Sierra Madre Occidental volcanic field of northern and central Mexico.

Following caldera collapse, smaller-volume eruptions continue during post-caldera volcanism. Similar to pre-caldera volcanism, this waning stage of volcanic activity is characterized by numerous eruptions of compositionally diverse magmas during a protracted period of thousands or millions of years. Dacite, andesite, and basalt lavas are erupted from volcanoes both within and outside the calderas, partially filling the calderas and further obscuring the original caldera geometry. Also during this time, thousands of cubic miles of magmas are emplaced into the middle and upper crust. Some of this magma never erupts but instead freezes and crystallizes to form plutons of intrusive granitic rock. Beneath individual calderas, and perhaps lying beneath the entire volcanic field, dozens to hundreds of plutons coalesce to form even larger intrusions called batholiths, likely resembling the types of rocks now exposed in the dramatic cliffs of Yosemite Valley in California. Not only do these intrusions mark the cessation of magmatic activity, but they also bring with them metal-rich fluids that alters the caldera-related rocks. Some of these zones of alteration were mined during the 1800s and 1900s for commodities such as gold, silver, tin, lead, and zinc.

Table of calderas, their ignimbrites, and their ages. Question marks indicate uncertainity concerning the location of that ignimbrite's caldera.

CALDERA (LOCATION)	IGNIMBRITE	AGE (million years)
Bear Trap (NW San Mateo Mountains)	Turkey Springs	24.6
Bear Trap (N San Mateo Mountains)	South Canyon	27.7
Hardy Ridge (N San Mateo & Magdalena Mountains)	Lemitar	28.4
Bursum (Mogollon Mountains)	Bloodgood Canyon & Apache Springs	28.4
Gila Cliff Dwellings? (Mogollon Mountains)	Shelly Peak	28.5
Nogal Canyon (S San Mateo Mountains)	Vicks Peak	28.9
Sawmill/Magdalena (Magdalena Mountains)	La Jencia	29.2
Gila Cliff Dwellings? (Mogollon Mountains)	Davis Canyon	29.4
Twin Sisters (Pinos Altos Range)	Upper Tadpole	31.8
Fall Canyon (W Mogollon Mountains)	Caballo Blanco	32.1
Socorro (Socorro Peak)	Hells Mesa	32.5
Schoolhouse (Schoolhouse Mountains)	Box Canyon	33.9
Sullivan Hole (N Black Range)	Blue Canyon	34.1
Mogollon? (W Mogollon Mountains)	Cooney	34.4
Skeleton Ridge (S San Mateo Mountains)	Rock House	34.9
Emory (Black Range)	Kneeling Nun	35.3
Sullivan (N Black Range)	Datil Well	35.3
Organ (Organ Mountains)	Squaw Mountain	36.0
Organ (Organ Mountains)	Achenback	36.2
Organ (Organ Mountains)	Cueva	36.5

Activity Since Volcanism Largely Ceased

During and following the caldera eruptions, the volcanic field was faulted by extension during both the Rio Grande rift and Basin and Range tectonic activity. During extension, some blocks of the crust were moved upward along normal faults to form mountain ranges; others were dropped downward to form large valleys adjacent to the mountains. The uplifted blocks are the mountains that we now know today, including the Magdalena, San Mateo, Black, Mogollon, Tularosa, and Organ mountain ranges; and some faulting continues even now. For example; the largest recorded earthquake in New Mexico, with an estimated magnitude of 6.5 or higher, occurred in 1906 near the town of Socorro, along a fault on the margin of the Rio Grande rift.

The Late Cenozcic (Miocene–Pleistocene) uplift of mountains caused major changes to meteorological patterns by blocking passing weather systems, which in turn caused increased precipitation, thereby increasing the rates of erosion. As a result, large volumes of material were eroded from the mountain blocks and deposited in the adjacent valleys. Silts, sands, cobbles, and boulders generated by millions of years of erosional processes are preserved as sedimentary deposits throughout the Mogollon–Datil region and are called the Gila Conglomerate. The valleys containing the Gila Conglomerate, such as the Mimbres and the Mangas, are filled with several thousand feet of sediment, much thicker than the adjacent mountains are tall. These incredible accumulations of sediment are both porous and permeable, making them valuable reservoirs for groundwater that support the hydrologic needs of communities throughout the region.

The youngest geologic activity in the Mogollon–Datil region includes the continued wearing down of the mountains by modern rivers, volumetrically minor volcanism, and geothermal activity. The Continental Divide, which separates river systems that drain into the Pacific Ocean (via the Gulf of California) from those that drain into the Atlantic Ocean (via the Gulf of Mexico), is located along a north-south trending irregular spine in the Mogollon–Datil volcanic field. Among the most significant of these drainages are the West, Middle, and

CLASSIFICATION & FLOW CHARACTERISTICS OF VOLCANIC ROCKS				
Volcanic rock name	Basalt	Andesite	Dacite	Rhyolite
Silica (SiO_2) content	48–52%	52–63%	63–63%	68–77%
Color	Dark ⟶			Light
Eruption temperature	1160°C			800°C
MOBILITY OF LAVA FLOWS	Low resistance to flow (thin, runny lava)			High resistance to flow (thick, sticky lava)

Decreasing mobility of lava ⟶

East Forks of the Gila River, one of the last free-flowing rivers in the western United States.

Although the majority of volcanism in the Mogollon–Datil volcanic field ended about 25 million years ago, some young vents and flows (many less than 5 million years old) are found within and along the margins of the field. These youngest volcanoes erupted dominantly basaltic lava flows with little explosive activity.

Economic Resources

The extensional faults that separate the basins and ranges provide a pathway for waters to circulate to great depths in the earth, where they are heated by the surrounding warm rocks, and then return to the surface along adjacent faults. Some thermal springs have been developed for heating and commercial use. However, most are undeveloped and provide relaxing soaks for those willing to hike to them.

As in many places throughout the western United States, mankind has left marks on the landscape. The Mogollon People built cliff dwellings in alcoves of the Gila Conglomerate between the 13th and 14th centuries. Petroglyphs and pictographs can be found on lava-capped mesas and canyon walls throughout the volcanic field. More recently, many of the calderas have been mined to extract their valuable mineral and metal resources. Some post-caldera intrusions developed large zones of adjacent hydrothermal fluid circulation during emplacement and deposited ore bodies of economic grade. One of the best-known mining districts is near the town of Mogollon, where in the 1870s gold was discovered on the western slope of the Mogollon Mountains, near the margins of the Mogollon and Bursum calderas. Mineral deposits were mined from almost all of the calderas in the Mogollon–Datil volcanic field at one time or another. Most of these mines are no longer active, as prices have driven mineral exploration to larger and more valuable districts elsewhere in the world. The only presently-active mines in the area are copper mines near Silver City. These are related to the older intrusions of the Laramide Orogeny rather than to magmatism of the Mogollon–Datil volcanic field.

—*Matthew J. Zimmerer and William C. McIntosh*

Historical photo of Mogollon Mining District.

Gila Cliff Dwellings National Monument

Gila Cliff Dwellings National Monument (pronounced Hee-luh) is located approximately 25 miles north of Silver City, within the heart of the Gila National Forest. The monument provides relatively easy access to the otherwise remote mountains and valleys of the Gila Wilderness interior. The cliff dwellings are located along the appropriately named Cliff Dweller Canyon, a tributary to the West Fork of the Gila River. In the 13th and 14th centuries, the Mimbres Mogollon people resided in these rock shelters built in alcoves above the ponderosas, junipers, and piñons of the canyon floor. These defensible shelters, with access to water and game, must have been a near-perfect setting for these early inhabitants.

Exposed along the valley walls and within the alcoves are rocks that record ancient volcanic activity, deposition of sediment following the cessation of eruptions, and classic examples of erosion and physical weathering processes.

OPPOSITE: **Cliff dwellings of the Mimbres Mogollon culture.**

Regional Setting

Gila Cliff Dwellings National Monument is located in a basin developed within the Gila Cliff Dwellings caldera, one of many ancient supervolcanoes of the Mogollon–Datil volcanic field. Although the field is widely known for spectacularly abundant volcanism between 40 and 25 million years ago, the area also is overprinted by several large sedimentary basins filled with detritus shed from the volcanic highlands. Toward the end of volcanic activity, the Earth's crust was stretched as part of the Basin and Range deformation. That extension created a series of fault-bounded basins, called grabens, and adjacent, uplifted mountain ranges. The particular basin in which the cliff dwellings are located is known as the Gila Hot Springs graben because it contains numerous thermal springs. Through-going drainages now dissect the area as exemplified by the West and Middle Forks of the Gila River that converge just below the Visitor Center, upstream of the confluence with the East Fork.

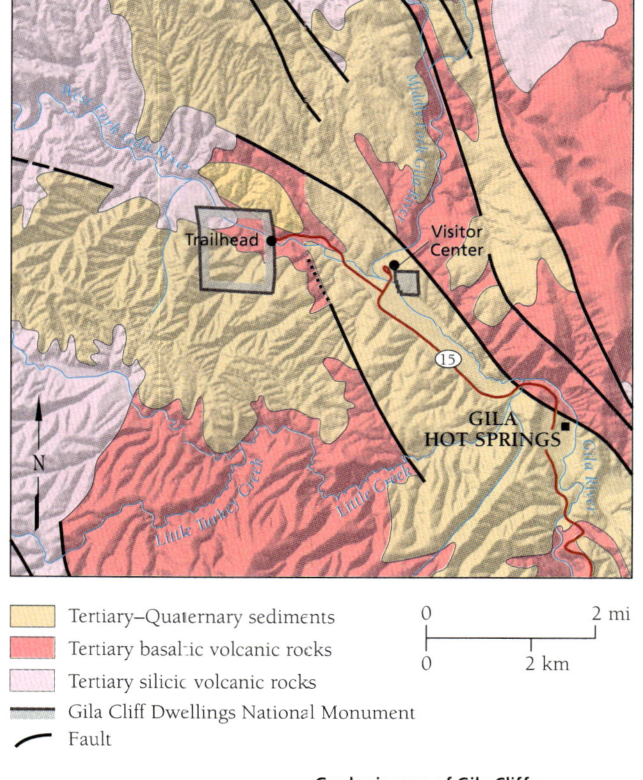

Geologic map of Gila Cliff Dwellings National Monument and adjacent lands.

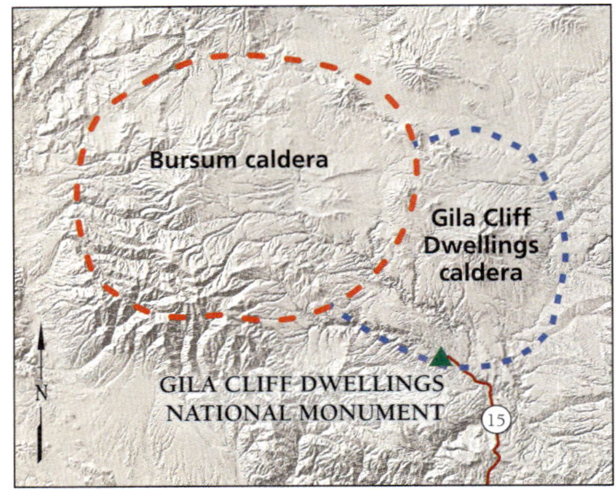

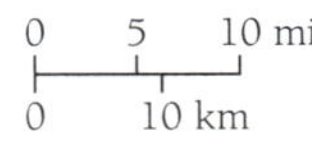

Shaded relief map showing the Bursum and Cliff Dwellings calderas with surrounding basins.

The Rock Record

The oldest rock unit exposed at Gila Cliff Dwellings National Monument is the Bloodgood Canyon Tuff. Outcrops of that unit can be seen at the base of the cliff to the north of the parking lot. This tuff was erupted about 28 million years ago during collapse of the 20- by 30-mile-wide Bursum caldera to the northwest of the national monument. The Bloodgood Canyon Tuff contains crystals of quartz and feldspar within a pumice- and ash-rich matrix. Some of the feldspar crystals are a gem variety known as moonstone. On a sunny day, the crystal faces show flashes of vibrant blue hues.

Dark reddish-brown lava flows exposed beneath the distinctive tan cliffs of Cliff Dweller Canyon are more evidence for volcanic activity of the Mogollon–Datil volcanic field. Unlike the highly explosive eruption of the Bloodgood Canyon Tuff, these flows were emplaced slowly as viscous lava. The lava flows contain iron-rich minerals, which give the lava flows their distinctive rust color. The lava flows belong to the Bearwallow Mountain Andesite, a unit composed of numerous flows that erupted from multiple vents throughout the volcanic field about 25 million years ago. The lava flows contain abundant vesicles, which formed around gas bubbles within the lava during solidification. Many of these eventually became amygdules, which are vesicles filled with light-colored minerals deposited by groundwater. Layers of reddish, highly fragmented lava are the brecciated bases or tops of individual flows within a stack of lavas.

Outcrops of volcanic rocks in the monument parking lot. The lower pinkish-brown cliffs are the Bloodgood Canyon Tuff. The upper slopes expose the lava flows of the Bearwallow Mountain Andesite.

The Gila Conglomerate caps the volcanic rock section and forms the prominent light-brown cliffs of Cliff Dweller Canyon. The age of the Gila Conglomerate ranges in age from about 15 to 5 million years. The conglomerate contains sand and gravel composed of lava and tuff fragments. The largest fragments are boulders more than a foot in diameter. The clasts are not very well rounded, indicating they were not transported very far from their original source. Some crystals in the conglomerate display the blue hues of moonstone, suggesting that the Bloodgood Canyon Tuff was a source of some of the sediment.

Although the cliffs of the Gila Conglomerate in the monument are impressive, looking up the canyon to the southwest indicates that the true thickness of the Gila Conglomerate is much greater. The cliff faces carved from the Gila Conglomerate have irregular, corrugated profiles that reflect slight differences in the strength of the rocks, where the weaker layers have eroded back faster than the more resistant, ledge-forming layers. Another characteristic of the cliffs is large, dark, vertical streaks. These streaks formed during repeated episodes of water flowing over the rim of the canyon and down the cliff face, leaving behind a very thin layer of manganese oxide.

Clasts in the Gila Conglomerate.

Vesicles in the Bearwallow lava. Some vesicles are filled with minerals and are known as amygdules.

Geologic History

Volcanism in the region surrounding the cliff dwelling began as early as about 35 million years ago. At approximately 30 million years ago, the Gila Cliff Dwellings caldera formed, most likely during the eruption of the Shelley Peak Tuff. About two million years later, the Bursum caldera to the west collapsed during the eruption of the enormous Bloodgood Canyon Tuff. As the Bloodgood Canyon Tuff flowed out of the Bursum caldera, it nearly completely filled the neighboring Gila Cliff Dwellings caldera much like water ponding in a boot print. Following the explosive caldera eruptions, the style of volcanism dramatically changed. Large stratovolcanoes and dome complexes erupted vast quantities of andesite and basaltic andesite lava, now known as the Bearwallow Mountain Andesite.

Following this volcanic activity, much of southwestern New Mexico, including the Cliff Dwellings area, was broken into basins and ranges by extensional faulting. Movement along large faults caused the Gila Hot Springs graben to subside within the southeastern part of what was

AGE (my)	ROCK UNIT
15–5	Gila Conglomerate
25	Bearwallow Mountain Andesite
28	Bloodgood Canyon Tuff

Exfoliation slabs (red arrow) along the back wall of an alcove.

formerly the Gila Cliff Dwellings caldera. Alluvial fans formed at the mouths of canyons draining the uplifted mountains, gradually filling the basins with sediments eroded from the now-extinct volcanoes. Continued burial, compaction, and cementation by flowing groundwater slowly turned the loose sediments into the solid rock of the Gila Conglomerate.

After basin filling concluded, erosion and river incision cut valleys and canyons in the headwaters of the Gila River. As Cliff Dweller Creek cut into the underlying bedrock, it also migrated laterally, preferentially eroding the soft layers of sediments in the canyon walls. When the weaker sedimentary rocks were worn away, the overlying layers were left unsupported and eventually collapsed to form arched alcoves. The cliff-dweller caves are all located at about the same level above the creek, indicating this level must have corresponded to a particularly soft layer within the Gila Conglomerate.

Continued downcutting of the canyon at no more than a fraction of an inch per year has now exposed the volcanic rocks beneath the Gila Conglomerate. Most recently, the combination of large wildfires, which destroy vegetation that stabilizes slopes, and massive monsoon rainfall events have created the perfect scenario for flooding and increased erosion. Log-jams of trees scattered throughout the monument and nearby river valleys testify to the ongoing geologic processes that continue to modify this diverse landscape.

Geologic Features

MONUMENT PARKING LOT—The hill to the north of the parking lot exposes the Bloodgood Canyon Tuff and Bearwallow Mountain Andesite. Only the top of the tuff is exposed at the cliff dwellings.

However, geothermal drilling at the nearby Gila Hot Springs indicates that unit may be as much as 600 feet thick in the area. Capping the hill are andesite lava flows, and each bench likely represents a different flow. At least three flows are present here.

DWELLER CAVES—The best examples of the Gila Conglomerate are exposed at the cave entrances. Some layers of the conglomerate are almost completely devoid of large rock fragments, whereas other layers appear to be dominated by coarse material. Within the caves, soot from ancient fires prevents identification of the rock types. Interestingly, all of the building material for the dwelling walls is composed of the Gila Conglomerate. Apparently, early inhabitants did not care much for lava and tuff blocks as building materials.

DESCENT TRAIL—The trail just below the caves takes you past the irregular contact between the lava and conglomerate. As sediment was deposited, it filled vertical fractures and fissures in the surface of the lava flow. Just below the top of the lava-flow sequence is a thin layer of sandstone and conglomerate interbedded between andesite flows. The top of the sedimentary layer was baked brick red from the heat of the overlying lava flow during its emplacement. This sandstone represents a brief pulse of deposition between eruptions of the lava. Upon exiting the canyon, the trail switchbacks through a slope of the West Fork burned by the Miller Fire in 2011. Grasses and small trees are slowly re-stabilizing the slope, which is still susceptible to rapid runoff and erosion.

THE MAIN VISITOR CENTER—The Visitor Center is located a short drive downstream of the cliff dwellings near the confluence of the West and Middle Forks of the Gila River. Visitors can sample the trails of the wilderness outside of the monument by hiking a half-mile up the Middle Fork from the Visitor Center to Light Feather hot springs. Numerous river crossings will eventually lead you to pools constructed on the east side of the river, which mix the 140°F thermal waters with the colder river waters to create a more pleasurable temperature.

—*Matthew J. Zimmerer*

Irregular contact (upper arrow) between the Bearwallow Mountain lava and the Gila Conglomerate. The lower arrow points to the contact with sediment that was baked brick red by the overlying lava.

Cliffs of the Gila Conglomerate with cliff dwellings built into a less consolidated horizon.

Additional Reading

Geology of the Gila Wilderness–Silver City Area, edited by Greg Mack, James Witcher, and Virgil W. Lueth, New Mexico Geological Society, 59th Annual Field Conference, Guidebook, 2008.

If You Plan to Visit

A joint National Park Service (NPS)/U.S. Forest Service Visitor Center and a NPS Trailhead Museum offer information, exhibits, and knowledgeable staff for both the park and the surrounding wilderness area. Only very limited food and primitive camping are available in or near the park. In addition, all visitors should remember that no collection of geological, archaeological, or any other materials is allowed within National Parks.

Regular passenger vehicles: Drive 43 miles north from Silver City New Mexico on NM 15. The trip can take up to two hours due to the narrow and curving nature of the road. Large RVs and vehicles pulling trailers should *not* use this route. Instead, take US 180 from Silver City to Santa Clara. Turn left onto NM 152 and drive to San Lorenzo. At San Lorenzo, turn left onto NM 35 and follow it to its termination at NM 15. Turn right on NM 15 and follow the signs to the park. For more information:

National Park Service
HC 68 Box 100
Silver City, NM 88061
(575) 536-9461
www.nps.gov/gicl

The Gila Wilderness
GILA NATIONAL FOREST

There are some who can live without wild things and some who cannot.

—Aldo Leopold

For those who cannot live without wild things, the Gila (pronounced Hee-luh) Wilderness is a paradise. In 1924 the Gila became the nation's (and the world's) first designated federal wilderness area, marking the birth of conservation through wilderness protection. The Gila Wilderness (commonly referred to simply as "The Gila") is vast. The Gila covers over 550,000 acres, stretching for almost 30 miles north to south and 40 miles east to west. The majority of the wilderness lies within the rugged Mogollon Mountains, where some of the highest peaks of southern New Mexico (Whitewater Baldy is 10,395 feet, Mogollon Baldy is 10,778 feet, Sacaton Mountain is 10,658 feet, and Black Mountain is 10,643 feet) tower over deep, isolated, and rugged canyons. These mountains catch the snows and rains that maintain water flow in the jewel of the wilderness—the Gila River. There are three forks of the Gila River within the wilderness—the east, west, and middle.

Bright-colored rocks near the center of this photo are altered rhyolite lavas of Alum Mountain. Everything from the foreground to the distant horizon is within the Gila Wilderness.

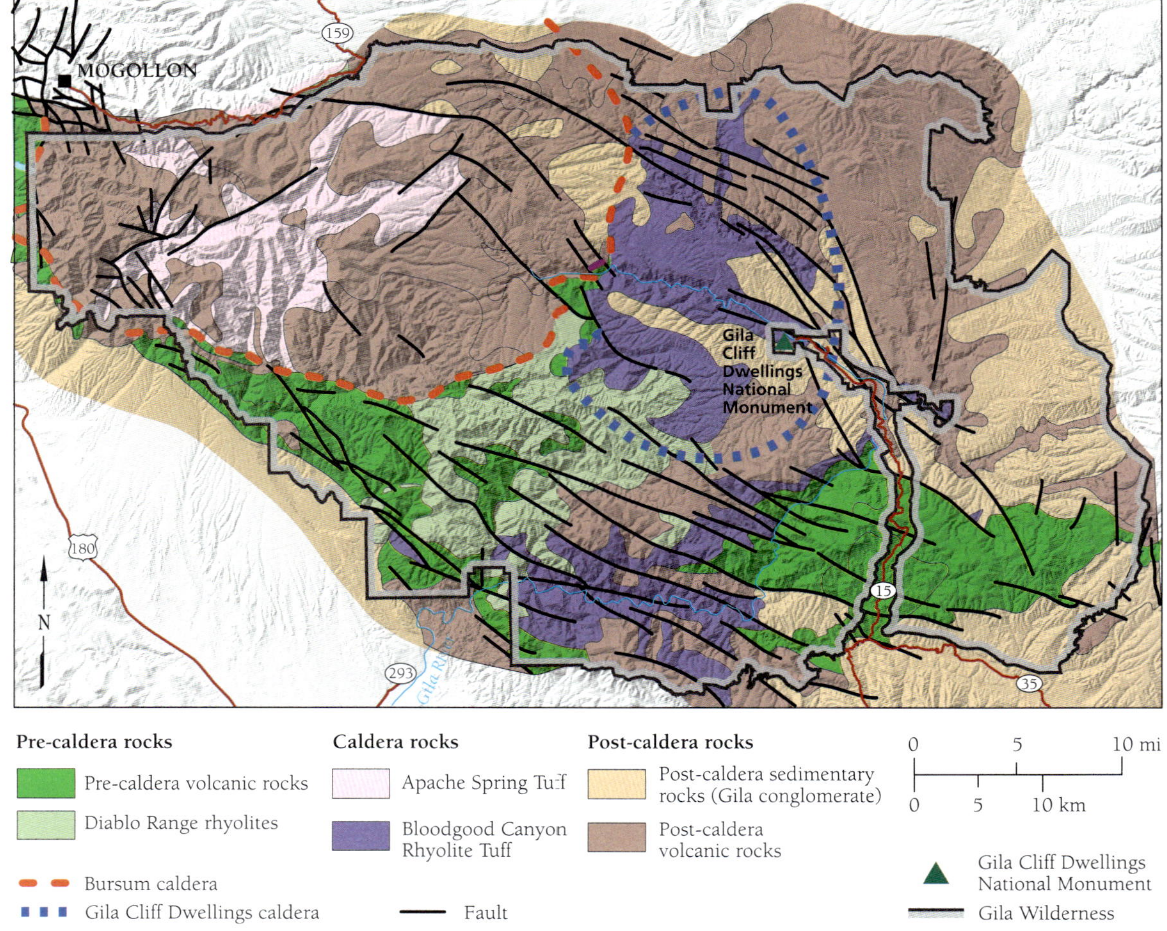

Pre-caldera rocks

- Pre-caldera volcanic rocks
- Diablo Range rhyolites
- ▬ ▬ Bursum caldera
- ▪ ▪ ▪ Gila Cliff Dwellings caldera

Caldera rocks

- Apache Spring Tuff
- Bloodgood Canyon Rhyolite Tuff
- ▬ Fault

Post-caldera rocks

- Post-caldera sedimentary rocks (Gila conglomerate)
- Post-caldera volcanic rocks
- ▲ Gila Cliff Dwellings National Monument
- Gila Wilderness

0   5   10 mi
0   5   10 km

Geologic map of Gila Wilderness and surrounding area.

Exploring this terrain is not for the faint of heart. The only road access to the interior of the wilderness is NM 15 leading to the Gila Cliff Dwellings. Otherwise, traveling through this landscape requires long hikes, horseback rides, or raft trips down the Gila River. The vast, relatively inaccessible landscape provides a sanctuary for animals including black bear, cougar, elk, mule deer, bighorn sheep, peccary (javelina), as well as the recently reintroduced Mexican gray wolf. Also found in the Gila Wilderness is the geological record of some of the largest volcanic eruptions ever to shake our planet.

Regional Setting

The Gila Wilderness is located in southwestern New Mexico, north of Silver City, in Catron and Grant County. The wilderness is bounded to the north by the Tularosa Mountains, to the east by the Black Range, to the south by the Pinos Altos Range, and to the west by the Mangas Basin. Within the wilderness is a nest of calderas and related volcanoes that peaked in activity between 30 and 25 million years ago. These volcanoes are only a few of the countless eruptive centers within the Mogollon–Datil volcanic field, which cover much of southwestern New Mexico. Abandoned mines and prospect pits throughout the region are part of the legacy of exploration of resources related to this prolonged episode of magmatism. The great topographic relief of the region is due to both the accumulation of enormous piles of volcanic rocks and to faulting within the Basin and Range province, which uplifted large blocks such as the Mogollon Mountains.

The Rock Record

Some of the best examples of rocks related to super-volcano eruptions are found in the Gila Wilderness. These explosive eruptions produced ground-hugging clouds of superheated material that moved across the landscape as pyroclastic flows, eventually stopping and cooling to form rocks known as ash-flow tuff or ignimbrite. If the tuff retains enough heat, the ash, pumice, and crystals fuse together, forming a welded tuff. Many of the cliffs within the wilderness, such as iconic spires of the Middle Fork, are composed of welded tuff. In contrast, smaller eruptions produce ash that falls from the sky, blanketing the landscape much like snow. Some of the seemingly endless whitish layers exposed on canyon walls are ash-fall deposits.

Another prominent volcanic rock within the wilderness is lava, which forms when magma erupts with very little explosivity. Lava flows in the Gila span a range in composition and viscosity. The more viscous lavas (rhyolite and dacite) typically formed thick flows, whereas less viscous lavas (andesite and basalt) typically formed thin, dark-colored flows.

Small dikes and intrusions are also scattered throughout the Gila. These features represent magma that was intruded into the crust and subsequently crystallized. These rocks hint at the enormous intrusions that must underlie the volcanic field.

The Gila Conglomerate records the relentless erosion of the volcanic field after volcanism ended about 25 million years ago. Great exposures of the Gila Conglomerate can be seen at the Gila Cliff Dwellings National Monument and within the adjacent Mangas and Mimbres basins. The Gila Conglomerate is further discussed in the previous chapter.

Hand samples of welded tuff. The light brown lenses were once pumice fragments that were later flattened during welding of the ignimbrite.

Spires of Bloodgood Canyon Tuff exposed along the Middle Fork of the Gila River. The tuff is over 1,000 feet thick with no exposed base.

Geologic History

Over a century of research in the Gila Wilderness has pieced together much of its complex history. Some of the earliest investigations were in the 1870's when Sergeant James C. Cooney (later killed by Apaches) first discovered gold and silver near Mogollon. A significant pulse of research was conducted between the 1970s and 1990s, when regional mapping was combined with advances in analytical techniques, most importantly high-precision geochronology, to tease out the details of the eruptive history.

Older ignimbrites ranging in age from about 34 to 30 million years are locally exposed in the Gila. Their source calderas have not yet been identified, likely because they are buried beneath the deposits of younger eruptions. The Bursum and Gila Cliff Dwellings calderas are the youngest and most well-studied supervolcanoes in the wilderness, and are described further in the Gila Cliff Dwellings National Monument chapter. Lava flows were erupted throughout the history of the field, and represent volcanic activity in the time intervals between caldera eruptions. Intense hydrothermal activity, common along fault zones and near intrusions, caused alteration and filled fractures in rocks, creating veins of minerals.

Almost as soon as the volcanism ended, faulting related to crustal stretching within the Basin and Range began to tear the region apart. Mountain blocks rose, and fault-bounded basins filled with thousands of feet of Gila Conglomerate, which is an accumulation of debris eroded from earlier volcanic deposits.

Geologic Features

BURSUM CALDERA—The Aldo Leopold Vista, along US 180 (south of Glenwood and Pleasanton), provides scenic views of the western side of the Gila Wilderness and the southwestern margin of the Bursum caldera. Horizontal beds of Gila Conglomerate dominate the rocks exposed in the valley. The canyons and ridges along the mountain front expose rhyolite lavas erupted after the Bursum caldera collapsed. The rounded peaks on the horizon define the resurgent dome within the caldera. Resurgent domes formed when hot, buoyant magma uplifts the caldera floor following collapse, similar to Redondo Peak at the Valles caldera near Los Alamos. The Bloodgood Canyon Tuff was erupted from the Bursum caldera about 28 million years ago; it is further described in the Gila Cliff Dwellings National Monument chapter. Educational signs at the vista provide additional information about the geography and geology of the area.

Contact between the Bearwallow Mountain Andesite and the overlying Gila Conglomerate. Much of the volcanic bedrock is covered by the Gila Conglomerate that fills the various basins within and surrounding the Gila Wilderness.

PRE-CALDERA ROCKS—Roadcuts along NM 15 from Sapillo Junction to the Cliff Dwellings expose a wide variety of pre-caldera rocks including lava flows, breccias, volcanic bombs, and the sediments derived from them. Most of these rocks are about 30 million years old and pre-date both the Gila Cliff Dwellings and Bursum calderas. The view from Anderson Overlook into the Gila drainage and wilderness includes unusual orange, red, and yellow rhyolite lavas. The coloration of these lavas was caused by intense alteration by hydrothermal fluids associated with Alum Mountain, a volcano that may have looked similar to present-day Mount St. Helens. Beyond Alum Mountain is the remote interior of the wilderness. Most of the high country adjacent to the

Gila River is composed of the Bearwallow Mountain Andesite, about 25 million years old. It is the youngest major volcanic unit in the area and caps thin tuffs erupted from various Gila calderas.

HOT SPRINGS—Numerous thermal springs and pools discharge within the boundaries of the wilderness. Two of the most popular are Jordan and Turkey Creek hot springs. Although it is easy to envision that these hot springs are related to volcanic activity, they are not. Chemical analyses of the thermal waters do not indicate any degassing from crystallizing magma. Instead, large faults allow water to circulate to great depths in the crust where it is heated before returning to the surface.

—*Matthew J. Zimmerer*

Additional Reading

The Gila Wilderness: A Hiking Guide, by John A. Murray, University of New Mexico Press, 1988.

If You Plan to Visit

Access to the southern Gila Wilderness is via NM 15 to the Cliff Dwellings National Monument. Most people enter the wilderness on trails along the west, middle, and east forks of the Gila River. There are numerous Forest Service roads along the west slope of the Mogollon Mountains. Access from the north is along FS 159 through the town of Mogollon. Visitors should bring plenty of water and watch out for rattlesnakes and bears.

Catwalk National Recreation Trail
GILA NATIONAL FOREST

The modern steel Catwalk Trail closely follows the course of the historic Catwalk pipeline through a narrow slot canyon cut into the Cooney Tuff.

The iconic Catwalk National Recreation Trail takes visitors along a dramatic trail up an historic and geologically dynamic canyon cut into welded Cooney Tuff of the western Mogollon–Datil volcanic field. Along with the Gila Cliff Dwellings National Monument, the Catwalk is one of the easiest ways for visitors to experience the rugged and remote Gila Wilderness and Mogollon Mountains area. The Catwalk Trailhead is located approximately 5 miles northeast of the village of Glenwood in Catron County, New Mexico.

The Catwalk Trail parallels a now-defunct pipeline constructed in 1893, which powered electric generators at Graham Mill, in what was then the town of Graham. Footings for Graham Mill can still be seen just

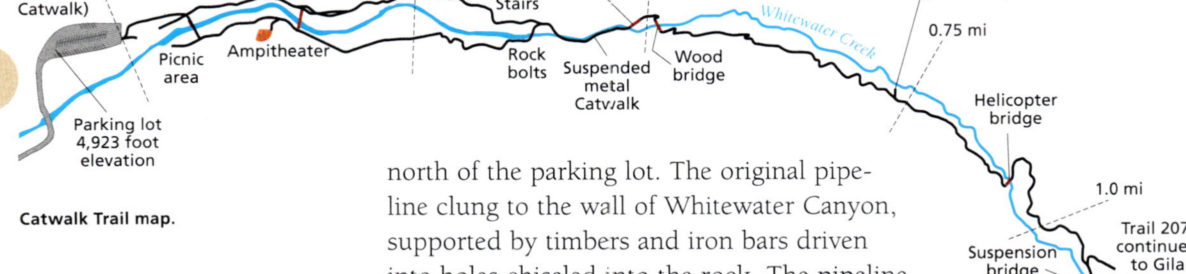

Catwalk Trail map.

north of the parking lot. The original pipeline clung to the wall of Whitewater Canyon, supported by timbers and iron bars driven into holes chiseled into the rock. The pipeline required frequent maintenance, and workers walking the pipeline named the route "The Catwalk." The pipeline operated intermittently until 1913. In the 1930s, a mile-long Scenic Trail was constructed by the Civilian Conservation Corps (CCC), featuring wood and steel walkways along the route of the original pipeline. The trail has been rebuilt and upgraded many times following destructive flash floods. A half-mile-long, wheelchair-accessible trail was added to the east side of the canyon in 2003, creating a loop with the original trail on the west side of the creek. The two trails join at a broad platform spanning the entire canyon, upstream of which the western trail continues another half mile to link to the extensive trail system of the Gila Wilderness.

Regional Setting

The Catwalk National Recreation Trail is located along the northwestern edge of a caldera cluster in the Mogollon–Datil volcanic field. The Mogollon–Datil field was active between about 40 and 25 million years ago, covering much of southwestern New Mexico with lavas, tuffs, and other deposits from a wide variety of small and large volcanoes. Later crustal extension and faulting produced the alternating basins and mountain ranges that characterize southwestern New Mexico today.

Eroded towers of Cooney Tuff south of Whitewater Creek, separated by columnar joints formed during cooling of the thick sheet of hot tuff.

The Rock Record

The main rock type exposed along The Catwalk National Recreation Trail is the Cooney Tuff, produced during a catastrophic supervolcano eruption about 34 million years ago. The Cooney Tuff is a light-purple welded tuff, in places interlayered with lavas and other deposits. Along much of the Catwalk's steel walkway, the Cooney Tuff has abundant cavities where rapid cooling trapped pockets of gas. Much of the Cooney Tuff exposed along the trail displays a horizontally layered texture defined by pumice clasts flattened into wispy shapes during welding and compaction of the hot material. Cliffs above the east side of the lower canyon expose dark layers of lava interlayered with lighter-colored welded tuff. Along the uppermost part of the trail, are thin dark interlayers of ash-fall deposits. Where the trail crosses faults, for example at the mouth of the canyon and along the lower half of the eastern trail, the rock is highly fractured and susceptible to landsliding, requiring rock bolts and metal nets to stabilize the slopes.

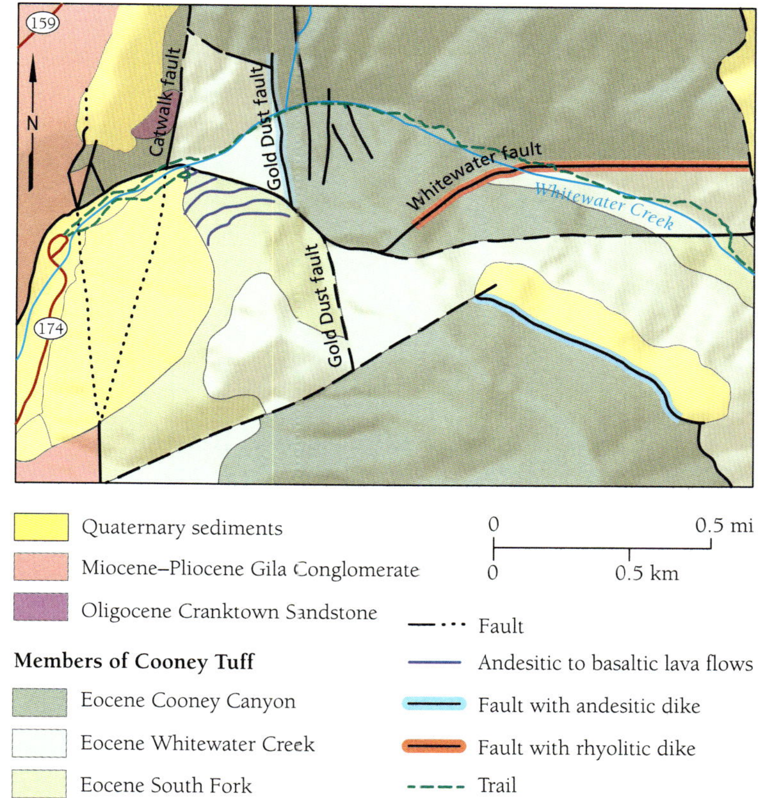

Geologic map of the Catwalk National Recreation Trail.

Geologic History

The Mogollon–Datil volcanic field in southwestern New Mexico contains many calderas (super-volcanoes) that erupted between about 36 and 25 million years ago, part of a volcanic flare-up that occurred throughout much of southwestern New Mexico. The Cooney Tuff was explosively erupted from one of these calderas in the western Mogollon Mountains about 34.4 million years ago. Flowing across the landscape as an incandescent mix of pumice, gas, and ash, the tuff retained enough heat to weld into solid rock when it came to rest. Numerous mines in the region have exploited mineral deposits that formed along and near caldera margins.

Whitewater Canyon documents the dynamic nature of canyon erosion along a faulted mountain front. In the relatively soft basin-fill sediments to the west, Whitewater Creek occupies a broad valley. In contrast, east of the mountain front, Whitewater Creek cut a narrow, steep canyon in the much harder Cooney Tuff. Storms and spring runoff can cause a rapid and dangerous rise of stream level. The power of floods in the canyon was on dramatic display following the 2012

Whitewater Baldy Complex Fire, which burned runoff-reducing vegetation in the headwaters of Whitewater Creek. During a subsequent rainfall event, the Catwalk bridge and trail system was severely damaged, requiring massive re-engineering of the bridge and trail.

Geologic Features

Outcrops adjacent to the trail expose three features characteristic of welded tuffs. Angular lithic fragments represent pieces of solid rock incorporated into the tuff during eruption. Gas cavities form when bubbles are trapped during welding of the tuff. Flattened pumice forms by flowage during welding. Fragments of flattened pumice along the western trail at the mouth of the canyon have been rotated to near vertical by movement along the mountain-front fault.

— *William C. McIntosh, James C. Ratté, and Matthew J. Zimmerer*

Additional Reading

Geology of the Gila Wilderness–Silver City Area edited by Greg Mack, James Witcher, and Virgil W. Lueth. New Mexico Geological Society Guidebook 59, New Mexico Geological Society, 2008.

If You Plan to Visit

The Trailhead area at the mouth of Whitewater Creek includes picnic facilities, a native plant garden, and graceful Arizona sycamores. Elevations along the trail range from 5,400 to 5,600 feet. During the summer, many visitors take advantage of swimming holes along Whitewater Creek. Only the recently reconstructed first ½ mile of the trail is open to the public as of the writing of this book.

To visit the Catwalk, take US 180 north from Silver City to Glenwood (approximately 60 miles) or NM 12 and then US 180 south from Reserve (about 37 miles). Turn west onto NM 174 from US 180 at the north edge of Glenwood and continue 5 miles to the Whitewater Picnic area. The route includes two river crossings—you should evaluate the water depth before proceeding. Before traveling to the Catwalk, check the trail closure status with the U.S. Forest Service **www.fs.usda.gov/alerts/gila/alerts-notices/?aid=25084.**

Gila Riparian Preserve
THE NATURE CONSERVANCY

The Gila Riparian Preserve, which encompasses more than 1,200 acres, lies along the Gila River, the last of the Southwest's major free-flowing rivers. At 649 miles, the Gila River is also one of the longest rivers in the West. The East Fork, Middle Fork, and West Fork of the river originate in the shadows of the Black Range along the Continental Divide and in the rugged Mogollon Mountains. These headwaters are part of the first designated American wilderness area, the Gila Wilderness. Periodic input from flows in Mogollon Creek, which drains the westernmost Mogollon Mountains and joins the Gila River in the preserve, helps to sustain consistent summer flow in the Gila. In Arizona, large irrigation diversions en route to its confluence with the Colorado River near Yuma have reduced most of the Gila River to a dry, sandy channel.

River Characteristics and Ecology

Extreme flow variability is a characteristic of the upper Gila River. Since 1928, when the stream gage at Gila River became operational, the

Canyon cut by the Gila River.

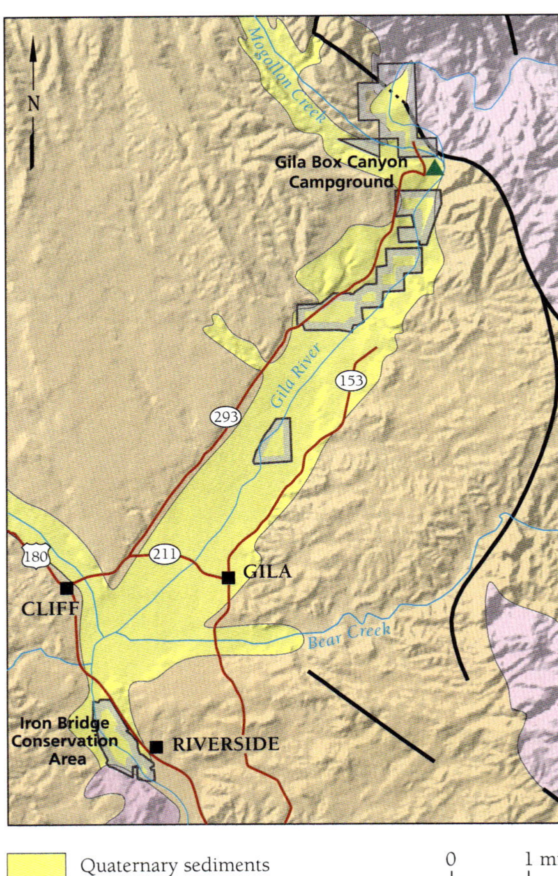

Quaternary sediments

Tertiary–Quaternary
Gila Conglomerate

Tertiary volcanic rocks

Gila Riparian Preserve

Fault

**New Mexico Chiricahua
leopard frog.**

highest recorded streamflow was an astonishing 35,200 cubic feet per second (cfs) in December of 1984. More recently, in September of 2013, 28,800 cfs thundered down the river. Such massive floods cause major reconfigurations of the river channel and its floodplain, but also support nutrient cycling, create new wetlands, and enhance vegetative biodiversity. In contrast, summer flows of less than 20 cfs are not unusual. Native plants and animals have evolved strategies and life cycles in response to this highly variable flow regime, with seasonal discharges playing distinct ecological roles that support aquatic and riparian ecosystems.

The Gila River supports an extraordinary assemblage of plant and animal life, including 40 at-risk species and over 300 types of birds. The river provides one of the most intact native fish and amphibian communities in the Lower Colorado River Basin, including populations of threatened or endangered loach minnow (a tiny member of the carp family), spikedace (a ray-finned fish), Gila trout, and the Chiricahua frog. The Gila has been referred to as "the nation's most variegated stream fishery," because, in places, wild trout, smallmouth bass, and catfish can occupy the same pool. The riverside cottonwoods and willows provide high-value habitat that supports one of the greatest concentrations of breeding birds in North America, including endangered Southwestern willow flycatchers and the threatened yellow-billed cuckoo. This Nature Conservancy preserve was established specifically to retain some undeveloped areas of this natural habitat.

Geologic Setting and Features

The Gila Riparian Preserve lies on the northern edge of the Basin and Range province and in the southern part of the Mogollon–Datil volcanic field. Views north of the preserve, toward the Mogollon Mountains, show the massive complex of volcanic rocks and associated sedimentary deposits of the Mogollon–Datil volcanic field. Most of the dark- and light-gray colored hills close to river level expose andesite lavas, tuffs, and breccias erupted from vents that predate the major calderas located within the wilderness. These moderately altered andesites, along with the quartz-fluorite-calcite veins that crosscut them, are collectively known as the Gila Fluorspar District. Mining in the district peaked in the 1940s, when fluorspar ore was needed for steel and aluminum production during World War II. Now the area is popular for hiking and mineral collecting. The laterally continuous cliffs in the higher country to the north are dominantly ash-flow tuffs erupted between 34 and 28 million years ago from calderas (supervolcanoes) in the Gila Wilderness. In contrast, the light tan rocks in the hills directly east of the preserve are not volcanic, but instead belong to a sedimentary unit called the Gila Conglomerate that was deposited in basins within and around the volcanic field. Similar rocks are exposed at the Gila Cliff Dwellings National Monument.

The landscape surrounding the Gila River changes dramatically near the Gila Riparian Preserve. Upstream from the preserve, the river flows over relatively resistant volcanic rocks and has incised into the bedrock a remarkably deep canyon with a small adjacent floodplain. As it exits the mountains, the river flows over relatively soft sediments that fill the basin downstream from the preserve. The meandering river has swept back and forth over these weaker rocks to create a broad valley with an extensive floodplain that is used for agriculture. Not only does the landscape change with distance along the river, so does the size of the material it

Photo showing grain size of sediments and the Mogollon Creek confluence of the Gila River.

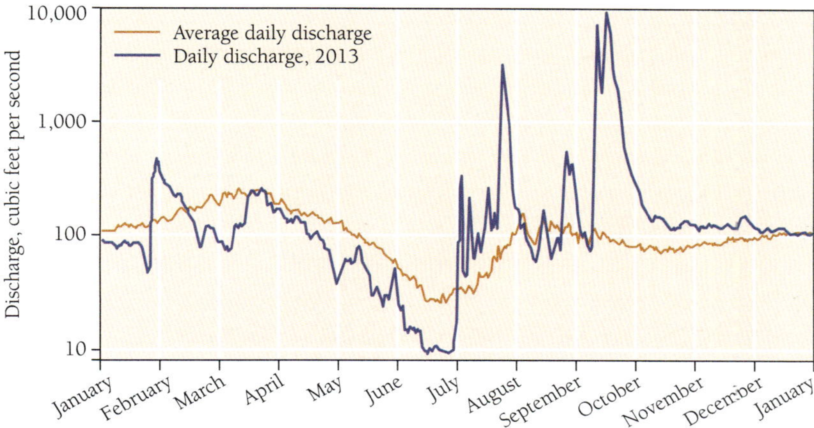

Hydrograph of the Gila River near Redrock, NM, showing the major flood events in 2013 in comparison to the mean discharge averaged over 54 years.

Wide river valley where the river emerges from the mountains near the town of Gila.

carries. The coarse pebbles and boulders within and adjacent to the Gila River at the preserve are typical for deposits of a steep stream near the mountains. Just a few miles downstream, where NM 211 crosses the river, lower stream gradients and less confined flow results in the deposition of much smaller pebbles and boulders with greater amounts of sand and silt.

—Paul W. Bauer and Matthew J. Zimmerer

If You Plan to Visit

There are three separate sites of Conservancy lands in this preserve. The main (northernmost) section includes lands directly downstream from the beautiful Gila Box Canyon and offers wonderful hiking and scenery along the river. Another section, termed the Gila River Farm, houses the Lichty Ecological Research Center, which supports research and environmental education activities. The third section, the Iron Bridge tract, provides walking access to an historic, 381-foot-long, Warren truss bridge across the Gila River (constructed in 1915). Trails also allow access to the river and riparian habitats along its banks and floodplain. Although not primarily managed as parkland, the Nature Conservancy allows year-round access to these three areas. Please observe the same rules and courtesy to the land that you would in a state or national park.

From Silver City, drive west and north on US 180 about 28 miles to Cliff. Then take a right on NM 211 and drive for about 1.5 miles. Follow the left fork in the road, which is NM 293 (also signed as Box Canyon Road) and continue up the valley about 7 miles until the road ends at Gila National Forest Box Canyon Campground. The preserve

begins on the north side of the green fence. Park here and walk north, across Mogollon Creek, to access the Gila Riparian Preserve. An old road parallels the river, or you can walk through the riparian forest.

The directions to Lichty Research Center are the same as above except that one only goes 4.1 miles up NM 293. After mile marker 4, take a right into the driveway, with a Gila River Farm sign and 426 on the fence post. Follow the driveway to the parking area. A trail map of the farm is posted on a kiosk in the parking area.

The Iron Bridge tract is located approximately 22 miles northwest of Silver City, and just south of Cliff, along US 180 between mile markers 86 and 87 (just downstream of where US 180 crosses the Gila River). Parking is located on both the east side of the river, next to a green ranch gate that can be seen from US 180, and on the west side of the river, off of Iron Bridge road. From either side, you can walk along the old highway roadbed that leads to the Iron Bridge. For more information:

Nature Conservancy of New Mexico
212 E. Marcy Street
Santa Fe, NM 87501
(505) 988-3867
www.nature.org/newmexico

The Iron Bridge crossing the Gila River.

Mimbres River Preserve
THE NATURE CONSERVANCY

The Mimbres River Preserve is located along a tranquil five-mile-long stretch of the Mimbres River in the northern Mimbres Basin. The Nature Conservancy established the 600-acre holding in 1994 with a mission to help re-establish the river's natural flow regime, restore fire to the watershed's uplands, and encourage the recovery of riparian forests and aquatic habitat lost to river channelization. Those goals are facilitated by the fact that much of the upland watershed is administered by the U.S. Forest Service, and the Conservancy lands extend those protections into areas that are mostly privately owned. The preserve supports riparian ecology and creates river habitat for the endangered Chihuahua chub and Chiricahua leopard frog, the imperiled Rio Grande sucker, the Desert Viceroy butterfly, and numerous other animals and plants.

OPPOSITE: Bear Canyon Falls just west of the preserve.

The River System

The Mimbres River flows southwestward from the 10,000-feet-high spine of the Continental Divide in the Black Range, through the mountains of the Gila National Forest, and into the Mimbres Basin (a 5,140-square-mile area). The lowest point in the basin (elevation 3,770 feet) is south of Deming, near the border with the Mexican state of Chihuahua.

The Mimbres is a lovely desert river with an intermittent upper reach, a perennial middle reach, and an ephemeral lower reach that normally vanishes into the desert sands near City of Rocks State Park. During very good water years, the river bed is wet even beyond Deming, where it sometimes fills a playa lake (Florida Lake) near the Florida Mountains. In an exceptional year, such as in 1905–1906, the Mimbres can flood the Chihuahuan Desert grasslands nearly to the international border.

Most precipitation in the Mimbres Basin falls either during the summer monsoon season as convective thundershowers or during large winter storms. Indeed, two of the four largest flows at the Mimbres gage during the interval 1978–2015 resulted from winter storms, with the highest flow of 6,360 cubic feet per second (cfs) recorded on

The Desert Viceroy butterfly is a mimic of the Monarch butterfly.

The Chihuahua chub can average 5–6 inches in length at maturity and may reach 12 inches.

December 28, 1984. In a typical water year, the Mimbres gage records a mean discharge of about 18 cfs, although it is not uncommon for the gage to read zero.

Although the modern river flow never leaves the Mimbres River basin, that has not always been true. In the past, the Mimbres has at times been connected to other river systems. As a result, the Mimbres River has evolved unique fauna and floral associations, including a handful of species, such as the Chihuahua chub, that are found nowhere else in the U.S. At least 37 species endemic to the Mimbres River are of great conservation concern, with 18 of those classified as "vulnerable, imperiled, or critically imperiled" statewide as well as nationally. Threats to the Mimbres habitat include surface and groundwater diversions, river channelization, parasites and pathogens, and non-native fish species.

The Mimbres River also has a long history of supporting human life in this arid region and was home to the Mimbres Mogollon culture particularly during the period from 200 to 1450 CE. That story of that culture is well represented at the Gila Cliff Dwellings National Monument.

Geologic Setting and Features

The Mimbres River upper stretches are in the Mogollon–Datil volcanic field, and the downstream stretches flow through the Basin and Range province. Nestled between the Black Range to the east and the Pinos Altos Range, the Mimbres Basin is a tilted, down-dropped block of the crust bounded by a fault along its western margin. Vast amounts of sediment shed from the adjacent highlands slowly filled the

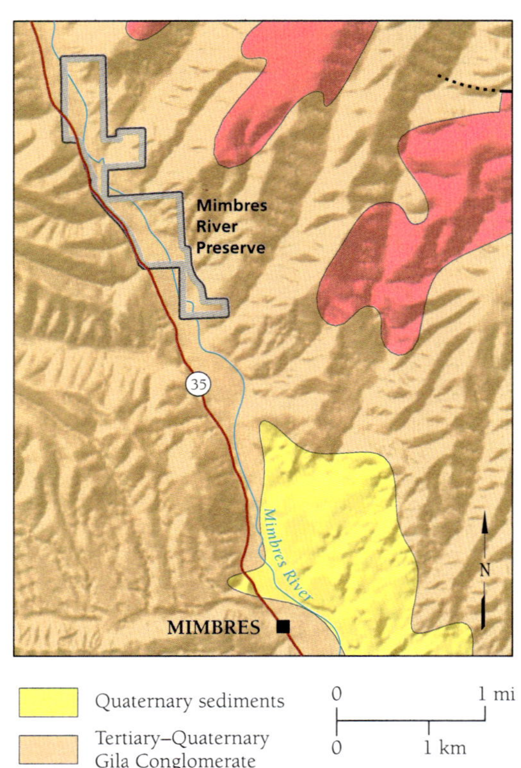

Outcrop of the Gila Conglomerate near the entrance to the preserve. Most of the cobbles and boulders are derived from volcanic rocks.

developing depression as it subsided. These sedimentary deposits, known as the Gila Conglomerate, are the buff-colored beds exposed at the preserve entrance and along most of the lower slopes surrounding the preserve. The adjacent highlands contain volcanic materials from supervolcanoes of the Mogollon–Datil volcanic field, although erosion and tectonic activity prevents easy identification of these calderas from topography alone. Basalt lavas that flowed from small local volcanic vents are found both below and above the Gila Conglomerate. Dating of these flows indicates that the Gila Conglomerate in this particular basin was deposited between 21 and 6.5 million years ago. After the basalt lava that caps the mesas was erupted, the Mimbres River began to cut downward through this lava and the underlying Gila Conglomerate until the present-day Mimbres Valley was formed.

—*Paul W. Bauer and Matthew J. Zimmerer*

Layered sediments of the Gila Conglomerate along NM 35 just north of the Nature Conservancy Mimbres Preserve.

If You Plan to Visit

Although not primarily managed as parkland, the Nature Conservancy allows year-round access to this riparian habitat area. Please observe the same rules and courtesy to the land that you would in a state or national park.

From Silver City take US 180 east for about 10 miles and turn left (east) onto NM 152. Take NM 152 east to NM 35/61 north near the town of San Lorenzo. Follow NM 35/61 north approximately 5 miles to

Mimbres River before and shortly after a monsoon storm.

the village of Mimbres Valley. Continue north about 3 miles. After mile marker 8, turn right into an unmarked driveway. There is a parking area on the left in front of an old barn. For more information:

Nature Conservancy of New Mexico
212 E. Marcy Street
Santa Fe, NM 87501
(505) 988-3867
www.nature.org/newmexico

Chino Mine Overlook
FREEPORT–MCMORAN INC.

The Chino Mine is part of the Santa Rita Mining District (one of the world's largest metal-bearing provinces) and is located in Grant County just east of Silver City. From this overlook, the spectacular open pits and waste-rock piles of the Chino copper mine are visible—the active pit is currently about 1.75 miles across and more than 1,350 feet deep. This overlook offers a view of the largest copper porphyry deposit in New Mexico. The mine is operated by Chino Mines Company, a subsidiary of Freeport–McMoRan. The Chino Mine's name was derived from the Spanish word for pyrite or chalcopyrite. The mine is also commonly referred to as the Santa Rita or Santa Rita del Cobre Mine, named for the former village of Santa Rita, which was removed in the 1950s as mining operations in the area expanded.

The Chino Mine is one of the oldest commercially mined mineral deposits in North America. Native Americans extracted copper for implements, jewelry, and weapons well before its "rediscovery" around 1799 when a local Apache revealed the locality to Colonel Jose Manuel Carrasco, of the Spanish military. Carrasco and Don Francisco Elguea obtained a grant from the Spanish government for the "Santa Rita del Cobre." In 1873, M.D. Hayes purchased and patented

View of Chino Mine looking southeast.

the claims. Primitive mining methods prevailed at the Chino Mine until the beginning of open pit mining in 1910. The first "modern" mill was constructed south of the mine at Hurley in 1911, rebuilt in 1982, and is still operating. Phelps Dodge Corporation acquired the mine, mill, and smelter in 1986. In November 2006, Freeport–McMoRan Incorporated acquired Phelps Dodge's assets.

Regional Setting

The Chino Mine lies at the southern edge of the Mogollon–Datil volcanic field, which is dominated by Tertiary volcanic rocks. However, the mine is also in the transition zone to the extensional terrane of the Basin and Range province. In fact, the mine is located on a northwest-trending fault block known locally as the Santa Rita Range or the Cobre (Copper) Mountains.

The Rock Record and Geologic History

The Cobre Mountains are composed of Proterozoic metamorphic and igneous rocks overlain by up to 4,800 feet of Paleozoic and Mesozoic sedimentary rocks. Cretaceous diorite sills subsequently intruded these older rocks. Shortly thereafter, dikes and other intrusive bodies were emplaced and roughly 2,000 feet of andesitic lavas and breccias were erupted onto the surface. This event was followed by the early Tertiary (60 to 58 million years) intrusion of large granodiorite plutons at Chino and at Hanover–Fierro to the northwest. Most of the mineralization formed during and shortly after intrusion, along the pluton margins. Later alteration further concentrated the ore. Polymetallic veins and skarns (stratabound, irregular to pod-shaped replacement ore deposits found in limestones) formed in proximity to the plutons.

The 35 million-year-old Kneeling Nun Tuff, the same unit that is found at City of Rocks State Park, postdates the intrusions and

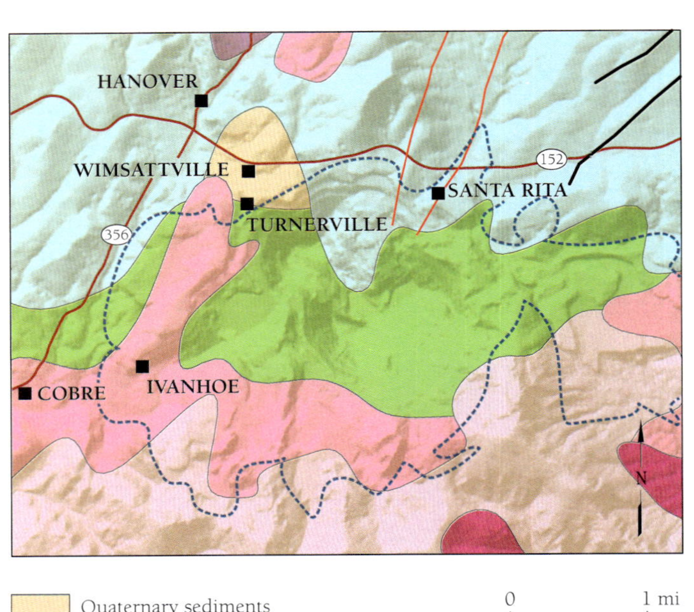

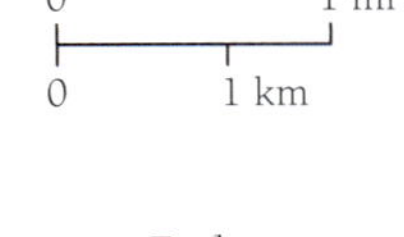

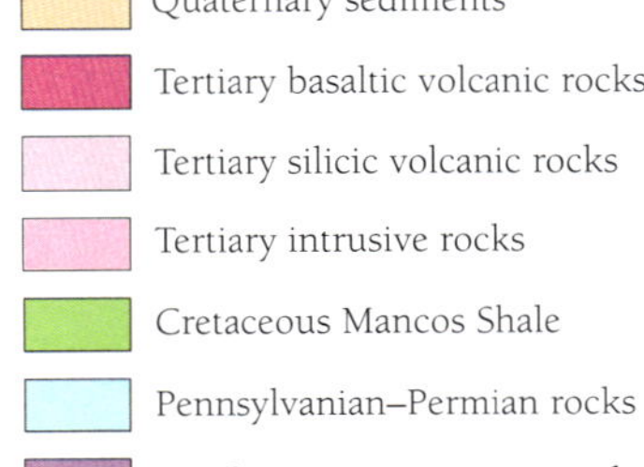

2016 satellite image of Chino Mine, Santa Rita district, Grant County.

mineralization and caps most of the mesas surrounding the mine. Its source was the Emory caldera in the Black Range to the east. The small isolated column or spire above the mine at the base of the tuff-formed cliffs, known as the Kneeling Nun, is a weathered block of the Kneeling Nun Tuff. That feature is the inspiration for the name of the tuff, and survival of this landmark during continued mining has been a source of local concern.

Mineral Deposits

The Chino district, a copper porphyry (with gold, silver, and molybdenum), is a large, low-grade (less than 0.8% copper) deposit that contains veinlets of copper and molybdenum sulfide minerals associated with mineralized porphyritic intrusions. Copper sulfide minerals are found in the upper, fractured granodiorite and adjacent sedimentary rocks. Several periods of alteration have further concentrated the ore. Much of the ore formed by high-temperature alteration of the sedimentary rocks adjacent to the granodiorite intrusion. This is especially true on the east side of the ore body where the Permian Abo Formation, the Cretaceous Beartooth quartzite and Mancos Shale were mineralized.

Early production from the Chino deposit was largely from disseminated chalcocite (copper sulfide and native copper) ores in granodiorite plutons, dioritic sills and chalcocite-pyrite skarn-type ores in adjacent Cretaceous and Pennsylvanian strata. Supergene minerals (mineral deposits formed in the near-surface by groundwater movement) include chalcocite, chrysocolla, covellite, cuprite, native copper, malachite, and azurite. Primary minerals in the porphyry ore bodies are chalcocite, bornite, and molybdenite, whereas those in the replacement ore bodies are chalcopyrite and magnetite.

The Chino Mine has produced more than nine billion pounds of copper, 500,000 ounces of gold, and 5.36 million ounces of silver plus some molybdenum and iron ore, worth more than two billion dollars at the time of production. Today, the Chino operation consists of open pits, a concentrator for copper and molybdenum ores, and a solvent extraction/electrowinning plant that yields copper cathode plates. The Chino Mine is the largest copper and gold deposit in New Mexico and is ranked ninth in New Mexico in silver production.

Features

At the Chino Mine overlook, active and inactive open pits are visible. Mining has excavated three large open pits that are more than 1,000 feet deep. The large piles of rock surrounding the open pits are mine dumps and composed of overburden that was mined from the open pits. Once the ore is blasted and removed from the open pit, it is trucked to either the nearby mill or the nearby heap-leach pile, where the ore is processed to form a copper concentrate (which is later refined to pure copper) or copper cathode plates. These copper plates are shipped to processing plants to manufacture copper wire, rod, tubing, and other products.

Other Nearby Features

At the road cut across the highway from the overlook, one can see dikes that have intruded the Cretaceous shale and sandstone. Approximately 18 miles southwest of the Chino Mine, the geologically

Postcard image of steam shovels and drill machines at Chino Mine about 1915.

Located in Bayard, NM, this steam boiler was used to run the compressor at Santa Rita Mine (ca. 1900).

similar Tyrone Mine on NM 90 in the Burro Mountains District provides views of the ongoing reclamation of mine-waste piles. Tyrone is the second largest copper porphyry deposit in New Mexico and is also operated by Freeport–McMoRan. As at the Chino Mine, Native Americans were the first to mine the area for turquoise and other trade goods from about 600 CE. Numerous small underground mines operated from the 1860s until, in 1968, open-pit mining commenced throughout the 35,200-acre mine site.

From 1969 to 1992, approximately 300 million tons of ore was produced from Tyrone. Subsequently, leaching technology has added approximately 425 million tons of copper ore to the production total. The Tyrone Mine is expected to cease mining in the near future due to the low ore grade, although leaching will continue for a few more years until the heap-leach pile is depleted of copper.

—*Virginia T. McLemore and Robert Eveleth*

The small column is called the Kneeling Nun. She sits above the colorful rock which is visible from the Chino Mine Overlook.

Additional Reading

A Copper Mining Community in New Mexico, Santa Rita del Cobre, by C.J. Huggard, and T.M. Humble, University Press of Colorado, 2011.

Potential for Laramide porphyry copper deposits in southwestern New Mexico, by V.T. McLemore, Geology of the Gila Wilderness–Silver City area, New Mexico, New Mexico Geological Society, 59th Annual Field Conference Guidebook, 2008.

If You Plan To Visit

Although not a public park or monument, the corporate Chino Mine Overlook has large educational displays and is accessible to visitors most of the year. It is located approximately 15 miles east of Silver City and can be reached by going east from Silver City on US 180 to Santa Clara, turning left on NM 152 and continuing just beyond the small town of Hanover. The overview site is along NM 152 and provides one of the best views of an active open-pit mining operation anywhere in the nation.

Although the Tyrone Mine is closed to visitors, a parking area with display boards explaining the mining and reclamation is found on the Tyrone Overlook Road off NM 90 south of Silver City.

The Gila River cuts through interbedded basaltic andesite, green volcaniclastic sandstone, and younger ash-flow tuffs at the mouth of White Rock Canyon (lower left to center) from Fisherman's Point Overlook.

The Gila Lower Box Wilderness Study Area (WSA) and the associated BLM Gila Lower Box Canyon Recreation area is one of the most scenic sections of the Gila River. Here, a lush riparian community with cotton-wood, sycamore, hackberry, and willow nestles in a narrow canyon surrounded by dramatic volcanic cliffs with the Chihuahuan Desert scrub and grasslands above and beyond. Raptors, including bald and golden eagles and peregrine falcons, are commonly seen soaring above the canyon. Coues (whitetail) deer are abundant. Nearly half of the vertebrate species found in New Mexico, including the Gila Monster,

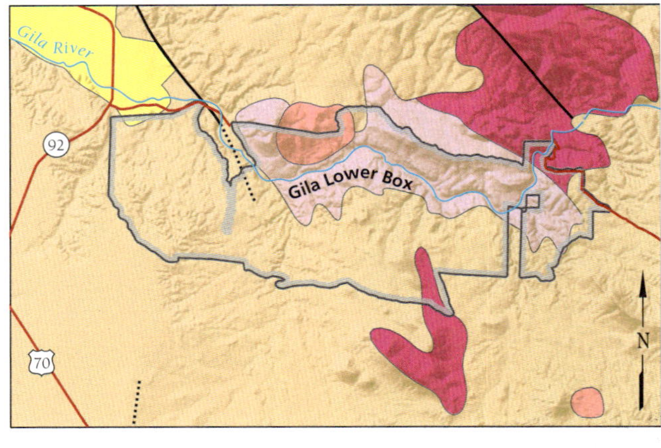

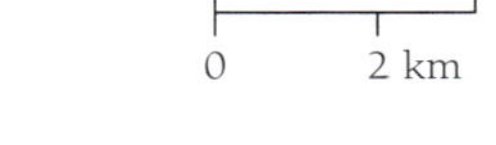

Quaternary sediments

Tertiary–Quaternary
Gila Conglomerate

Tertiary basaltic
volcanic rocks

Tertiary intrusive rocks

Tertiary silicic
volcanic rocks

Fault

Gila Lower Box WSA

reside here. Collectively, these make the Gila Lower Box one of the most ecologically rich regions in southwestern New Mexico. Recreational enticements also abound—swimming holes, birding opportunities, floating the mostly non-technical Class 1+ river during spring runoff; fishing for smallmouth bass and catfish in these relatively warm waters; and hiking along fishing trails. Not surprisingly, this rich habitat was well known to ancestral groups and has abundant archaeological resources including prehistoric dwellings and large petroglyph panels.

Geologic Setting

Part of the Basin and Range province, the Gila Lower Box is also located in the southwest corner of the extensive middle Eocene–Miocene Mogollon–Datil volcanic field, just north of the vast Lordsburg plain. The Mogollon–Datil field dominates the geology of southwestern New Mexico. Nevertheless, Mogollon–Datil volcanic rocks are not seen until reaching the canyon. On any of the roads leading to the Lower Box from the south, one travels over sands and gravels of the Lordsburg Plain. These young sediments bury the volcanic rocks (lavas and ash-flow tuffs), which are only exposed in the canyon cliffs and on the rocky uplands north of the Gila River along Deadman Ridge.

Geologic History

Initial activity in the Mogollon–Datil volcanic field began about 40 million years ago with eruptions of andesite and basaltic andesite lavas. This was followed by eruptions of vast rhyolitic ash-flow tuffs from supervolcanoes, along with minor basaltic andesite lavas, from about 36 to 25 million years ago. Extensive radiometric dating of volcanic rocks has determined that eruption of ash-flow tuffs from Mogollon–Datil calderas was highly episodic. Eruptions in the Mogollon–Datil field clustered into four distinct time intervals (pulses 1–4), each lasting less than about 2.5 million years and separated by quiescent periods lasting 1.5 to 3 million years. Within the Gila Lower Box, three ash-flow tuffs are found: Box Canyon Tuff from the initial pulse, Caballo Blanco Tuff from the second, and the Bloodgood Canyon Tuff from the third.

The Rock Record and Geologic Features

The rock record is best described via a backwards journey through time, beginning with the youngest rocks at the top of the canyon, then down to the river just east of the WSA, then downstream along the river. The desert uplands surrounding the Lower Box consist of Miocene to Pliocene sands and gravels eroded from the older volcanic rocks. These stream-deposited sediments are known as the Gila Conglomerate. On the road to Nichols Canyon, one passes through

Silica-filled vesicles in an outcrop of basaltic andesite.

The lush oasis of the Gila Lower Box. The Gila River flows through andesites, basaltic andesites, and ash-flow tuffs from right to left. On the skyline are Canador Peak, Deadman Ridge, and Caprock Mountain (left to right).

the Caprock Mining District. At the Consolidation Mine on Caprock Mountain, some 5,000 tons of manganese oxide was produced from the Gila Conglomerate in the 1950s. At the mouth of Nichols Canyon, a 21 million-year-old basalt flow within the Gila Conglomerate forms the eastern cliffs, as well as the bluffs across the river valley.

Turning downriver, the first of two northwest-trending faults is crossed at the mouth of Nichols Canyon, exposing an older, dark green andesite downstream. Shortly thereafter, the river crosses a second fault, exposing two ash-flow tuffs interbedded with basaltic andesites and a volcaniclastic sandstone. The upper ash-flow tuff is the 28 million-year-old Bloodgood Canyon Tuff that erupted from the Bursum caldera, approximately 50 miles to the north in the heart of the Gila Wilderness. The lower Caballo Blanco Tuff is about 32 million years old. The source of this tuff is uncertain, but evidence suggests that it erupted from a caldera that was subsequently obliterated by the Bursum caldera, which was centered northwest of

The Gila River flows past tan outcrops of the Caballo Blanco Tuff below Fisherman's Point.

Gila Cliff Dwellings National Monument. These strata, all tilted about 10 degrees to the northeast, can be seen just east of Fisherman's Point, at the mouth of White Rock Canyon—perhaps the most scenic locality in the Lower Box.

Downstream from Fisherman's Point, the river continues to cut into older rocks. From just east of the mouth of Box Canyon to slightly upstream of the access trail at the end of Spring Bluff Road, the cliffs expose the 33.5 million-year-old Box Canyon Tuff, erupted from the Schoolhouse Mountain caldera 20 miles to the north, and andesite lavas of similar age. Approximately a mile downstream from Spring Bluff Road, the river bends to the southwest and a younger volcaniclastic sandstone is seen along the south bank. Shortly thereafter, a small andesitic sill has intruded the sandstone. Rising above the Box Canyon Tuff to the north is Canador Peak, the eroded remnants of a 35 million-year-old intrusion.

—*David J. McCraw*

If You Plan to Visit

From Lordsburg, take US 70 north to NM 464. Take NM 464 14.2 miles north and turn left on Fuller Road, an unsigned, unpaved (but maintained) county road just south of the Grant County line. Three Lower Box access roads take off from Fuller Road to the north: Caprock Ranch Road at mile 3.4, White Rock Canyon Road at mile 4.7, and Spring Bluff Road at mile 7.9. Caprock Ranch Road enters into Nichols Canyon and provides the only direct access to the river. On this road, take the right fork at mile 3.6 at Caprock Mountain; you will reach the river at mile 5.9. As of this writing, this is the roughest of the three access roads. Four-wheel drive and high clearance are mandatory.

On the White Rock Canyon Road, Fisherman's Point overlook is 4.2 miles from Fuller Road. The overlook is on a high point between White Rock and Box Canyons. At mile 3.9 on this approach, a rough, four-wheel drive track provides access down to a bench above the river, where a foot trail leads to the mouth of White Rock Canyon.

The terminus of Spring Bluff Road is 14 miles from Fuller Road, and reaching it involves taking a right fork at mile 3.5 and the left fork 0.1 miles later. A hiking trail about one mile long begins at 13.8 miles and is very scenic.

To float the river through the entire Gila Lower Box WSA during spring runoff, boaters put in at the NM 464 bridge at Red Rock, New Mexico. A 30-mile shuttle is required from put-in to take the take out, a point just upstream of the NM 82 bridge at Virden, just below the diversion dam. For more information:

Bureau of Land Management
Las Cruces District Office
1800 Marquess Street
Las Cruces, NM 88005-3370
(575) 525-4300
www.blm.gov/visit/gila-lower-box-canyon
www.recreation.gov/camping/gateways/10052

Gila Monster

PART 3

BASIN AND RANGE

The semiarid southwestern corner of New Mexico, also known as the Bootheel, lies within the Basin and Range province, a huge geologic region that also includes parts of west Texas, southern Arizona, western Utah, most of Nevada, southern Idaho, eastern California, and southeastern Oregon. The Basin and Range province extends far into northern Mexico as well, occupying sizable portions of the states of Sonora, Chihuahua, and Baja California. The Rio Grande rift can be considered to be a prong of the Basin and Range that bisects New Mexico and southern Colorado, although the rift is broken out as a separate region in this book. The New Mexico section of the Basin and Range province is characterized by northerly- to northwesterly-trending mountain ranges separated by broad basins filled with debris eroded from adjacent highlands. The ranges are fault blocks that were uplifted by movement along the steep faults that bound the range on one or both sides. Drainages from the mountains debouch onto alluvial fans that merge to form broad piedmont slopes at the base of the mountains.

Through-flowing streams are rare in the Bootheel region. An exception is the Gila River, a tributary to the Colorado River, which crosses a part of the Basin and Range north of Lordsburg. Most of the drainages in the Bootheel have no outlet to the ocean, so runoff collects in adjacent basins forming ephemeral lakes called playas. The Playas, Hachita, and Mimbres basins to the east of the Continental Divide generally drain southeast into playa lakes in Mexico. West of the Continental Divide, playa lakes in the Animas and Lordsburg basins contribute to groundwater systems that drain northward to the Gila River.

The timing and cause of faulting varies considerably across the vast Basin and Range province, but researchers agree that the distinctive topography is the result of crustal stretching that followed close on the heels of regional volcanism. This volcanism heated and weakened the crust, allowing it to be easily deformed. The amount of crustal stretching varies from as much as 200 percent in parts of Nevada to approximately 50 percent in southwestern New Mexico, where the crust is now half again as wide as it was when extension began about 25 million years ago. Sensitive GPS measurements show that Nevada is currently widening at rates of 2 to 6 millimeters (0.08 to 0.24 inches) per year. In contrast, measured extension rates across the Basin and Range in southwestern New Mexico are less than 1 mm per year (0.04 inches).

OPPOSITE: **Algal bioherms ("reefs") and associated limestones of Pennsylvanian–Permian age form the sheer cliffs on the western face of Big Hatchet Peak.**

Map of the Basin and Range and Rio Grande rift in the western United States and Mexico.

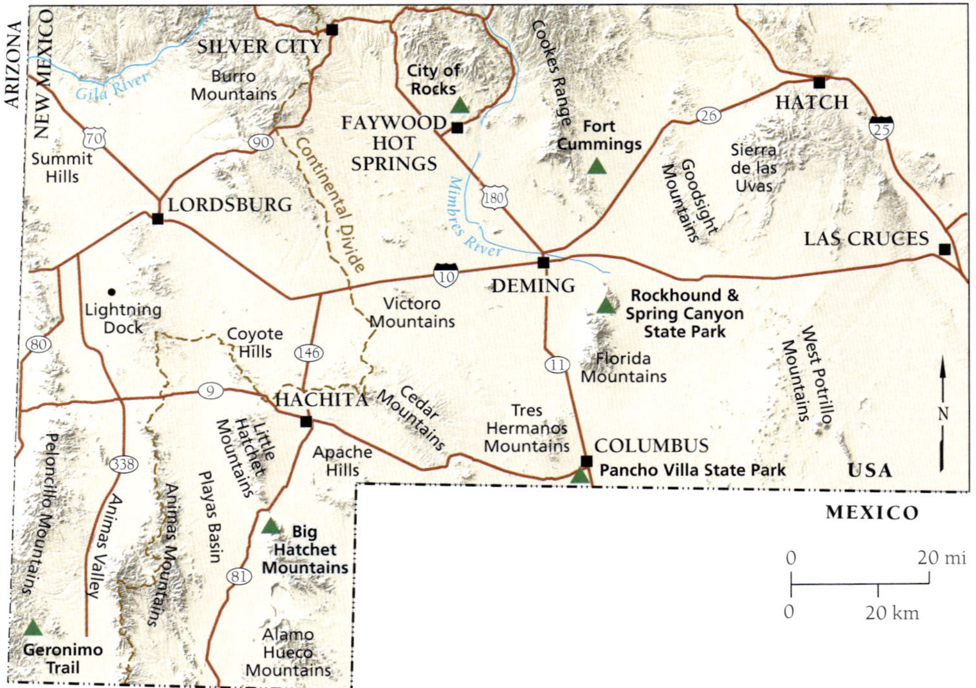

Map showing the recreational areas of interest in the Basin and Range province of southwestern New Mexico.

Geologic History

Rocks exposed in the desolate ranges of southwestern New Mexico record multiple episodes of mountain building, sedimentation, and volcanism. Major mountain building events include continental assembly during Proterozoic time, Late Paleozoic Ancestral Rocky Mountain deformation, middle Mesozoic Bisbee Basin extension, development of Laramide compressional highlands and basins during the Late Cretaceous to Early Cenozoic, and the creation of Late Cenozoic basin-and-range topography. Shallow marine sedimentation dominated in the area through Early Paleozoic time and during Late Cretaceous time, and eruptions from many volcanic centers in the Bootheel field and from the Mogollon–Datil field to the north blanketed the landscape with tuffs and lavas during middle Cenozoic time.

THE FORMATION OF A CONTINENT (PROTEROZOIC)—Southwestern New Mexico is blessed with extensive exposures of ancient rocks that preserve one of the longest records of sedimentation, plutonism, and mountain building during Proterozoic time in the state. A few of these rock outcrops are present in the parks and monuments discussed in this section, and larger exposures can be viewed in roadcuts along NM 90 through the Burro Mountains between Lordsburg and

Silver City. The earliest part of the story is recorded in small outcrops in the Burro Mountains that started their history as mud and basalt laid down on a sea floor. The sediment and lava were buried and then caught up in a mountain-building event about 1.7 billion years ago. Because the mudstone and basalt were subjected to great heat and pressure deep in the crust, they became metamorphic rocks. In addition, plutonic rocks were intruded during this mountain-building cycle, with examples preserved in the Burro and Florida Mountains. About 1.4 billion years ago, a broad band of granitic intrusions developed across the southwestern and central United States. Plutons of this age are exposed in the Burro, Peloncillo, Animas, Little Hatchet, and Big Hatchet mountains. Renewed deformation about 1.2 billion years ago was associated with the assemblage of the Rodinia supercontinent, which resulted in the deposition of sediments and intrusion of plutons and dikes. As Rodinia began to break apart (510 million years ago), a granitic pluton was emplaced in the Florida Mountains as a broad zone of extension formed across parts of New Mexico and Colorado.

Stratigraphy of the Basin and Range area of southern New Mexico. Numerals indicate the age of igneous intrusions in millions or billions (Ga) of years. Ss denotes sandstone.

AGE		UNIT	ROCK TYPES	DEPOSITIONAL SETTING	TECTONIC EVENT
Cenozoic	Quaternary	Alluvium	Silt, sand, gravel	Playa, wind-blown dunes, alluvial fan	Basin and Range extension and mafic eruptions
		Basalt flow (0.3 Ma)			
	Pliocene-Miocene	Gila Conglomerate	Sand, gravel, and silt interbedded with mafic lavas	Alluvial fans, streams, playas	
	Oligocene	Younger silicic tuffs	Ash flow tuffs, lava flows, and sediments	Calderas	Ignimbrite flare-up, extension begins
		Older silicic tuffs			
	Eocene	Rubio Peak Formation	Lava and debris flows	Stratovolcanoes	Start plate rollback
	Paleocene	Lobo Fm. 57	Conglomerate interbedded with andesite	Erosion of Laramide uplifts and volcanoes	Late Laramide deformation, volcanism, copper mineralization
Mesozoic	Late Cretaceous	Hidalgo Fm. 72	Andesite flows and breccia	Laramide volcanoes	Early Laramide deformation
		Ringbone Fm.	Conglomerate, ss., shale	Laramide basin	
	Early Cretaceous	Bisbee Grp. — Mojado Fm.	Sandstone and shale	Marine to nonmarine	Bisbee Basin rifting
		Bisbee Grp. — U-Bar Fm.	Limestone, shale, ss.	Marine	
		Bisbee Grp. — Hell-to-Finish Fm.	Limestone, ss., mudstone	Marine to nonmarine	
	Jurassic	Bisbee Grp. — Broken Jug	Conglomerate, mudstone, carbonates, basalt	Marine	
Paleozoic	Pennsylvanian-Permian	Naco Group	Limestone, dolomite, mudstone, redbeds	Marine	Pedregosa basin Gondwana collides with Laurentia
	Mississippian	Paradise Formation Escabrosa Limestone	Limestone	Marine	Passive margin
	Devonian	Percha Shale	Shale	Marine	
	Silurian	Fusselman Dolomite	Dolomite	Marine	
	Ordovician	Montoya Formation	Dolomite	Marine	
		El Paso Group	Dolomite, limestone	Marine	
	Cambrian	Bliss Sandstone	Sandstone	Marginal marine	
	Proterozoic	510 / 1.2 Ga / 1.4 Ga	Granite, gneiss, schist, phyllite, amphibolite, carbonate	Passive margins and island arcs	Continental assembly

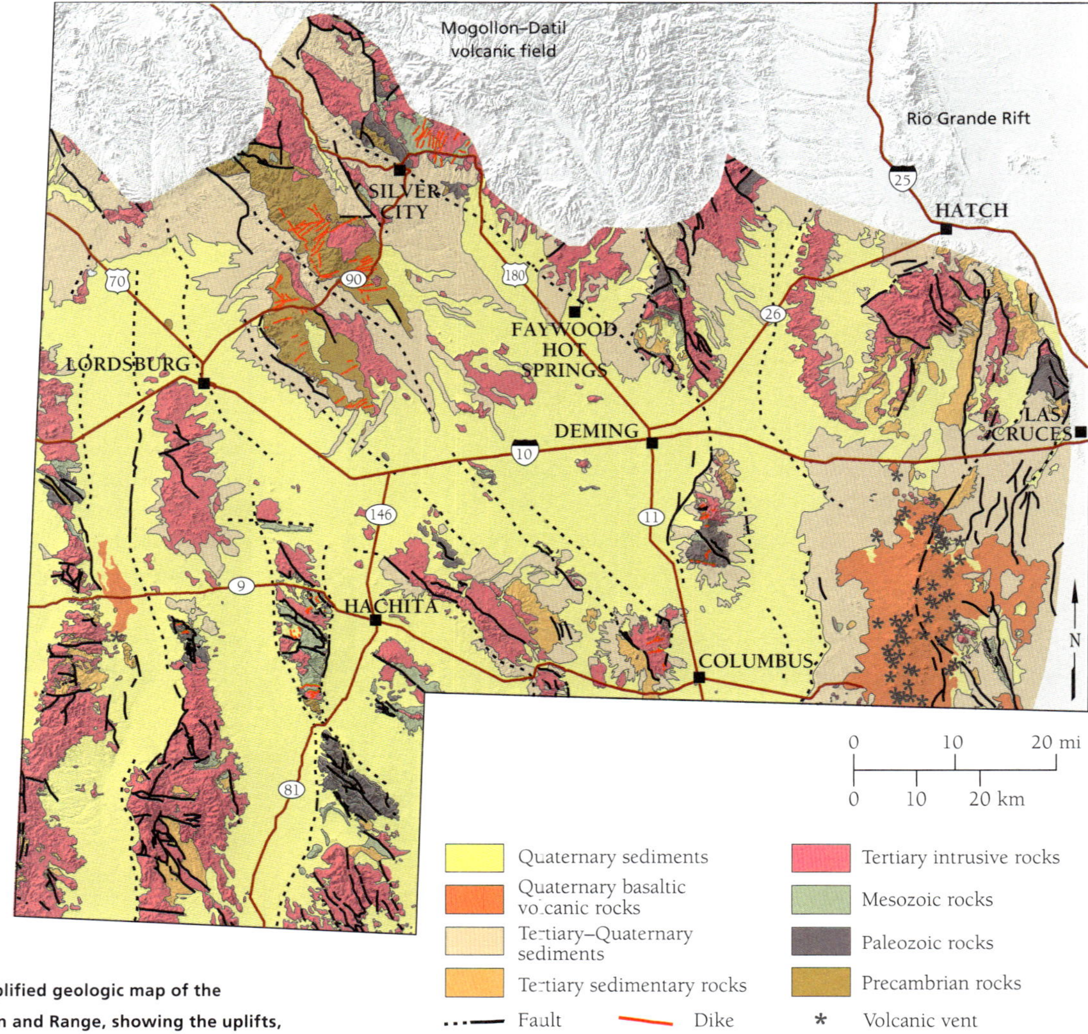

Simplified geologic map of the Basin and Range, showing the uplifts, called ranges, and the sedimentary basins in between them.

CONTINENTAL MARGIN DEPOSITION (EARLY PALEOZOIC)—After a long period of erosion, a shallow ocean covered the region starting about 500 million years ago. Shallow seas advanced and retreated across the area several times between about 500 and 340 million years ago, first depositing sandstone (the Cambrian and Ordovician Bliss Sandstone), then limestone and dolomite (the Ordovician El Paso and Montoya formations and the Silurian Fusselman Dolomite), and finally shale (the Devonian Percha Shale). This sedimentary succession forms cliffs and slopes on the north side of the Big Hatchet Mountains, the south side of the Florida Mountains, and the north side of Cookes Peak. These lower Paleozoic units are preserved only in southern New Mexico and pinch out northward.

UPLIFT AND EROSION OF THE ANCESTRAL ROCKY MOUNTAINS (LATE PALEOZOIC)—Late Paleozoic Ancestral Rockies deformation began at about 310 million years ago, following the deposition of the Mississippian Escabrosa Limestone. The northwest-trending Florida uplift rose during Pennsylvanian time in the general vicinity of Silver City and Deming. The Pedregosa basin, a NW-trending basin that extended from Chihuahua, Mexico through the southwest corner of New Mexico, and into southeastern Arizona, subsided starting in the Early Pennsylvanian. The area between the Pedregosa basin and the Florida highland was occupied by a shallow marine shelf. Superimposed on the tectonic changes affecting the landscape were episodes of sea-level rise and fall of 100–250 feet in amplitude that were driven by the recession and advancement of continental glaciers in the Gondwana supercontinent.

About 290 million years ago, during the transition from Pennsylvanian to Permian time, renewed uplift of earlier Pennsylvanian highlands led to deposition of red fluvial and less colorful marginal marine units near the highlands, while open marine deposition persisted to the south. All rock units deposited during Pennsylvanian and Permian time in southwestern New Mexico are included in the Naco Group. The algal mounds (bioherms) that form the spectacular cliffs in the Big Hatchet Mountains were deposited during this time interval.

Topographic evolution of a basin-and-range landscape. The top panel shows the landforms that developed when the area was actively extending. Coarse-grained alluvial fans (stippled gray in cross section) form along the mountain fronts and finer grained sediment is carried to the middle of the basin to be deposited in playa lakes (pink in cross section). The bottom panel shows pediment surfaces that form on the mountain blocks as erosion wears down the ranges after extension wanes.

BISBEE BASIN RIFTING AND WESTERN INTERIOR SEAWAY DEPOSITION (LATE JURASSIC TO LATE CRETACEOUS)—A northwest-trending rift basin, known as the Bisbee Basin, developed in the Bootheel area during Late Jurassic to Early Cretaceous time. The trend of the Bisbee Basin followed the general trend of the earlier Ancestral Rocky Mountain structures. Deposits within this basin are known as the Bisbee Group, which form low hills on the southwest side of the Big Hatchet Mountains. Triassic and Jurassic rocks were either never deposited on, or were eroded from, the northern shoulder of the rift basin, where a highland formed in the Silver City–Las Cruces area. Upper Cretaceous sediments then lapped onto and buried the Bisbee Basin and its rift shoulder. These Cretaceous sediments preserve the earliest and most southerly transgression of the Western Interior Seaway.

LARAMIDE DEFORMATION (LATE CRETACEOUS TO EARLY CENOZOIC)—
Starting in the Late Cretaceous and reaching a peak in Eocene time, Laramide deformation produced basins and uplifts that followed the northwest trend of the previous Jurassic rift basin. Many of the faults associated with Jurassic crustal stretching were reactivated with the opposite sense of motion during Laramide crustal shortening. Important copper porphyry intrusions, such as those at the Tyrone and Chino mines, were emplaced during this time.

Typically dry river bed of the lower Mimbres River looking north from US 180.

REGIONAL VOLCANISM (MIDDLE CENOZOIC)—Following the end of Laramide deformation, upwelling hot mantle rocks caused widespread volcanism in southwestern New Mexico and adjacent areas (see Part 2). Large andesitic stratovolcanoes formed beginning about 40 million years ago, burying older Laramide highlands and basins beneath lavas and volcanic sediments. Starting about 36 million years ago, caldera eruptions covered the area with rhyolitic ash-flow tuffs and associated sediments, while basaltic lavas from shield volcanoes flowed across the region. Remnants of nine calderas are preserved in the Bootheel region, forming part of a once-continuous volcanic field that stretched from the Socorro–Quemado area on the north to Guadalajara in Mexico to the south. Two tuffs from calderas in the Mogollon-Datil volcanic field spread southward to form the geologic features at Fort Cummings and City of Rocks. Geronimo Trail in the Peloncillo Mountains crosses tuffs erupted from calderas in the Bootheel volcanic field.

BASIN AND RANGE EXTENSION (LATE CENOZOIC)—Following volcanism, the crust in southwestern New Mexico began to stretch and break up into innumerable fault blocks. At first, during latest Eocene to late Oligocene time, faulting was relatively minor. This phase gave way to rapid stretching during late Oligocene to late Miocene time (about 25 to 10 million years ago), creating the landscape that we see today. Although extension has waned since the late Miocene, recent deformation is recorded by numerous fault scarps that cut Quaternary deposits in the Animas Valley, along the southwestern margin of the Burro Mountains, and on both sides of the Florida Mountains. Small-magnitude (approximately 3.0) earthquakes also indicate continued crustal unrest in this area. Young volcanism is recorded by a cinder cone on the west side of the Animas Valley that erupted a basalt that flowed north across the basin floor about 320,000 years ago.

Resources

Several deep oil and gas wells were drilled in the basins of southwestern New Mexico, especially targeting the sediments within the now-buried Pedregosa basin, but no significant reservoirs have been found. However, these wells provide important scientific information about rocks and geologic structures buried beneath the younger alluvial basins. A significant geothermal resource was inadvertently discovered by a rancher drilling a stock well in 1948. Hot water from the Lightning Dock geothermal field south of Lordsburg in the Animas Valley has since been used to grow roses, raise fish (tilapia), and generate electricity. Other geothermal resources, like Faywood Hot Springs, are used for recreational soaking.

Numerous mines are located in the ranges of southwestern New Mexico. Many mineral deposits are associated with volcanism during late-stage Laramide deformation and during middle Cenozoic time. Metallic ores and agate form as a result of hydrothermal alteration during the waning stages of volcanism. Other mineral deposits are caused by cooling hydrothermal fluids migrating along fault zones during Basin and Range extension. Resources include lead, zinc, copper, fluorite, gold, silver, manganese, asbestos, clay, cinder, and agate.

—Shari A. Kelley

Fort Cummings National Historic Site
BUREAU OF LAND MANAGEMENT

Cookes Peak looking west from Fort Cummings.

Fort Cummings is located at Cookes Spring near the southern entrance to Cookes (or Massacre) Canyon in Luna County. The spring is nestled among volcanic rocks along the south flank of the Mogollon–Datil field. In 1858, the Butterfield Overland Mail Company began using Cookes Spring as a stage stop along the Butterfield Trail. Although the mail company was forced to close in 1861, the trail remained a popular but perilous route for travelers. Cookes Canyon became notorious as an ambush site used by hostile Apaches, and the fort was established to protect travelers in that area.

Fort Cummings was established on October 2, 1863 by Captain Valentine Dresher of the California Volunteers during the Civil War and was relocated at least twice in attempts to find the best defensive position. The facility was a base for the 125th Regiment of the U.S. Colored Infantry, a contingent of the Buffalo Soldiers who fought with great distinction in the Apache Indian War during the late 1860s. The fort was abandoned in 1873 but was reoccupied in 1880 when Victorio and his band of Warm Springs Apaches began raiding in southern New Mexico. The fort was permanently abandoned in 1886 following Geronimo's surrender. Today, there is little left of Fort Cummings except for some adobe walls and earthen mounds where buildings once stood.

Stone marker at Fort Cummings.

The Rock Record

Rocks in the Cookes Range span nearly the entire extent of the geologic record, from Precambrian (Proterozoic) to late Cenozoic, but most of the rocks in the vicinity of Fort Cummings are early Tertiary basalt and andesite flows along with sediments derived from them. The fort walls were mostly made of adobe, derived from the local soil and alluvial fan deposits, but some of the walls were made of boulders of gray andesite and orange iron-oxide-banded sandstone. The surrounding hills are mostly composed of the Bear Springs basalt and the rhyolitic Sugarlump Tuff, the product of an explosive eruption early in the history of the Mogollon–Datil volcanic field, about 35 million years ago. These rocks also form the hills in the vicinity of Cookes Spring and Cemetery Ridge. Cookes Spring flows from a fault within the Sugarlump Tuff. Today, an elongated ridge of Sugarlump Tuff ends at a rock wall that was once part of the Butterfield stage station, adjacent to a pump house built in 1881 by the Atchison, Topeka and Santa Fe Railway Company. The pump house roof was reconstructed in 1987 and now has a solar panel to pump water.

The entrance to Cookes Canyon is accessible by a rough road beyond the spring house, but much of the canyon is private land. Volcanic flows and interlayered sediments of the Rubio Peak Formation, about 36 million years old, are exposed within the canyon. Massacre Peak (elevation 5,667 feet), the high point of the southern Cookes Range just south of the canyon, is an granodiorite intrusion of similar age.

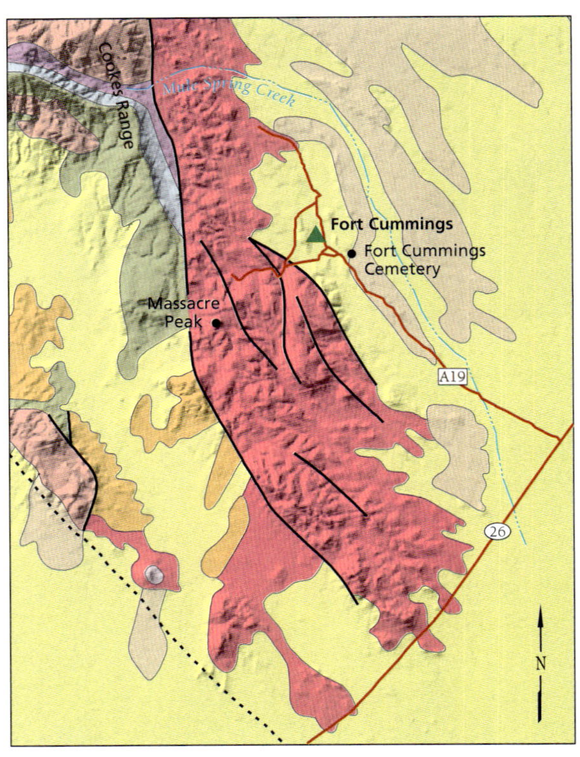

Quaternary sediments

Tertiary–Quaternary sediments

Tertiary sedimentary rocks

Tertiary intrusive rocks

Tertiary basaltic volcanic rocks

Tertiary silicic volcanic rocks

Cretaceous Mancos Shale

Pennsylvanian–Permian rocks

Mississippian–Cambrian rocks

Fault

Geologic Features

Cookes Peak Mining District was discovered in 1876 by Ed Orr and by 1900 had produced approximately $3 million worth of ore from carbonate-hosted vein deposits. Large-scale mining ceased in 1905, but in 1951, H.E. McCray reopened the district and produced lead, copper, zinc, silver, and gold until 1953.

Chalcedony, white to gray agate nodules, and geodes can be found in the rhyolitic to andesitic flows exposed in the hills surrounding Fort

Cummings, and agate is common throughout the Cookes Peak Mining District. Rock and mineral collecting is allowed on the BLM land surrounding the ruins, but not near the ruins or on private property.

Adobe ruins of Fort Cummings.

—*Virginia T. McLemore*

Additional Reading

Annals of Old Fort Cummings, New Mexico, 1867–8, by William T. Parker, M.D., Northampton, Massachusettes, 1916.
babel.hathitrust.org

If you Plan to Visit

To get to Fort Cummings, take NM 26 northeast about 7 miles from Deming or 13.2 miles southwest from Nutt and the southern terminus of the Lake Valley Backcountry Byway. Turn onto Cookes Canyon Road (CR A19) at the railroad water tank at Florida Station. Set your trip odometer here. After traveling 1 mile north, the gravel road crosses a cattle guard; turn left. At mile 1.9, take the right fork and continue straight. A viewpoint of the fort to the north (with an accompanying plaque) is on the left, and the old cemetery on the right is encountered at mile 5.1. Fort Cummings is located at mile 5.6, and Cookes Spring is a few hundred yards further to the west. Both are located on BLM land west of the barbed wire fence; private ranch lands are east of the fence. High clearance vehicles are recommended, but the road is generally passable with two-wheel drive vehicles.

City of Rocks State Park, located in Grant County about halfway between Silver City and Deming, was established in 1952 and encompasses one square mile (680 acres) of land at an elevation of about 5,000 feet. The park features the largest of several intricately eroded exposures of ash-flow tuff in southern New Mexico. The park is truly a geologic monument with large rock columns and pinnacles as much as 40 feet tall separated by eroded areas that resemble city streets. Prehistoric people likely camped by these rocks for shelter from the elements and predators. The Mimbres Mogollon culture settled the area about 200–1450 CE (Mimbres is Spanish for willow), and the Apaches were present from 1400–1886. Their arrowheads and pottery sherds are still found today. Mortar holes, small smooth-sided, cylindrical depressions formed by grinding seeds on the rock surface with stone manos over many years, are visible in the rocks along the trails, especially in the northern part of the park. These features are sometimes called "Indian wells" because water collects in the holes.

OPPOSITE: **Columns of ash-flow tuff at the City of Rocks State Park.**

Grinding holes, small smooth-sided cylindrical holes, are found in the rocks along the trail in the northern part of the park.

Regional Setting

The park lies within the Mimbres Valley of the Chihuahuan Desert in the Basin and Range physiographic province, on the southern margin of the Mogollon–Datil volcanic field and near the western margin of the Rio Grande rift.

After the Mexican War of 1846–1847, the Mormon Battalion under Captain Philip St. George Cooke blazed a trail south of the park to link the newly acquired territories of New Mexico and Arizona with the eastern United States. The mountains visible southeast of the park are named Cookes Range after Captain Cooke. Cookes Range consists of rocks ranging in age from Precambrian (Proterozoic) to Late Cenozoic. The prominent peak in the southern part of the range is named Cookes Peak and is formed by a gray granodiorite intrusion that was emplaced approximately 38 million years ago. Younger Basin and Range extensional faulting has since brought this intrusion to the surface, and subsequent erosion has formed the notable landmark that we see today.

At the top of Observation Point, south of the park entrance and accessible by car, a much younger alkali basalt flow is exposed, which is only 2–3 million years old. From Observation Point, you can see the

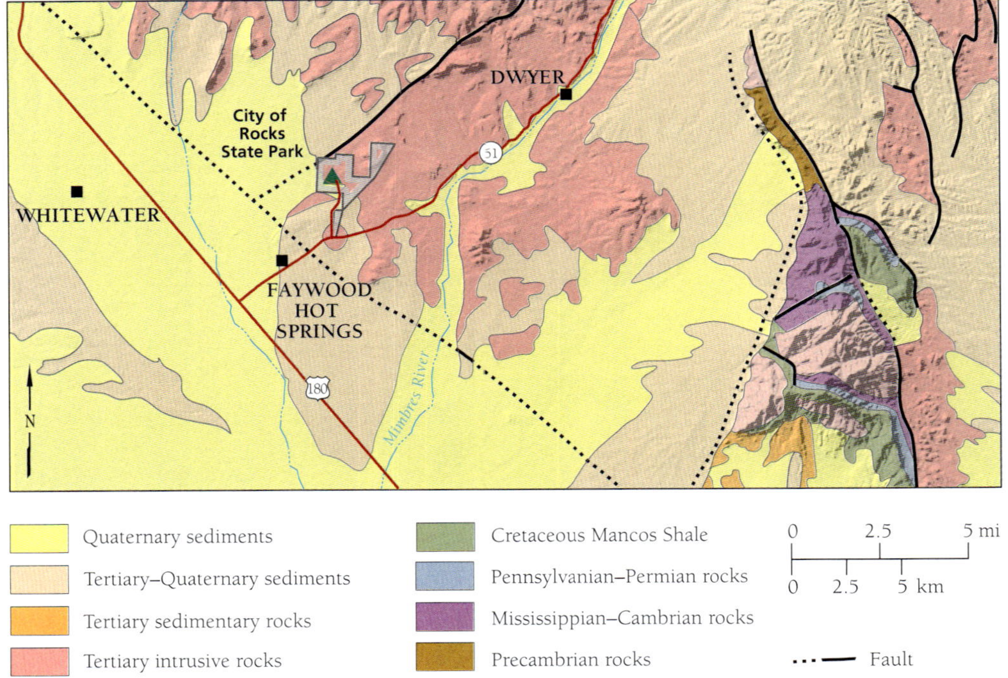

Quaternary sediments

Tertiary–Quaternary sediments

Tertiary sedimentary rocks

Tertiary intrusive rocks

Cretaceous Mancos Shale

Pennsylvanian–Permian rocks

Mississippian–Cambrian rocks

Precambrian rocks

Fault

0   2.5   5 mi

0   2.5   5 km

Geologic map of the City of Rocks State Park and surrounding area.

Tyrone Mine dumps (tan to orange) to the northwest and the Burro Mountains far to the west. The whitish dumps north of the park, near Hurley, are Chino Mine tailings (the residues left after removal of valuable ore materials from uneconomic ones).

Faywood Hot Springs lies southwest of City of Rocks State Park (along NM 61). The private resort, then known as Hudson Hot Springs, once was one of the most famous spas in the west; the water was even bottled and shipped out of state. The hotel burned down in 1891, and the new owners, J.C. Fay and William Lockwood, rebuilt and renamed the complex Faywood, after both men. Patronage declined after WWI and the hotel was demolished in 1952. In 1966, the land was owned by Kennecott Copper, then later sold to Phelps Dodge Company. Today Faywood Hot Springs is, once again, a private resort designed for camping and bathing in the geothermal waters.

The Rock Record

City of Rocks is made up of an ash-flow tuff (named the Kneeling Nun Tuff) emplaced during a violent volcanic eruption about 35 million years ago from the Emory caldera in the Black Range to the northeast. During the eruption, dense, ground-hugging clouds of nearly molten volcanic ash flowed rapidly outward from the caldera, filling the surface topography. After emplacement, the ash compacted and welded together, forming a continuous layer of ash-flow tuff. As the ash cooled and contracted, vertical shrinkage cracks called columnar joints formed. Erosion and weathering eventually widened the columnar joints and other fractures, leaving the free-standing columns we see today. Columnar joints can be seen in the cliffs flanking nearby Table Mountain and along the Table Mountain hiking trail, northeast of the campground. Table Mountain is a slightly younger tuff from an eruption around 34 million years ago.

Ash-flow tuffs in southwestern New Mexico are typically rhyolites, a volcanic rock similar in chemical composition to granite. A close look at the rhyolites in City of Rocks reveals phenocrysts (large crystals) surrounded by a finer groundmass or matrix. The shiny black needles are hornblende crystals, the platy black crystals are biotite, and the grayish-white to clear rounded crystals are quartz. White to clear, glassy laths or rectangles are feldspar crystals. The matrix is too fine grained to be seen with the naked eye, but under a microscope one can see that it is composed mostly of tiny bits of pumice, a frothy volcanic glass.

Geologic History

The volcanic rocks exposed in the park are part of the Mogollon–Datil volcanic field, one of several large volcanic fields active in southwestern North America between about 40 and 25 million years ago. These volcanic fields hosted many super-volcanoes (such as the Emory caldera)—huge, highly explosive volcanoes similar to the more recent examples in the Jemez Mountains and Yellowstone. The deeply eroded Emory caldera occupies much of the Black Range northeast of City of Rocks State Park. Beginning about 25 million years ago, crustal extension broke much of southwestern New Mexico into a series of fault blocks, forming the basin-and-range topography we see today.

Satellite image of the southern part of City of Rocks State Park on 23 February 2016.

Panoramic view of City of Rocks.

Geologic Features

The "streets" in City of Rocks are formed by fractures that trend north-northeast, east-northeast, and northwest. Some fractures may be columnar joints formed as a result of cooling; others may be associated with the release of load pressure as overlying rocks were eroded away. A combination of freeze-thaw cycles, soil processes, wind, and surface and/or subsurface weathering over millions of years enlarged the fractures to produce the rock columns and pinnacles that we see today. Differences in shape and color of the pinnacles from an upper, steep-sided, dark gray to a lower, recessed cream to reddish brown suggest that most of the weathering occurred in the subsurface. Subsurface weathering by water and humic acids in the soil horizon slowly weakened and dissolved the rocks, forming the recessed bottoms. Several periods of weathering and exposure occurred, and

Table Mountain on the skyline, north of the campground, is a slightly younger tuff from an eruption around 34 million years ago.

surface runoff has stripped away the former soil mantle and revealed the bare rock below. Most of the columns and pinnacles at City of Rocks have weathered in place and have not moved. They are not boulders or detached blocks.

Climbing around among the rocks is a favorite pastime that can be enjoyed by people of all ages and skill levels. The sound-baffling effect of the rocks preserves the solitude and quietness of the park and because many of the campsites are isolated from each other by pinnacles, there is a sense of privacy even when the park is filled with people.

—Virginia T. McLemore

Additional Reading

City of Rocks State Park, by V.T. McLemore, New Mexico Bureau of Geology and Mineral Resources, New Mexico Geology, v.19, no.2,1997.

City of Rocks, by R.H. Weber, New Mexico Bureau of Geology and Mineral Resources, New Mexico Geology, v.12, no.1, 1980.

If You Plan to Visit

City of Rocks State Park offers camp sites, numerous hiking trails, excellent mountain biking, wildlife viewing, birding, picnic areas, and a desert botanical garden. The uniquely designed Visitor Center includes a display area and modern restrooms with hot showers.

To reach the park from I–10 in Deming go northwest on US 180 for 24.1 miles. Turn right (northeast) on NM 61 and go 3.2 miles; turn left onto a paved road signposted for the park and go north for 1.8 miles to the Visitor Center. From Silver City, follow US 180 east and then southeast for about 27 miles; turn left (northeast) on NM 61 and go 3.2 miles; turn left onto a paved road signposted for the park and go north for 1 8 miles. For more information:

City of Rocks State Park
327 Hwy 61
Faywood, NM 88034
(575) 536-2800
www.emnrd.state.nm.us/spd/cityofrocksstatepark.html

Rockhound State Park and Spring Canyon Recreation Area

Ash-flow tuffs form the top of Little Florida Mountains, looking east from the park.

Rockhound State Park lies southeast of Deming, in the Little Florida (pronounced Flo-ree-da) Mountains, and Spring Canyon Recreation Area is located in the nearby northern Florida Mountains. Rockhound State Park was established in 1966 as one of the few parks in the U.S. that allows collecting of rocks and minerals for personal use. Each visitor is allowed to collect a small amount of rocks and minerals from the 1,100-acre park; mineral dealers are not allowed to collect for the purpose of sale. Spring Canyon is a quiet, sheltered, steep-sided canyon (day-use only) with picnic areas, hiking trails, and wonderful views.

While water is scarce and generally limited to wells and hidden springs, a diverse flora and fauna makes its home here. Lizards, snakes, deer, pronghorns, coyotes, and small mammals such as prairie dogs, rabbits, and badgers are common, along with many birds. Mountain lion and desert bighorn sheep may be seen at the higher elevations of the Florida Mountains. A variety of plants thrive in this environment, including yucca, prickly pear cactus, barrel cactus, ocotillo, creosote bush, mesquite, and hackberry; juniper and scrub oak are common in the canyons.

Regional Setting

The main park provides excellent views of surrounding desert and mountains with rugged topography reflecting their location in the Basin and Range province. The Cobre Mountains form the far northern horizon, the Burro Mountains lie to the west-northwest, the Victorio Mountains lie to the west-southwest, and the Cedar Mountains lie to the south-southwest. Florida Gap separates the Little Florida Mountains (Rockhound) and the Florida Mountains (Spring Canyon). The dark mountain north of Deming is Black Mountain, which is composed of volcanic rock.

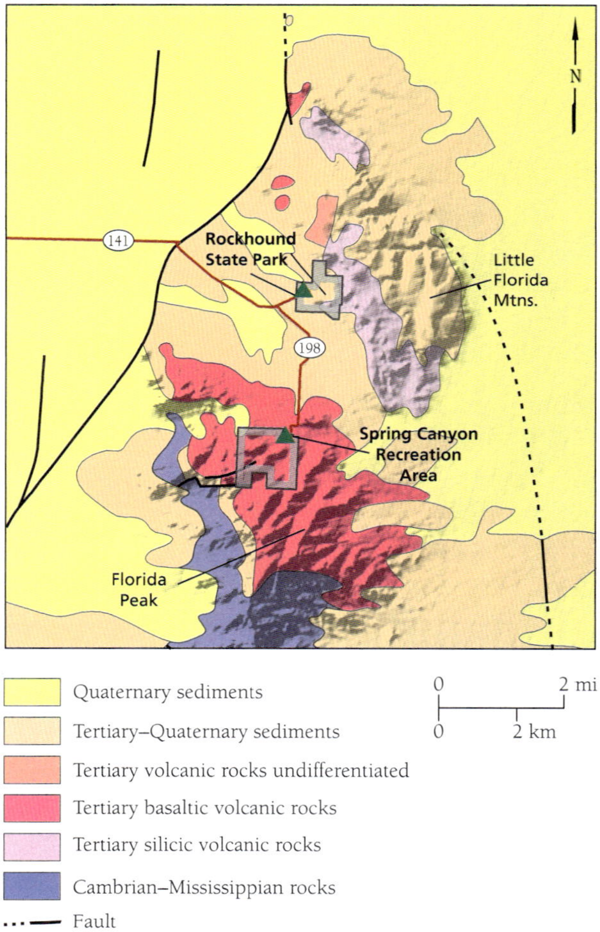

The Rock Record

The Little Florida Mountains predominantly consist of interbedded mid-Tertiary rhyolitic ash-flow tuffs and andesitic to dacitic lavas, which were later intruded by rhyolite domes and dikes between about 33 to 28 million years ago. Some of these volcanic rocks were produced by very explosive volcanic eruptions, with source vents as far away as Lordsburg in southwestern New Mexico. The ash from these eruptions was generally very hot when deposited, and then was fused, or welded, into very dense, hard layers of rock called ash-flow tuff. In places, these welded tuffs contain flattened clasts of pumice, termed fiamme. The overlying andesite lava flows were less explosively erupted, probably from shield volcanoes or stratovolcanoes.

The volcanic breccias, conglomerates, and andesitic lavas found at
Spring Canyon Recreation Area belong
to the Rubio Peak Formation and
were emplaced about 46 to 36 million
years ago (middle Eocene to early
Oligocene). A rhyolite dike, which
cuts the rocks near the head of Spring
Canyon, is about 25 million years old.

Geologic History

During and after the period of
volcanic activity, block faulting related
to crustal stretching uplifted the Little
Florida and Florida Mountains. Erosion
has since worn the mountains down
to their present elevation above the surrounding desert. The basins
around the Little Florida and Florida Mountains subsided as the
mountains were uplifted. Wind, water, and freeze-thaw cycles continue
to break up the rocks through time and to carry fragments downslope,
forming the gently sloping bajadas and alluvial fans at the base of the
mountains. The campground at Rockhound State Park lies on one of
these bajadas.

Volcanic breccia in Rockhound State
Park. This rock may have formed in
a mud flow.

Geologic Features

Rockhound State Park is aptly named for the variety of rocks
and minerals that, with some work, can be found by visitors. While
mineral deposits such as manganese and fluorite exist in the area,
all mining ceased after 1959. Today, the most popular activity at the
park is collecting perlite, thundereggs, geodes, jasper, onyx, agate,
crystalline rhyolite, Apache tears (obsidian), and quartz crystals. Agate
(microcrystalline quartz) is present in a wide range of colors, and some
thundereggs and geodes contain zones of multicolored agate in addition
to well-formed quartz crystals.

Geodes, thundereggs, and spherulites may have been formed by
similar geological processes. Geodes are hollow or near-hollow, with
crystal-lined cavities that form in igneous and sedimentary rocks.
Spherulites are round to oval masses of radial crystals that grew outward from a central nucleation point within cooling volcanic rocks.
Thundereggs, a type of spherulite, are solid or near-solid nodules that
commonly appear to have expanded and shattered, creating a hollow

Looking north from the picnic area in Spring Canyon Recreation Area.

center that is filled late by multicolored silica minerals. Geodes, spherulites, and thundereggs are typically rounded to slightly oval in shape.

Agate, chalcedony, and quartz in veins and open-space-fillings in thundereggs, geodes, and spherulites formed by multiple episodes of late-stage hydrothermal and surficially-derived fluids moving through the volcanic rocks as they cooled and were buried. Hydrothermal fluids were formed from fluids escaping deeply-buried magma bodies and mixing with local groundwaters. The fluids commonly contain a variety of dissolved material derived from the magma and from the surrounding rock. As these hydrothermal fluids move through fractures in the rocks, veins or bands of agate, chalcedony, and quartz may precipitate. In larger open pores or pockets (like thundereggs and geodes), concentric bands of crystals precipitate along the walls of the cavity, gradually decreasing the amount of open space available after each generation.

The different colors of the bands in the silica void fillings are mainly a result of minute fluid and crystal inclusions of a variety of minerals. For example, minerals containing iron result in red-colored phases and manganese in black- or pink-colored phases. Occasionally, small amounts of elemental iron, aluminum, and a few other elements within the void filling can change the color of silica phase. Perfectly formed quartz crystals indicate that the crystals precipitated slowly under relatively slow-changing conditions. The presence of calcite in some thundereggs and geodes indicates another episode of fluid movement through the rock, likely related to groundwater movement.

An interesting form of thundereggs found at Rockhound State Park are ones in which the filling shows evidence of tilting during formation. These originally horizontal fillings record either local landslides or tilting of fault blocks within the Little Florida Mountains during the time in which the crystals were precipitating from the fluid.

Thundereggs, geodes, and spherulites cannot be distinguished from one another until they have been broken apart. The beauty of a thunderegg or geode is locked in the void space inside the hard, rough textured outer shell. Spherulites, although geologically interesting, do not contain these colorful cores. All it takes to look for thundereggs and geodes is a rock hammer, pick, shovel, chisel, collecting bag or pack, persistence, and patience. Ask the park staff for suggestions on the best collecting sites.

—Virginia T. McLemore and Nelia W. Dunbar

Variations in angles of internal layering indicate that this thunderegg rotated at least two times during crystal growth. The white layers were deposited before the precipitation of the bands of blue chalcedony filled the thunderegg.

Additional Reading

Geology of the Florida Mountains, southwestern New Mexico, by R. Clemons, New Mexico Bureau of Mines and Mineral Resources, Memoir 43, 1998.

The Formation of Thundereggs (Lithophysae), by R. Colburn, 1999. www.zianet.com/geodekid/thndregg.htm

If You Plan to Visit

The Rockhound State Park offers covered shelters for camping and picnicking, a group shelter, restrooms with showers, an RV dump station, a playground, hiking trails, wildlife viewing areas, and rock and mineral collecting areas. Facilities at Spring Canyon include picnic tables, a group shelter, restrooms, and hiking trails. Both parks are open year round.

To get to Rockhound State Park from I–10 in Deming, take Exit 85. Go south on Business I–10 for less than a ½ mile and turn east on

NM 540 for 6.6 miles. Turn right on NM 140 (Stirrup Road) and go south for 6.2 miles to the park entrance. Alternatively, take NM 11 south for five miles, and then go east on NM 141 for about nine miles to Stirrup Road and the park entrance. To get to Spring Canyon Recreation Area, follow either of the routes described above but turn south on NM 198 just before reaching Rockhound park and follow that road for about 3 miles to Spring Canyon. For more information:

Rockhound State Park
9880 Stirrup Road SE
Deming, NM 88030
(575) 546-6182
www.emnrd.state.nm.us/spd/rockhoundstatepark.html

Pancho Villa State Park
NEW MEXICO STATE PARKS

On March 9, 1916, Pancho Villa with approximately 500 men raided and burned part of the small town of Columbus, New Mexico, on the border with Mexico. By the end of the raid, ten civilians and eight U.S. soldiers had been killed. At least 80 of Villa's men were killed, and many more were wounded. Six days later, General John J. (Black Jack) Pershing led a punitive force of 10,000 men into Mexico in an unsuccessful, eleven-month pursuit of Villa. Soldiers used Coote's (or Villa) Hill, capped by a basaltic flow, as a lookout before and after the raid.

When Pancho Villa State Park was established in Columbus in 1959, it became the only park to be named after an invader of the United States. The park occupies the site of abandoned Camp Furlong (1912–1924), a former Army post.

Exhibit of uniforms and equipment used by Pershing's men.

Regional Setting

The park lies on a nearly flat alluvial plain extending southward from the Tres Hermanas Mountains (Spanish for three sisters) in the transition between the Rio Grande rift and the Basin and Range province.

The Rock Record

Pancho Villa State Park rests upon an alluvial plain that consists of gravel, sand, and mud that were transported from the adjacent mountains, mostly by streams. Many of the boulders now found throughout the park also were transported by streams during major

Pancho Villa in 1914.

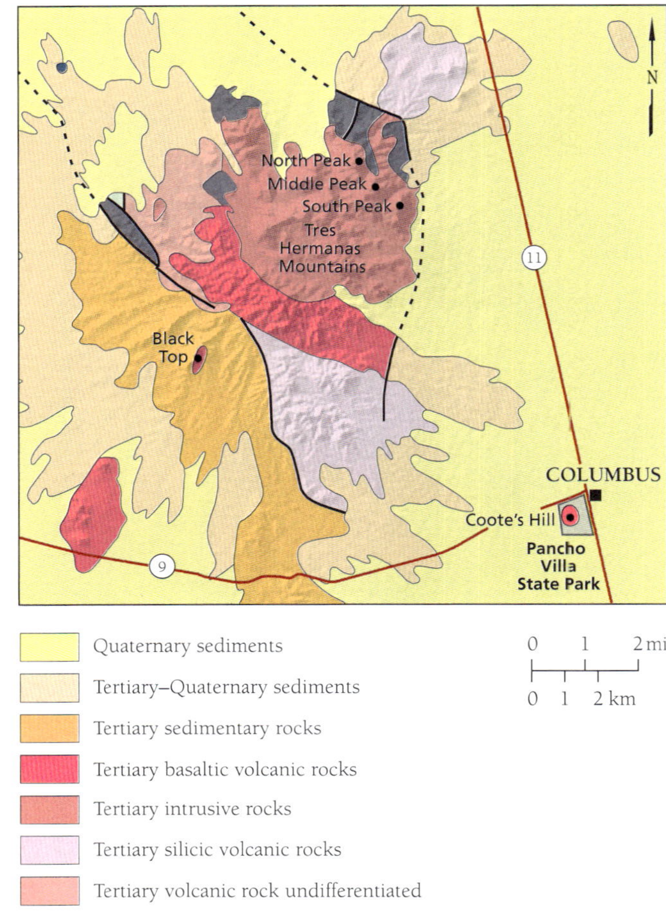

Generalized geology of the Tres Hermanas Mountains northwest of Pancho Villa State Park.

floods. Small sand dunes, testaments to the efficacy of the wind in transporting sediment in dry climates, are common throughout the park. Groundwater contained in the alluvial fan deposits is pumped to irrigate fields in the Deming and Columbus areas.

The most prominent feature within the northwestern part of the park is Coote's Hill, located near the park entrance. This hill rises approximately 25 feet above the surrounding desert floor. It is composed of reddish-brown to black basalt that erupted nearly 4 million years ago. Similar basalts were erupted southwest of Columbus and in the Potrillo volcanic field south of Las Cruces and east of Columbus. These basalts were derived from the mantle and were extruded during the late stages of formation of the Rio Grande rift. Evidence for the deep-seated origin of these magmas comes from the recognition of fragments of crust and mantle materials that were incorporated into the magma as it rose. Another

prominent feature in the basalts, the abundant gas bubbles or vesicles, formed as the magma came to the surface and cooled.

Geologic History

The broad alluvial plain and the mountains that rise above it are part of the Rio Grande rift as it merges southward and westward with the Basin and Range province. The crustal stretching associated with rifting produced down-dropped fault blocks that form basins and are flanked by uplifted mountain blocks. The Rio Grande rift consists of a series of such linked basins extending from central Colorado southward to northern Mexico. The Mimbres Basin east of Columbus and Deming is one of these down-dropped basins.

Much of the alluvial plain was deposited by the Mimbres River, which has its headwaters in the Pinos Altos Mountains and Black Range to the north. During the wetter climate of the Pleistocene, the Mimbres flowed past Deming and the Florida Mountains before passing east of Columbus into Mexico to fill playa lakes in Chihuahua. Today, the Mimbres River sinks into the sand of the Mimbres Basin near Deming.

Basalt outcrop on Coote's Hill.

Tres Hermanas Mountains northwest of Pancho Villa State Park.

Geologic Features

Looking south from Coote's Hill, the small village and port of entry of Palomas, Mexico, and the desert of northern Chihuahua beyond are visible. The Tres Hermanas Mountains are three prominent peaks

The reddish-brown 4 million-year-old basalts that make up Coote's Hill.

Visitors Center at Pancho Villa State Park. The Florida Mountains are in the background.

visible 7 miles northwest of the park. They consist mostly of volcanic rocks and a granitic intrusion that is about 35 million years old. The Tres Hermanas mining district (1885–1957) lies in these mountains. It produced an estimated $600,000 worth of zinc, lead, silver, gold, and copper. Along the far western skyline, the jagged peaks of the Big Hatchet Mountains consist mostly of faulted and tilted Paleozoic limestones and Cretaceous shales and sandstones. The Florida Mountains, an eastward-tilted Basin and Range fault block, are visible to the north and northeast. That range includes Precambrian basement rocks, Cambrian granites, a thick Paleozoic sedimentary section, and extensive Cenozoic volcanic rocks and alluvial fan deposits.

—Virginia T. McLemore

Additional Reading

Pancho Villa, by F.E. Kottlowski, New Mexico Bureau of Geology and Mineral Resources, New Mexico Geology, v.2, no.3, 1980.

Pancho Villa State Park, by V.T. McLemore, New Mexico Bureau of Geology and Mineral Resources, New Mexico Geology, v.22, 2000.

Pancho Villa and his men.

If You Plan to Visit

The Visitors Center was originally the U.S. Customs House, built in 1901. It has extensive displays related to Pancho Villa's raid and the punitive expedition into Mexico by Pershing. Across the street from the park at the old railroad depot is the Columbus New Mexico Historical Society Museum.

Pancho Villa State Park with its campground is located in the village of Columbus, New Mexico. From I–10 in Deming go 35 miles south on NM 11. The park lies just south of the junction with NM 9.

Pancho Villa State Park
400 West Highway 9 (P.O. Box 450)
Columbus, NM 88029
(575) 531-2711
www.emnrd.state.nm.us/spd/panchovillastatepark.html

Big Hatchet Mountains Wilderness Study Area

The Big Hatchet Mountains Wilderness Study Area (WSA) covers about 66,000 acres in Hidalgo County (southwestern New Mexico), near the United States/Mexico boundary, approximately 70 miles south-southeast of Lordsburg. The WSA encompasses Big Hatchet Peak, which, at an elevation of 8,366 feet, is the highest mountain in the "bootheel" of New Mexico. The Big Hatchet Mountains expose the thickest successions of Paleozoic and Early Cretaceous sedimentary rocks in southwestern New Mexico. The Continental Divide National Scenic Trail passes through the area, and a diverse flora and fauna can be found, which include mountain mahogany, desert bighorn sheep, and bat colonies.

Geography and Geology

The Big Hatchet Mountains are a fault-block mountain range within the southern Basin and Range physiographic province. They are a complexly faulted block of relatively old sedimentary rocks (the "range") surrounded by nearly flat-lying valleys (the "basin") filled with geologically young, Tertiary and Quaternary sediments. Rocks in the mountain range are Paleozoic and Cretaceous in age, and most of them are limestones. The nearly flat basin floor to the west of the Big Hatchet Mountains is the Playas Basin, whereas the basin to the east (and extending into Chihuahua, Mexico) is the Hachita Basin.

The Rock Record

The bedrock succession of mostly limestone and dolomite has a total thickness of approximately 20,000 feet, about half of it Paleozoic and the other half Cretaceous. The oldest rocks in the Big Hatchet Mountains are Precambrian granite and quartzite seen at small outcrops along the northern and northeastern periphery of the range.

The oldest Paleozoic strata are separated from the Precambrian rocks by a major erosional break (The Great Unconformity). Paleozoic rocks range in age from Late Cambrian (about 490 million years old) to Early Permian (about 280 million years old). Another significant break in the rock record exists between Permian and Cretaceous time. The Cretaceous rocks are mostly of Early Cretaceous age, about 125 to 100 million years old. Fault blocks of Paleozoic rocks make up the high topography and peaks of the Big Hatchet Mountains, whereas the Cretaceous rocks are exposed in the lower hills and ridges of the range.

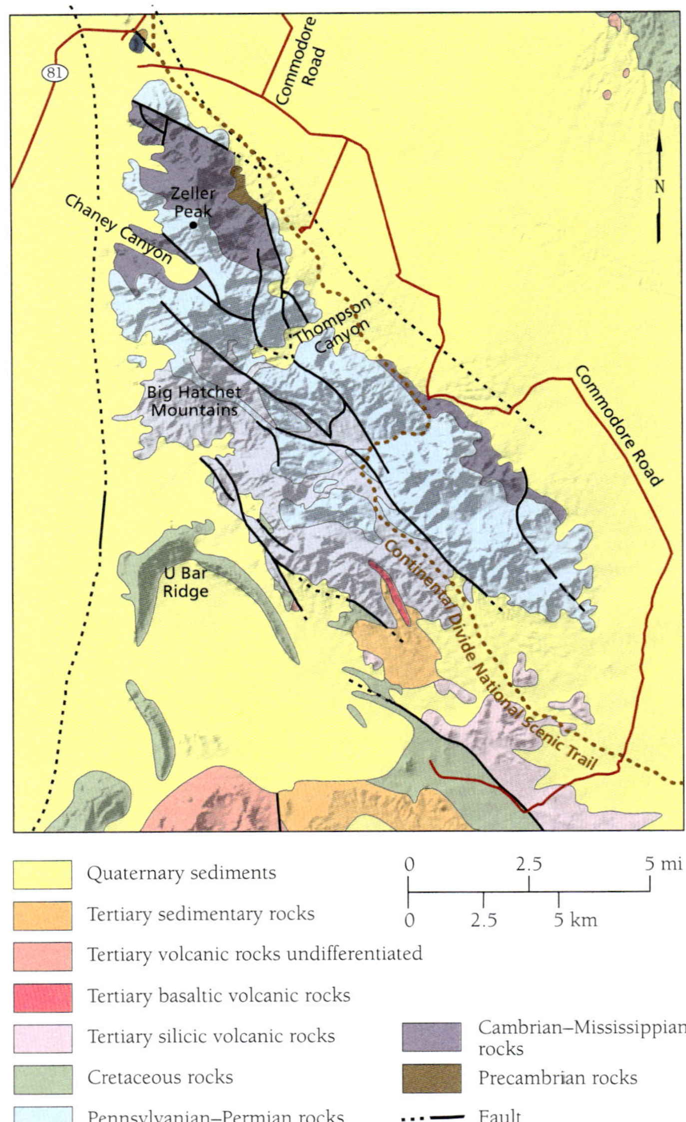

Geologic map of the Big Hatchet Mountains and vicinity. Several private inholdings lie within the wilderness boundary. Please refer to the BLM website for current land status.

Geologic History

In the Big Hatchet Mountains, Paleozoic sedimentary strata are mostly the deposits of shallow tropical seas that covered this part of New Mexico during the Late Cambrian, Ordovician, Devonian, Mississippian, Pennsylvanian, and Permian. The earliest (Late Cambrian) marine deposits are shoreline sandstones of the Bliss Formation—the oldest Phanerozoic rocks in New Mexico. During the Ordovician (about 470 to 445 million years ago), marine deposition of carbonate muds produced the limestones and dolomites of the El Paso Group and the overlying Montoya Formation.

After a period of nondeposition or erosion that lasted through the Silurian and into the Devonian, sea levels rose again during Late Devonian time (about 380 to 360 million years ago), producing the organic-rich black shales that are characteristic of the Percha Shale. During the Mississippian (about 360 to 320 million years ago), marine carbonate sediments of the Escabrosa Limestone and the overlying Paradise Formation were deposited. The Horquilla Formation of Pennsylvanian and early Permian age (about 320 to 290 million years old) contains thick limestone beds deposited as huge algal mounds (bioherms, or "reefs") that formed on the shallow, sunlit Pennsylvanian/Permian seafloors The sediments of overlying Earp Formation were deposited on tidal flats and in shallow marine embayments, also during early Permian time (about 290 to 280 million years ago). The tidal flat deposits yield fossil foliage impressions of extinct conifers and fossil footprints of amphibians and reptiles. Both the Horquilla and Earp formations are within the Naco Group.

The overlying strata of the Colina, Epitaph, Scherrer, and Concha formations (in ascending order, also in the Naco Group) are mostly shallow-water marine limestones deposited during the late part of the Early Permian (about 280 to 270 million years ago). The exception is the Scherrer Formation, an interbedded sandstone composed of wind-blown sand derived from the Coconino–Glorieta desert that extended across northern Arizona and northern and central New Mexico during a drop in sea level.

Following a major episode of exposure and erosion, lasting more than 130 million years, another incursion of the seas caused renewed deposition in the Big Hatchet Mountains during the Early Cretaceous. These strata belong to the Bisbee Group, which consists of three formations in the Big Hatchet Mountains. Within the Bisbee Group, the initial river and alluvial fan deposits of the Hell-to-Finish Formation are overlain by marine limestones and shales of the U-Bar Formation and those, in turn, are capped by river-deposited sandstones and marine shales of the Mojado Formation. The striking, thick limestone ridges of the U-Bar Formation represent reefs formed by rudistids, a group of unusual, reef-building clams of the Cretaceous that went extinct along with the dinosaurs.

Geologic Features

Paleozoic limestones and dolomites form bold cliffs and all of the higher topography in the Big Hatchet Mountains WSA. These rocks are full of marine fossils, mostly shells of brachiopods, bryozoans, and

Cliff-forming Paleozoic rocks flank the hills along the northwestern edge of the Big Hatchet Mountains.

At 8,366 ft elevation, Big Hatchet Peak (left-center) is the tallest mountain in New Mexico's bootheel.

crinoids, typical inhabitants of the Paleozoic seafloors. Big Hatchet Peak, particularly its north face, features tall cliffs of light-colored limestone. These are algal bioherms (or "reefs") of the Horquilla Formation that formed on shallow, tropical seafloors during the Pennsylvanian.

A hike up Mescal Canyon, beginning at the Continental Divide Trail entrance to the Big Hatchet Mountains WSA, allows you to examine the entire Cambrian, Ordovician, Devonian, and Mississippian succession of sedimentary rocks. Bugle Ridge and Newell Peak in the eastern part of the WSA provide relatively accessible outcrops of the Pennsylvanian–Permian Horquilla Formation. The Early Cretaceous rudist reef can be seen on U-Bar Ridge on the southwestern periphery of the WSA.

—*Spencer G. Lucas*

Brachiopods

Additional Reading

Mineral Resources of the Big Hatchet Mountains Wilderness Study Area, Hidalgo County, New Mexico: USGS Bulletin 1735-C, by Harald Drewes, H.N. Barton, W.F. Hanna, and D.C. Scott, 1988.

If You Plan to Visit

The Big Hatchet Mountains WSA is one of the more remote sites described in this book. To reach it, drive west from Deming or east from Lordsburg on I–10 to Exit 49 (the NM 146/Hachita/Antelope Wells Exit). Follow NM 146 south to the small village of Hachita, drive east on NM 9 and go south from the east side of Hachita on

NM 81. Between mileposts 35 and 34, turn left onto
CR C11 (Hatchet Road). In about a o.5 miles, the road
veers right. At approximately 2.1 miles, turn right on
to Commodore Road and follow it approximately
3.3 miles to a T-intersection. Turn right and drive
approximately 1.75 miles to where the Continental
Divide National Scenic Trail crosses the road. The
Trail is marked by cairns; there is no trailhead. Hike
southeast to go into the WSA. Note that once you begin
traversing CR C11, all roads are unpaved and will, in
places, require high clearance and four-wheel drive.
Before you go double check public land access online.
For more information:

Las Cruces Field Office, Bureau of Land Management
1800 Marquess Street
Las Cruces, NM 88005
(575) 525-4300
**www.blm.gov/programs/national-conservation-lands/
new-mexico/hatchet-mountains**

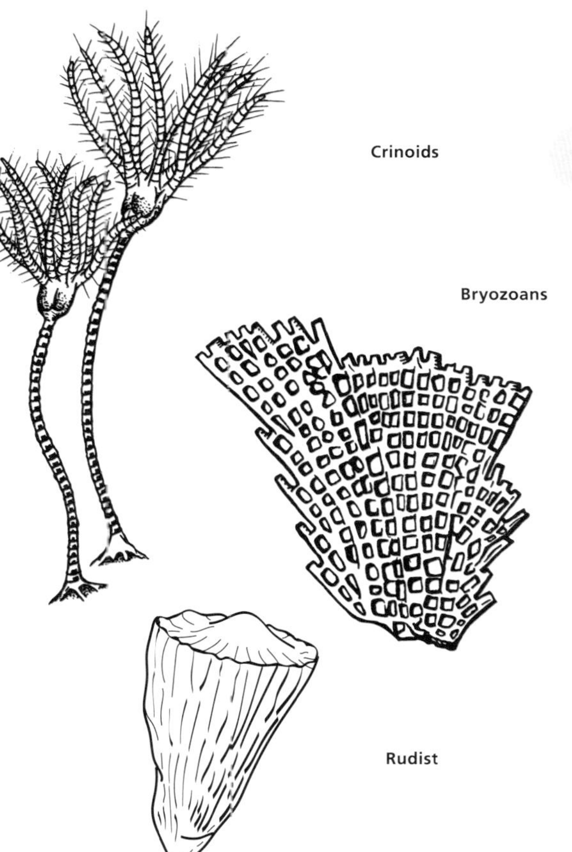

Geronimo Trail in the Southern Peloncillo Mountains

Many public lands in New Mexico's vast Bootheel region are difficult to visit because public roads are few and access may be blocked by locked gates. A notable exception is the historic Geronimo Trail, which bisects the rugged Peloncillo tract of the Coronado National Forest. Not to be confused with the Geronimo Trail National Scenic Byway, with its Visitor Center located in Truth or Consequences, this route follows Clanton Draw up to the Continental Divide, through the southernmost pass in the Peloncillo Mountains, and then into the Cottonwood Creek drainage, and on into Arizona. In addition to a scenic drive, there are several hiking trails, and the Geronimo Trail is a popular mountain biking destination.

There is a rich history in this area. The Apache warrior Geronimo led many raids from here and finally surrendered in nearby Skeleton

Trailhead of the Geronimo Trail just west of the divide heads north into the Whitmire Canyon Wilderness Study Area.

Canyon just to the north. The Clanton gang, outlaw cowboys of Tombstone notoriety, owned a ranch in the Animas Valley and ran rustled cattle through there. The Mormon Battalion, the only religion-based unit in U.S. military history, passed through in 1846, establishing a southern route through to California.

Today, an historical marker at the divide commemorates the Mormon Battalion and its travails in this rugged area, which involved road-building and the lowering of wagons over cliffs on ropes. Visitors can enjoy scenic views to the northeast of the Animas Valley and to the southwest into southern Arizona from the parking area here as well as hiking trailheads. A pack trail leads south into the Bunk Robinson Wilderness Study Area and further into the Guadalupe Canyon Wilderness Study Area beyond. Just west of the divide, a trail goes north into the Whitmire Canyon Wilderness Study Area. Beyond the divide, the Geronimo Trail drops steeply down switchbacks and passes a state line marker a few miles there after en route to Douglas, 34 miles further to the southwest.

Ecologically, the Peloncillos are a "sky island" that exhibits an amazing diversity of plants and animals due to a higher, cooler, and wetter local environment. As a result, the Peloncillo Mountains support Madrean pine and oak woodlands from the south that intermingle with elements of the chaparral of the Mogollon Rim country to the north, Chihuahuan Desert scrub from the east, and upper elevation Sonoran desert flora from the west. The region supports several endangered species, including Gila monsters and pincushion cacti. A confirmed U.S. sighting of a jaguar in the area occurred in 1997.

Geologic Setting

The Peloncillo Mountains lie in the Mexican Highlands section of the southern Basin and Range province, which is characterized by long north-south mountain ranges separated by low-lying desert basins. This fault-block topography is the result of east-west crustal stretching during the middle Cenozoic. The Geronimo

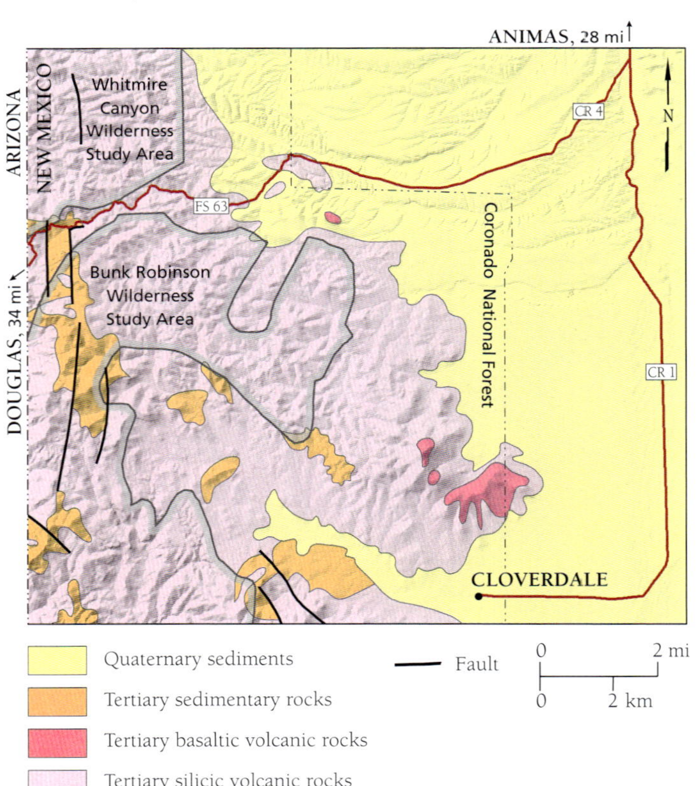

The Geronimo Trail through Coronado National Forest in the southern Peloncillo Mountains, gateway to the Whitmire Canyon and Bunk Robinson Wilderness Study Areas.

The southern Animas Valley and the northern Animas Mountains from the eastern side of the Coronado National Forest.

Trail leaves upper Quarternary fan surfaces and Neogene basin fill just beyond the Forest Service boundary, where it then winds through volcanic rocks from the massive explosive eruptions of two calderas, or supervolcanoes, within the southern Peloncillo area. These calderas, along with seven others in the neighboring Chiricahua, Pyramid, and Animas Mountains, are part of the Bootheel volcanic field. While volcanic eruptions and caldera formation ceased long ago here, Basin and Range crustal stretching continues today.

Geologic History

From the Late Cretaceous through the early Eocene (roughly 75 to 45 million years ago), the Laramide Orogeny deformed western North America, creating the Rocky Mountains to the north. At the end of the Laramide, the southwestern margin of North America began a transition from crustal shortening to crustal stretching. At the same time massive melting began in the upper mantle, setting off one of the largest episodes of volcanism in the planet's history.

Major volcanism impacted the Peloncillo region between about 33 and 27 million years ago. The oldest caldera in the southern Peloncillo Mountains, named the Geronimo Trail caldera, erupted lavas and huge volumes of ash-flow tuff that were deposited from rapidly flowing, incandescent clouds of gas and volcanic ash. About 5 million

132

Warping and folding of flow bands within Clanton Draw rhyolite.

years later, more ash-flow tuff was erupted from the Clanton Draw caldera, which developed on the northern flank of the older Geronimo Trail caldera.

The Rock Record and Geologic Features

Travelling west on the Geronimo Trail, one passes from rocks related to the Clanton Draw caldera into the older rocks erupted from the Geronimo Trail caldera. Upon entering the forest, the road drops from a high Quaternary gravelly surface into Clanton Draw, and within the first mile passes through hills of poorly exposed, water-lain tuffs and sandstones. These quickly give way to exposures of Clanton Draw rhyolite. In places the rhyolite is banded with quartz; further along the rhyolite has been greatly warped and contorted due to flowage. Approximately six miles west of the forest boundary, rocks related to the Geronimo Trail caldera appear, including the Gillespie Tuff, lavas, and breccias consisting of fragments of broken volcanic rock. These rocks are exposed all the way to the divide. Beyond the divide, more breccia is exposed before the road again descends, near the state line, into the Gillespie Tuff.

—*David J. McCraw*

From near the bottom of the switchbacks looking southeast, deposits of both calderas are apparent. Ash-flow tuffs of the Geronimo Tuff (GT) comprise the hills, while near the skyline are the Clanton Draw (CD) ash-flow tuffs.

If You Plan to Visit

To get to the east (New Mexico) end of the trail, take NM 338 south from I–10 near Lordsburg, New Mexico. Continue south from Animas on Hidalgo CR 1 for approximately 28 miles and turn right (west) at the sign for the Geronimo Trail. The road (shown on most maps as Hidalgo CR 4) crosses private land for 7 miles before reaching the National Forest (FS 63).

The region is beautiful, but very isolated and very near the Mexican border. Be aware that the town of Animas is the last opportunity for services unless you follow the entire Geronimo Trail to Douglas, Arizona. Be aware also of highly changeable weather and road conditions—high-clearance or four-wheel vehicles are generally recommended. For more information:

Douglas Ranger District
Coronado National Forest
1192 West Saddleview Rd.
Douglas, AZ 85607
(520) 364-3468
www.fs.usda.gov/coronado

Outcrop of volcanic breccia, just west of the divide.

THE RIO GRANDE RIFT

The formation of the Rio Grande rift has been one of the most influential tectonic events to have affected New Mexico during the Cenozoic era and is intimately linked with the state's landscape, wildlife, scenery, distribution of resources, and cultural history. The Rio Grande rift extends well beyond the state of New Mexico, and spans an impressive north-south distance of over 625 miles from Colorado, through New Mexico, and into western Texas and Chihuahua, Mexico. The rift bisects New Mexico and forms a spine of mountains and basins down the central axis of the state. This region hosts many of the iconic landscapes associated with New Mexico, ranging from tall peaks with extensive pine forests to arid lowlands of the Chihuahuan Desert.

The Rio Grande rift is located at the nexus of five main physiographic provinces of New Mexico, including the Rocky Mountains to the north, the Great Plains to the east, the Basin and Range to the southwest, the Mogollon–Datil volcanic field to the west, and the Colorado Plateau to the northwest. The rift also cradles the Rio Grande (or the Rio Bravo del Norte as it is known in Mexico), a river which has sustained human life for millennia. Its flanking mountains expose the oldest rocks found in New Mexico, yet its basins are still being deposited with sediments. This rich landscape also contains numerous and diverse wilderness areas, state parks, and national monuments aimed at preserving the geologic wonders, cultural heritage, and wildlife found within the Rio Grande rift.

What is the Rio Grande Rift?

A rift is a linear zone within Earth's crust that has been stretched and thinned due to extensional stresses, resulting in faulting and magmatism. In the Rio Grande rift, extensional stresses are primarily oriented east-west to produce a north-south belt of uplifts and basins throughout central New Mexico. The rift is composed of many individual basins, and in central and southern New Mexico, these include the Albuquerque, Socorro, Engle, Palomas, Mimbres, Mesilla, Jornada del Muerto, and Tularosa basins. Not coincidentally, some of these basins also contain the Rio Grande, a major river with headwaters in the San Juan Mountains of southern Colorado.

Extension in the Rio Grande rift began roughly 35 million years ago during the Eocene, although this early episode of crustal stretching was relatively minor and the region was mainly affected by widespread

OPPOSITE: Pennsylvanian limestones cap the summit ridge of the Caballo Mountains, an uplifted flank of the southern Rio Grande rift.

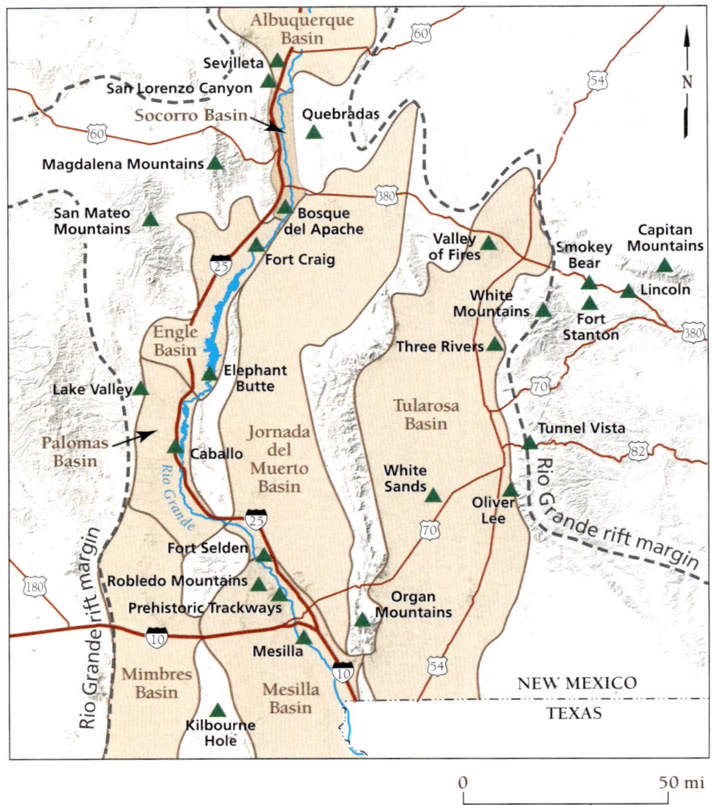

Index map showing location of the Rio Grande rift area parks included in this volume. Also shown are basins occupied by the Rio Grande during its evolution.

outpourings from volcanoes scattered throughout central and southern New Mexico. From 25–10 million years ago, extension along the entire length of the rift intensified. It was during this time period that the majority of the sediment was deposited within rift basins, most of which was derived from mountain ranges that rose along the flanks of the rift.

Extension in the rift has since slowed, although there is evidence that it is still active. Geologists have used extremely accurate global positioning systems (GPS) to measure movement of Earth's crust in New Mexico over a time period of several years. They found that the rift is currently spreading about 1–2 millimeters/year (0.04–0.08 inches) in southern New Mexico. This means that every year the western edge of the rift moves away from the eastern edge of the rift by about the thickness of a penny. These GPS studies tell us that the Rio Grande rift is still active, and while 1–2 millimeters/year is not very much over a human lifetime, it can still add up to a significant distance if the rift continues to open for thousands to millions of years.

Although each basin in the rift preserves its own unique geologic history, there are several general attributes of the Rio Grande rift as a whole that have a profound impact on the landscape in southern New Mexico. Within the deep roots of the rift, the lower crust is thinned by stretching, allowing mantle material to rise to shallower levels because it is hot and buoyant. As it rises, the mantle begins to partially melt, creating pools of molten rock called magma chambers. These melts usually cool and solidify within the crust of the Earth, but occasionally they are able to make their way to the surface and erupt, producing mainly basaltic lava flows.

As the upper, shallow part of crust stretches and breaks along faults, some crustal blocks drop down along the fault lines to create depressions called basins. In each rift basin, the faulting is asymmetric, with a major fault developing along only one margin of the basin. As one side of the fault is down dropped, the other side is raised to form a rift-flank uplift. These rift-flank uplifts include many of the rugged, iconic mountain ranges of central and southern New Mexico, including the Sandia and Manzano mountains, the Sierra Ladrones, the Magdalena Mountains, the San Mateo Mountains, the Sacramento Mountains, the White Mountains, the San Andres Mountains, the Caballo Mountains, and the Organ Mountains (as well as the Franklin Mountains in west Texas, just south of the New Mexico border).

Streams and rivers carried mud, sand, and gravel from adjacent highlands and deposited them in the developing rift basins. Many of these deposits are porous and provide excellent reservoirs for groundwater. Groundwater aquifers within rift basins are essential to human habitation in our semiarid environment, and this accounts for why the majority of the state's population resides within the rift.

Faulting in the Rio Grande Rift

As noted earlier, the crust in the rift zone stretches mainly through movement along faults. As this happens, older rocks are brought to shallower levels in the crust within the uplifted blocks and are eventually exposed at the surface. Some of the oldest rocks found in southern New Mexico include 1.4 to 1.7 billion-year-old Precambrian igneous and metamorphic rocks that are exposed within most of the uplifted ranges along the rift flanks. These uplifts provide opportunities for geologists to investigate these otherwise-buried rocks that record some of the early geologic history of New Mexico.

Although the majority of faults in the central and southern Rio Grande rift are millions to tens of millions of years old, many are still active. By studying these faults, geologists have determined that many of them are younger than about 15,000 years old. This is evident in the topography of the land, where young faults commonly form steps in the landscape called fault scarps. These scarps are an indication that the rift in New Mexico is still active today.

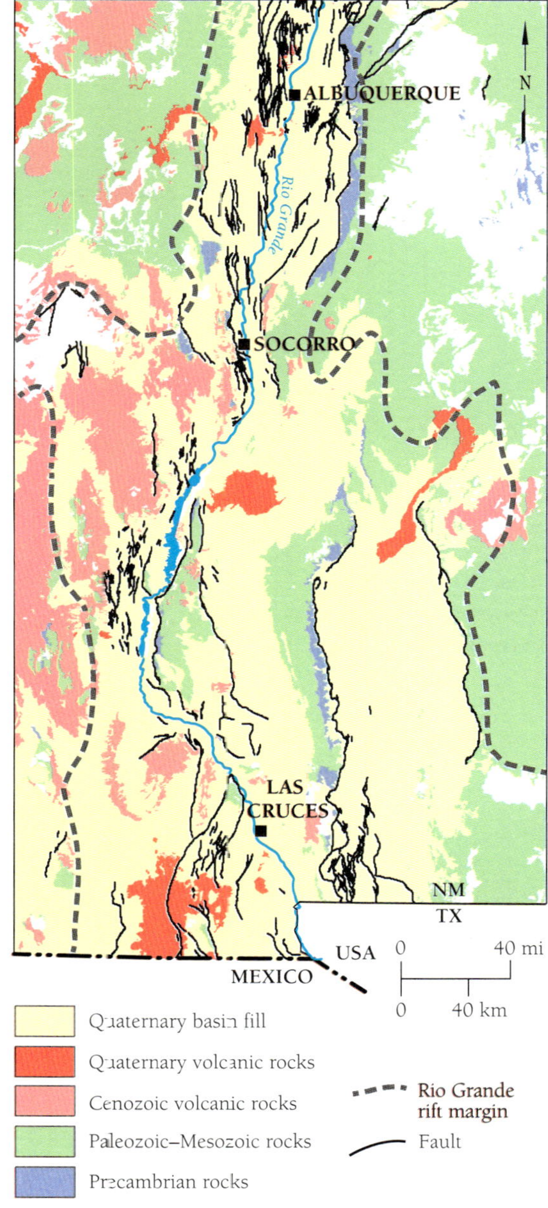

Simplified geologic map of the southern Rio Grande rift. Note the multitude of faults that mark the rift margins.

The north-south orientation of mountains produced by rift faulting imparted a topographic "grain" to the landscape that has influenced the locations of trade routes and settlements for hundreds and perhaps thousands of years. The longest and one of the most heavily traveled trade routes was El Camino Real de Tierra Adentro (the Royal Road of the Interior Lands), which extended roughly 1,600 miles from Ohkay Owingeh (San Juan) Pueblo in northern New Mexico to Mexico City. Except where it crossed the Jornada del Muerto, the Camino Real in central and southern New Mexico closely followed the Rio Grande. This route was initially a trade corridor for Native Americans between what is now New Mexico and central Mexico. During the 16th and 17th centuries, it was heavily utilized by Spanish explorers and colonists, and it was later exploited by Confederate soldiers during the Civil War. I–25, running north-south through central and southern New Mexico, largely stays within the basins of the rift and shows that the rift-flank ranges continue to influence human travel and commerce.

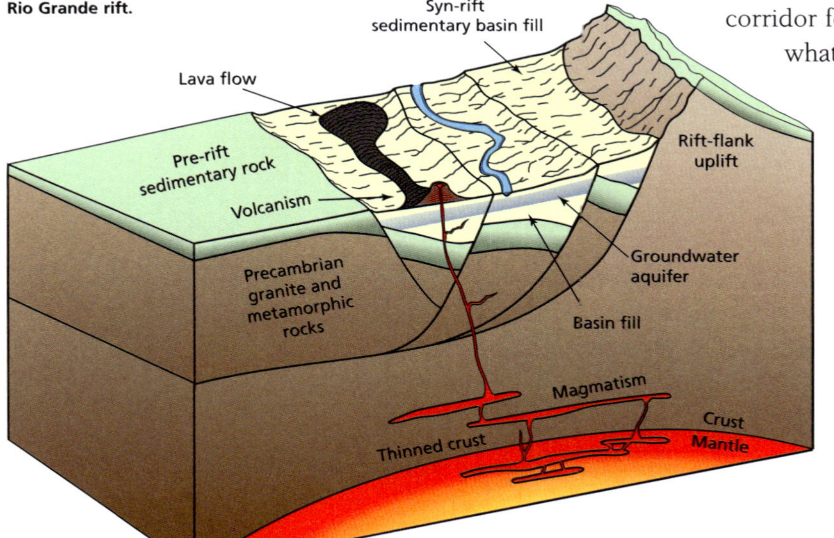

Schematic diagram showing the generalized structure of the Rio Grande rift.

The Manzano Mountains that form the eastern flank of the Rio Grande rift near Belen.

Volcanism in the Rio Grande Rift

Although faulting in the rift influenced both trade routes and the location of settlements, volcanism has also significantly influenced human cultural history in the region. Volcanic rocks are a natural byproduct of continental rifting. As the crust stretches and thins, hot rock in Earth's mantle moves upward and partially melts. These melts rise and pool within Earth's crust as magma chambers, and one of these magma chambers is still active today. Termed the Socorro magma body, it lies about 12 miles deep beneath the central Rio Grande rift and covers an area of over 1,300 square miles between the towns of Belen and San Antonio, New Mexico.

Many other magma chambers have existed and solidified in the crust during the development of the Rio Grande rift. Some of these magma chambers erupted at the surface to form volcanoes. Volcanic eruptions have occurred throughout the life of the Rio Grande rift, including some very recent ones. The Albuquerque volcanoes produced a series of basaltic lava flows within the Albuquerque Basin about 220,000 years ago. This is the location of Petroglyph National Monument, one of the largest petroglyph sites in North America (described in the companion volume to this book, The Geology of Northern New Mexico's Parks, Monuments, and Public Lands). Other significant outpourings of lava that have shaped the landscape include the Cat Hills volcanoes in the southern Albuquerque Basin (about 140,000 years old), Mesa del Contadero (820,000 years old), the Jornada del Muerto volcano (78,000 years old), Kilbourne Hole (47,000 years old), and the Potrillo

Maps showing major stages of
evolution of the Rio Grande system
over the past 5.3 million years.

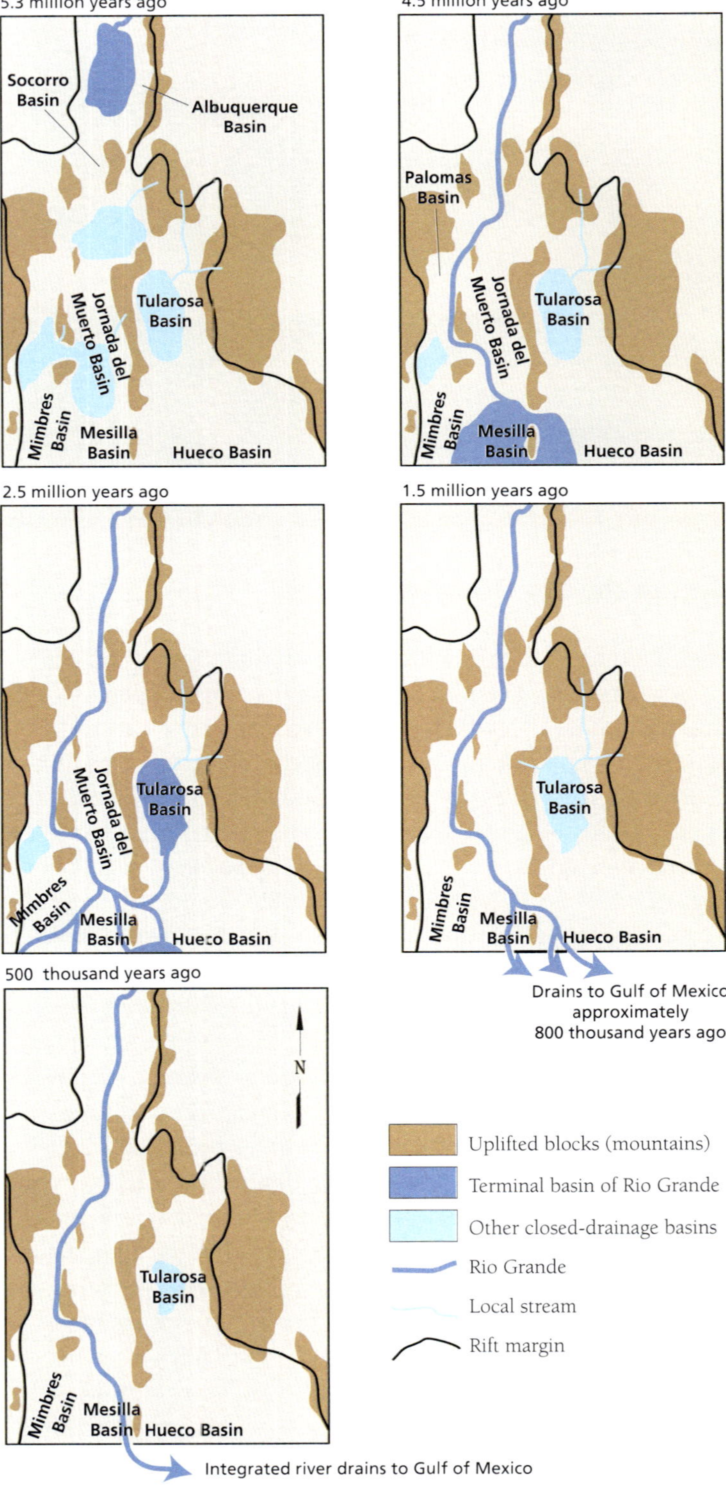

volcanic field of southern New Mexico, which was the site of hundreds of basaltic eruptions between 1.2 million and 20,000 years ago. Valley of Fires National Recreation Area is the location of the extremely young Carrizozo lava flows, which erupted a mere 5,200 years ago and flowed southward into the Tularosa Basin. These youthful eruptions are yet another testament to an active Rio Grande rift.

The Rio Grande

The Rio Grande is the fourth longest river in the United States and flows nearly 1,900 miles from the San Juan Mountains in Colorado to the Gulf of Mexico. Along that distance, it drains more than 260,000 square miles of the southwestern United States and northern Mexico. The river bisects the state of New Mexico, providing a lifeline of water and habitat within an otherwise semiarid region. This river system evolved over a span of millions of years, localized by the north-south grain of basins within the Rio Grande rift.

Interestingly, the Rio Grande is not a typical river that grew by "headward" erosion from coastal areas toward the mountains. Rather, it is a river that spilled southward from its headwaters to eventually reach the Gulf of Mexico. Beginning about 25 million years ago, crustal stretching in the Rio Grande rift had begun to form individual basins bounded by uplifts. These basins were not initially connected by a through-flowing river, but instead contained ephemeral playa lakes (similar to, but smaller than, the Great Salt Lake of Utah). The lakes mostly received runoff from local flanking uplifts. Runoff carried sediment into the basins, eventually producing accumulations (now known as the Santa Fe Group) as much as several miles thick in some basins. The ancestral Rio Grande initially drained into rift basins in southern Colorado and northern New Mexico. It eventually filled these basins and spilled southward, reaching the southern Albuquerque Basin about 5 million years ago. By 4.5 million years ago, the growing river system had spilled over into the Socorro and Palomas basins, and by 3.5 million years ago, it had reached the Mesilla Basin in the Las Cruces area. About 2 million years ago, it crossed the southern border of New Mexico, where it terminated within playa lakes in northern Mexico.

Satellite image of the Socorro area showing a section of the Rio Grande rift with relatively well-defined flank uplifts. The green irrigated areas along the river stand in sharp contrast with the otherwise arid landscape.

Through this process of downward integration, the ancestral Rio Grande grew in length, eventually reaching the Gulf of Mexico about 800,000 years ago to form the Rio Grande as we know it today.

As the Rio Grande rift developed, surface waters from other smaller rivers and streams percolated into underlying sandy sediment and began to accumulate, forming aquifers that are now found within every basin of the Rio Grande rift. Rainfall, surface water, and groundwater are naturally interrelated through the hydrologic cycle. However, groundwater in aquifers can take decades to millennia to replenish, and much of the present groundwater accumulated during the wetter climates of the ice ages some 10,000 or more years ago.

142

View north from Mesilla during the fall showing a nearly dry Rio Grande river bed. Picacho Mountain (left) and the Robledo Mountains are visible in the background.

In New Mexico, groundwater is being used more rapidly than it is being replenished. Although this water is hidden from view, it represents one of the greatest natural resources of New Mexico and provides clean drinking water for nearly four out of five New Mexicans. This groundwater, along with surface water diverted from the Rio Grande through an extensive system of dams and ditches (acequias), also remains a vital resource for agricultural production and ranching within the state. Sustainable use of these precious resources will remain a formidable challenge in our semiarid environment as population grows and the demand for clean, fresh, drinking water increases.

—*Jason W. Ricketts*

Cotton, pecans, and chile peppers are three high-value crops grown with irrigation water from the Rio Grande.

Sevilleta National Wildlife Refuge
U.S. FISH & WILDLIFE SERVICE

Sevilleta National Wildlife Refuge (NWR) is located in central New Mexico about 15 miles north of Socorro. The refuge spans the Rio Grande valley from the Sierra Ladrones on the west to the Los Pinos Mountains on the east, an area of approximately 30 miles by 15 miles, encompassing 230,000 acres. The area was designated a Spanish land grant, Sevilleta de la Joya, in 1819. The land grant was sold in 1928 to Socorro County and then, in 1936, to General Thomas D. Campbell, under whom it became a ranch for sheep and cattle. The Campbell Family Foundation donated the land to the Nature Conservancy in 1973, which then passed it on to the U.S. Fish & Wildlife Service to become a national wildlife refuge dedicated to preservation and enhancement of the integrity and natural character of the land. The transfer stipulated that the land should undergo natural processes of succession, including floods and fires, without human interference. Thus, most of the refuge has very limited public access, but the Mesa View Trail west of the Visitor Center is open for hiking at least five days per week.

OPPOSITE: Water-filled depressions carved in Pennsylvanian marine limestones were formed by waterfalls due to spring flow and flooding events. This photo location (Cibola Springs) is generally not open to the public.

Geological Overview

Sevilleta NWR straddles the southern edge of the Albuquerque Basin of the Rio Grande rift where it narrows into the Socorro Basin. Pre-rift rocks (older than 25 million years and ranging in age back to 1.6 billion

Block diagram of Sevilleta National Wildlife Refuge illustrating the generalized stratigraphic units and fault blocks of the Rio Grande rift.

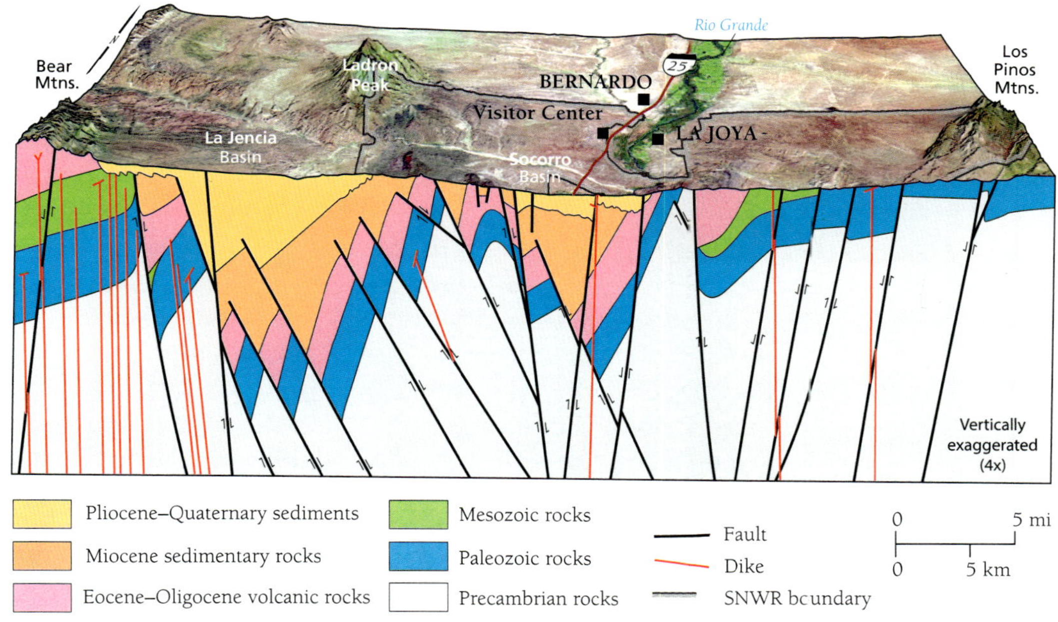

years) are exposed on both sides of the Rio Grande valley in the southern part of the refuge. Rift-flank uplifts to the west are the Sierra Ladrones and the southeastern margin of the Colorado Plateau. To the east are the Los Pinos Mountains, where a complex zone of faults related to Laramide crustal shortening 75–45 million years ago is located. The Rio Grande rift is still expanding as western New Mexico moves very slowly away from eastern New Mexico, resulting in east-west stretching and north-south rending of the continent. Hot mantle rocks have risen to fill the space beneath the stretched crust, resulting in episodic volcanic eruptions within the rift. The most recent manifestation of this is the Socorro magma body, a pancake-shaped igneous intrusion located at a depth of about 12 miles. The Rio Grande itself is a relatively recent addition to the geologic story. It began to flow through the southern Albuquerque and Socorro basins about 5 million years ago.

Geologic Features

The Mesa View Trail starts and ends at Sevilleta NWR Visitor Center and forms two loops ascending, crossing, and descending the top of the mesa to the west, where panoramic views are stunning. The trail rises 0.8 miles and more than 200 feet to the mesa at the top of the cliffs and branches there. The north loop (3.8 miles) parallels the top of the cliffs along the edge of the mesa to a spectacular view of an eroded bowl exposing the downthrown and warped sediments west of the north-trending Cliff fault. The route continues to the north and east on the downthrown side of the fault, then crosses through a narrow gap cut into the upthrown side, and continues south to the Visitor Center over gentle ridges and drainages. The south loop, newly called the Ladrones Vista Trail (1.9 miles) follows the Cliff fault and adjacent eroded bowls cutting the downthrown block of sediments. Before turning east, the trail crosses a high terrace of the Rio Salado, today 3 miles farther south. The route descends across tilted sediments and crosses the fault onto deeply eroded sediments on the upthrown side of the fault. The trail then continues east and north to the Visitor Center.

The Mesa View Trail traverses weakly-consolidated sedimentary rocks, uplifted by the Cliff fault, that are about 2 to 4 million years old. Gravels exposed in the cliffs were carried here by the Rio Salado and

Conglomerate fill in a former river channel is exposed along the Mesa View Trail. Pebble orientations and compositions indicate that this channel flowed to the southeast and was related to the Rio Salado, now 3.6 miles south of the trail.

147

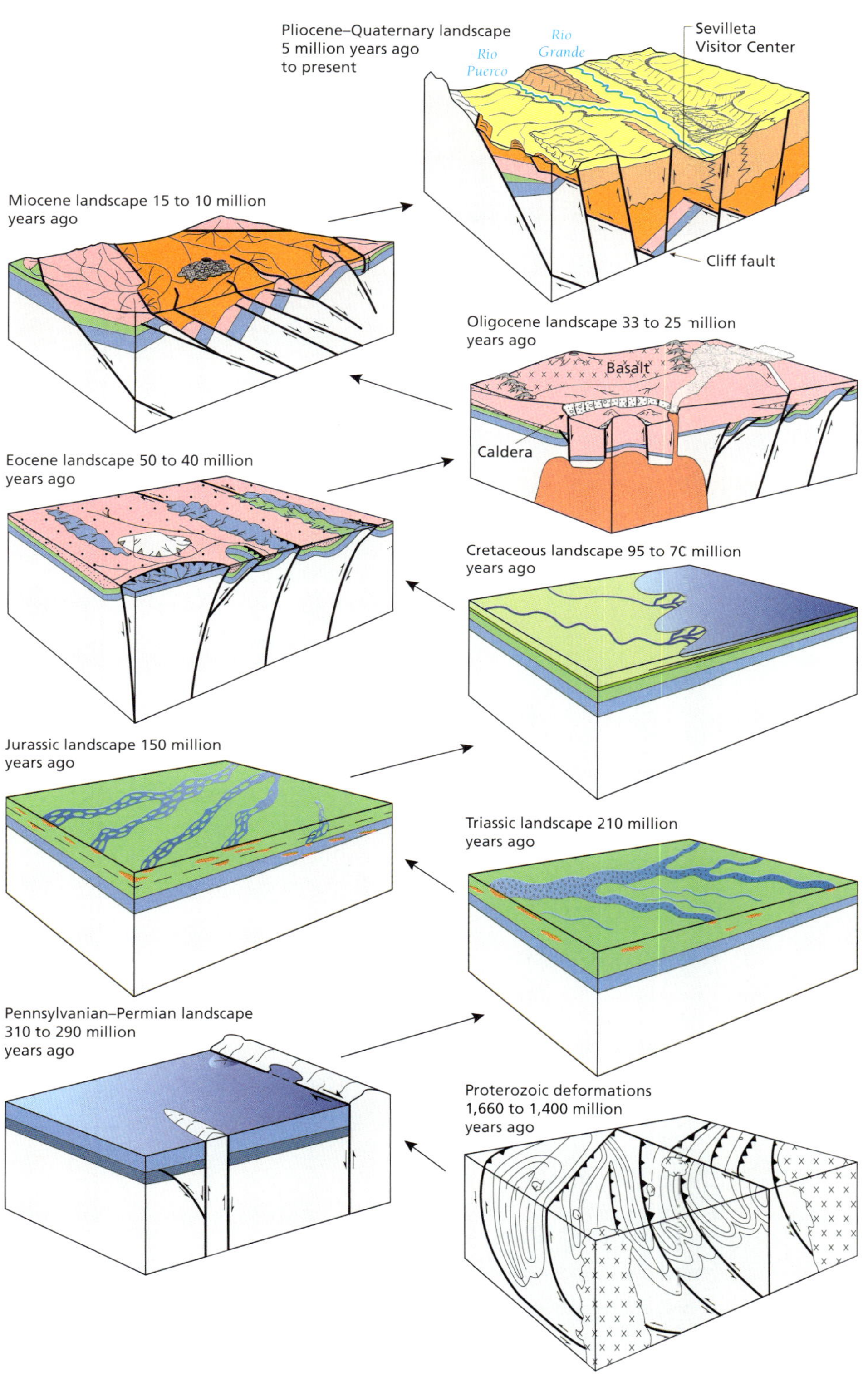

Block diagrams showing a progression of different landscapes through geologic time from the oldest rocks at the lower right, to the present landscape near Sevilleta NWR Visitor Center at the top. Many intervening geologic episodes are not illustrated.

Panorama of sedimentary deposits cut by the north-trending Cliff fault. The trace of the fault (between the white lines) follows the dark rock and steep grassy slopes. To the west (left) are the pinkish sediments of the downthrown and warped block. To the right of the fault are the more cliff-forming cemented sedimentary rocks of the upthrown block.

the Rio Puerco. From the top of the trail, magnificent vistas of the vast surrounding landscape show the extent of the rift across Sevilleta NWR and the many fault blocks forming the distant hills and mountains. Much of the area between the Ladrones Vista Trail and San Lorenzo Canyon is currently being uplifted by the Socorro magma body, but so slowly that it is not noticeable without careful measurement.

—*David W. Love and Richard M. Chamberlin*

Additional Reading

Block diagrams and cross sections illustrating geologic and tectonic evolution of the Sevilleta National Wildlife Refuge, Rio Grande rift, central New Mexico, R.M. Chamberlin and D.W. Love, New Mexico Bureau of Geology and Mineral Resources, Open-File Report 579, 2016.

If You Plan to Visit

The Sevilleta Visitor Center is 0.4 miles west of I–25 at Exit 169, 56 miles south of Albuquerque and 19 miles north of Socorro. The trails and a Nature Loop (1.1 miles) can be accessed behind the Visitor Center. Check their website for hours. For more information:

Sevilleta National Wildlife Refuge
P.O. Box 1248
Socorro, NM 87801
(505) 864-4021
www.fws.gov/refuge/Sevilleta

San Lorenzo Canyon Recreation Area
BUREAU OF LAND MANAGEMENT

San Lorenzo Canyon is a hidden recreational and geological gem located near the axis of the Rio Grande rift in central New Mexico, about 15 miles north-northwest of Socorro. A spectacular angular unconformity, rift faults, and basin-fill deposits are beautifully exposed in the walls of San Lorenzo and adjacent canyons. The canyon is a primitive recreation area primarily managed by the Bureau of Land Management with collaboration from the adjacent Sevilleta National Wildlife Refuge.

Rock pillar (hoodoo) in middle Popotosa conglomerate beds at the Narrows. Hoodoos and cliffs in San Lorenzo Canyon are the expression of geologically rapid uplift and erosion over the last 125,000 years. Like a giant seesaw, San Lorenzo Canyon is now on the high side of an active west-tilting 12-mile-wide fault block.

Regional Setting

The intermittent stream in San Lorenzo Canyon cuts across Oligocene lava flows and Miocene conglomerates and sandstones exposed in variably tilted fault blocks on the north flank of the Lemitar Mountains, where domino-style extension has greatly stretched (2x) the brittle upper crust within the Rio Grande rift. As shown in the block diagram, two large rift faults (the Lemitar and Silver Creek faults) intersect an older northeast-trending shear zone within the crystalline basement rocks underlying San Lorenzo Canyon. Between the north end of the Lemitar fault and the south end of the Silver Creek fault, the shear zone has been reactivated as a broad southeast-down

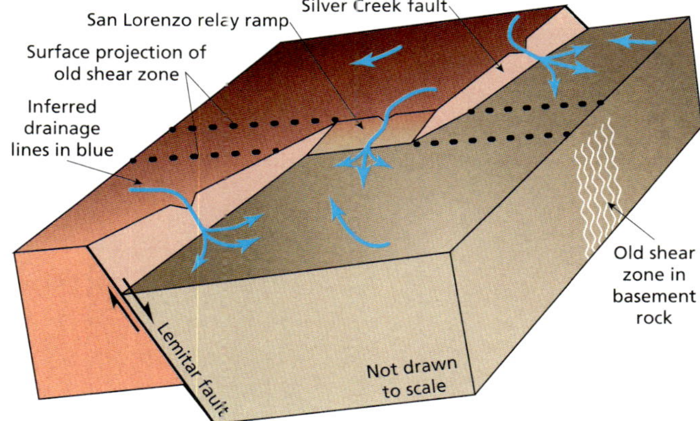

Block diagram illustrating regional westward block tilting (rifting) and initial growth of the San Lorenzo relay ramp in early Miocene time.

fault zone and relay ramp. Along lower San Lorenzo Canyon, moderate-displacement rift faults have been deflected near the shear zone, in much the same way as a board splits around a knot. Many other structural complexities occur in this area but are beyond the scope of this book. Interested readers should consult the free online publication listed at the end of this chapter.

Erosional remnants of ice-age gravels (less than 125,000 years old) stand about 180 feet above the floor of the narrows of San Lorenzo Canyon and attest to the relatively recent cutting of the canyon. Satellite measurements indicate the ground surface here is rising about 0.75 inches every 10 years. At this rate of uplift, the cliffs could have formed during the past 30,000 years. This ongoing uplift of the San Lorenzo Canyon region is a result of expansion of the seismically detected Socorro magma body, a broad pancake-shaped body of molten rock at 12 miles depth.

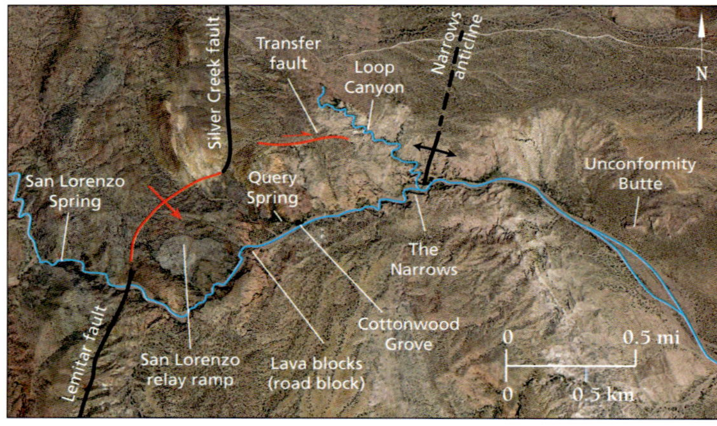

Satellite image showing geologic sites of interest and major structural features along San Lorenzo Canyon.

Rock Record and Geologic History

Beginning about 33 million years ago, explosive eruptions of huge volumes of ash-flow tuff from calderas near Magdalena, and many eruptions of thin basaltic lavas (29–28 million years ago) from fissures, formed a volcanic plateau over 2,600 feet thick in the region. A 28 million-year-old dacite lava flow and ash beds locally cap the volcanic plateau near San Lorenzo Spring. Disruption of this volcanic plateau by block faulting began about 20 million years ago.

During Miocene time, gravels and sands were shed from fault-block volcanic highlands to fill the adjacent, subsiding rift basins. The early basin-fill deposits are up to 4,000 feet thick and consist of volcanic-rich stream gravels and playa-lake muds, now assigned to the Popotosa Formation. The Rio Grande first flowed into the San Acacia area about 5 million years ago. At this time, renewed faulting caused the Lemitar Mountains and San Lorenzo Canyon area to be uplifted and eroded. At least 1,000 feet of upper Pliocene and lower Pleistocene sediments (Sierra Ladrones Formation) accumulated in the basin south of San Acacia. During the Pleistocene ice ages, large streams flowing to the Rio Grande deposited wide ribbons of coarse gravels across what is now the north flank of the Lemitar Mountains and San Lorenzo Canyon.

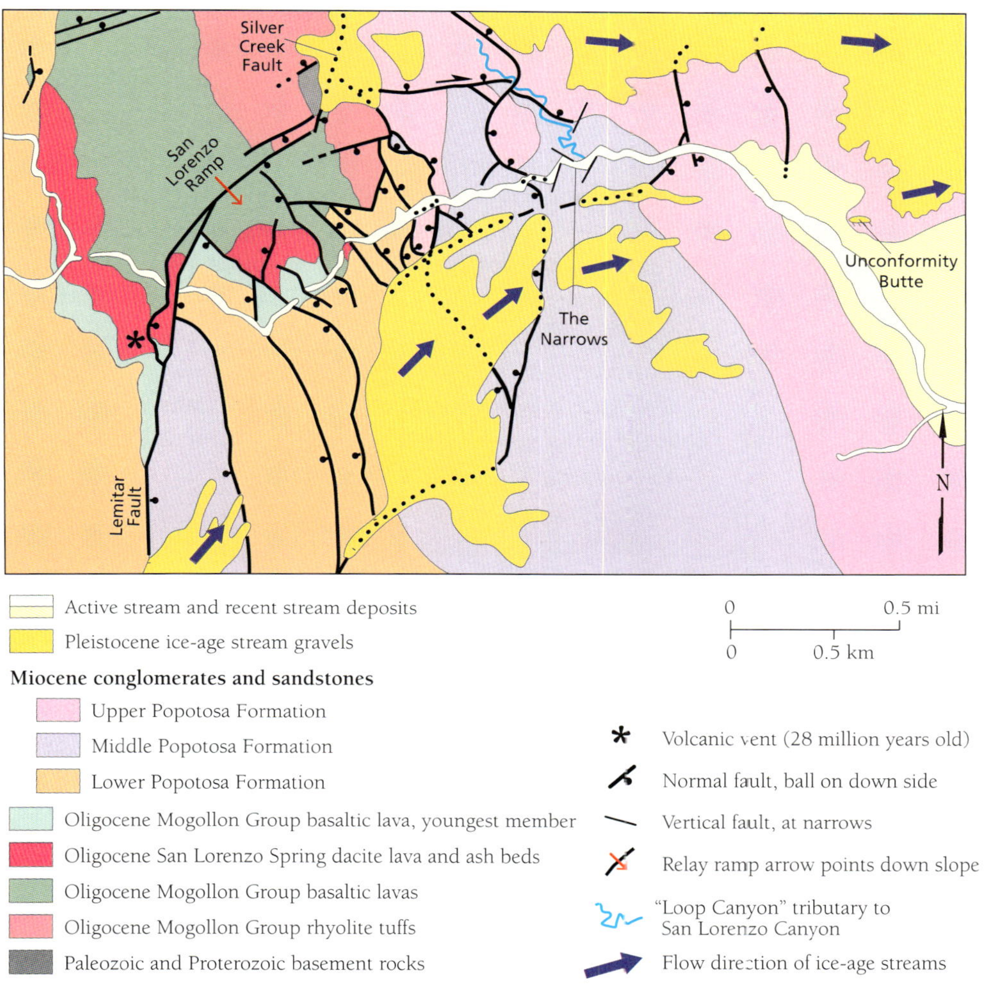

Active stream and recent stream deposits

Pleistocene ice-age stream gravels

Miocene conglomerates and sandstones

Upper Popotosa Formation

Middle Popotosa Formation

Lower Popotosa Formation

Oligocene Mogollon Group basaltic lava, youngest member

Oligocene San Lorenzo Spring dacite lava and ash beds

Oligocene Mogollon Group basaltic lavas

Oligocene Mogollon Group rhyolite tuffs

Paleozoic and Proterozoic basement rocks

0   0.5 mi

0   0.5 km

✱   Volcanic vent (28 million years old)

Normal fault, ball on down side

Vertical fault, at narrows

Relay ramp arrow points down slope

"Loop Canyon" tributary to San Lorenzo Canyon

Flow direction of ice-age streams

Geologic Features

Driving up San Lorenzo Canyon, one encounters several unique geologic features. The easternmost feature is the photogenic "Unconformity Butte," which has been used as a prime example of an angular unconformity in several geologic textbooks. This angular unconformity represents nearly 10 million years of missing rock record. Above the unconformity, horizontal stream gravels were deposited by ice-age streams about 125,000 years ago. Below the unconformity are east-tilted sandstone beds of the Miocene Popotosa Formation. These were originally horizontal when deposited in a subsiding rift basin about 10 million years ago. The Popotosa beds were subsequently tilted to the east by later rift faulting (5 to 3 million years ago), then beveled by erosion and buried by the younger gravels.

Generalized geologic map of the San Lorenzo Canyon area. Lower San Lorenzo Canyon parallels an east-northeast trending boundary between west-tilted blocks on the north and east-tilted blocks on the south.

Most visitors consider the nearly 180-feet-high vertical cliffs of the Narrows area to be the heart of San Lorenzo Canyon. It is a popular spot for picnics and hiking up a tightly wound tributary slot canyon, known by some as "Loop Canyon." The crooked course of the main canyon here is guided by near-vertical faults and fractures. Several erosional rock spires, called hoodoos, are present at the Narrows.

Cottonwood trees near "Query Spring" on the west end of lower San Lorenzo Canyon were planted by the BLM in the early 1990s as part of a riparian restoration project to eradicate non-native salt cedars. Looking west from the cottonwood grove, there is a good view of the south-tilted lavas and tuffs that form the San Lorenzo relay ramp.

A 1-mile hike along upper San Lorenzo Canyon provides some solitude and a good look at mafic (dark colored) lava flows and coarse conglomerates of the lower Popotosa Formation that contain abundant cobbles of dark lava and light-gray tuffs.

Three perennial springs are present in upper San Lorenzo Canyon. The final one, San Lorenzo Spring, is at a 3-foot-high waterfall that usually produces about 1 gallon of water per minute. Do not drink from the springs since cattle graze the ranchland upstream from here.

—*Richard M. Chamberlin*

Additional Reading

Block diagrams and cross sections illustrating geologic and tectonic evolution of the Sevilleta National Wildlife Refuge, Rio Grande rift, central New Mexico, R.M. Chamberlin and D.W. Love, New Mexico Bureau of Geology and Mineral Resources, Open-File Report 579, 2016.

South-facing cliff of "Unconformity Butte" shows east-tilted, sandstone beds of the upper Popotosa Formation. These beds were initially horizontal. The tilted beds were truncated by a horizontal conglomerate bed (cemented gravel of ancestral San Lorenzo Canyon) that was deposited by ice-age streams about 125,000 years ago. The erosion surface between the tilted beds and horizontal conglomerate is called an angular unconformity. It represents about 10 million years of missing geologic record.

If You Plan to Visit

To get to San Lorenzo Canyon, exit I-25 at the Lemitar (Exit 156). Take the west frontage road 4.5 miles north to the end of the pavement and the BLM sign to San Lorenzo Canyon. Head west on the dirt road 2 miles to the welcome sign for San Lorenzo Canyon. Turn north and then drive west up the channel of San Lorenzo Wash. The lower part of San Lorenzo Canyon is accessible by motor vehicle, but the upper part is accessible only by hiking. Four-wheel drive is recommended, because deep, loose sand is common; two-wheel and all-wheel drive vehicles can generally be used, but call the BLM Socorro Office to ascertain current conditions. Do not enter the channel when water is flowing or after a major rainstorm. No water, food, or other amenities are available. For more information:

Bureau of Land Management
Socorro Field Office
901 S Old US Hwy 85
Socorro, NM 87801
(575) 835-0412
**www.blm.gov/nm/st/en/prog/recreation/socorro/
san_lorenzo_canyon.html
www.recreation.gov/recreationalAreaDetails.do?contract
Code=NRSO&recAreaId=1840**

These two small faults cut Popotosa conglomerate beds in north wall of San Lorenzo Canyon about 0.1 mile west of the narrows.

Faults at Unconformity Butte are shown as they formed in Miocene time, when Popotosa sandstone beds were horizontal (the photograph has been rotated).

Quebradas Back Country Byway
BUREAU OF LAND MANAGEMENT

The Quebradas (Spanish for "breaks," a rugged or cliffy area) is a region of splendid scenery and easily accessible geologic exposures. Rocks from some of the most eventful times in New Mexico's geologic past, especially the Pennsylvanian and Permian are exposed along the Quebradas Back Country Byway. Triassic, Upper Cretaceous, Tertiary, and Quaternary strata are also present. The Byway is a 24-mile-long unpaved road with many stretches of nearly continuous geologic exposures. A free 21-page brochure is available online as well as at the New Mexico Bureau of Geology and BLM offices in Socorro (citation at end of chapter). It provides far more detail than this chapter can, especially to the 10 sign-posted stops along the Byway.

OPPOSITE: **View of the Yeso Formation (lower two thirds of the photo), Glorieta, and San Andres section.**

Regional Geology

The Quebradas Byway passes through the eastern flank of the Rio Grande rift, crossing numerous large and small faults in the process. A remarkable diversity of rocks are visible in the Quebradas area (see stratigraphic chart). Starting from the north end, the initial east-trending section of the road climbs in elevation on a ramp of young (less than 5 million years old) rift-filling detritus mainly eroded from Paleozoic rocks. The northern half of the road passes through exposures of Pennsylvanian and Lower Permian rocks, and the southern part of the road traverses more open country with younger Permian and Triassic deposits.

Geologic History

Most of the major geologic events in New Mexico's past are reflected in the Quebradas region. First, granitic and metamorphic basement rocks formed roughly 1.6 billion years ago at a depth of about 10 miles within the crust. This was followed by a prolonged episode of uplift and deep erosion that brought these rocks to the surface, producing the

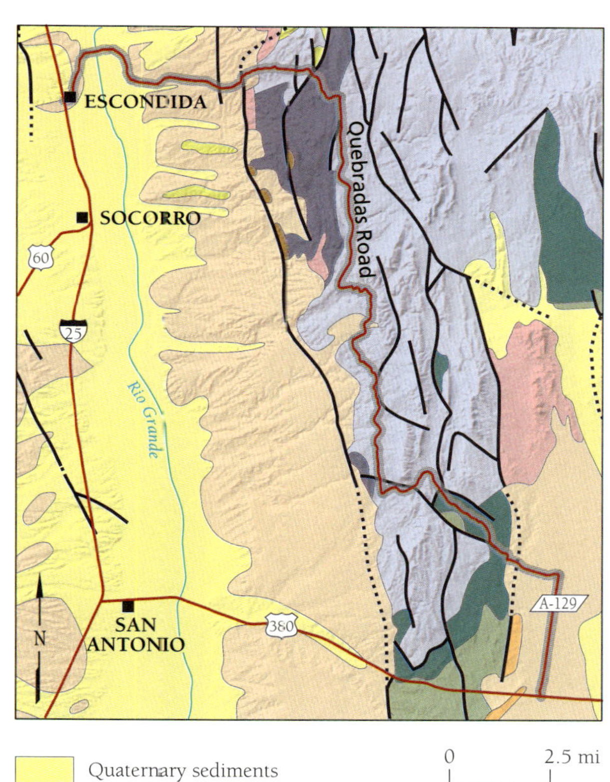

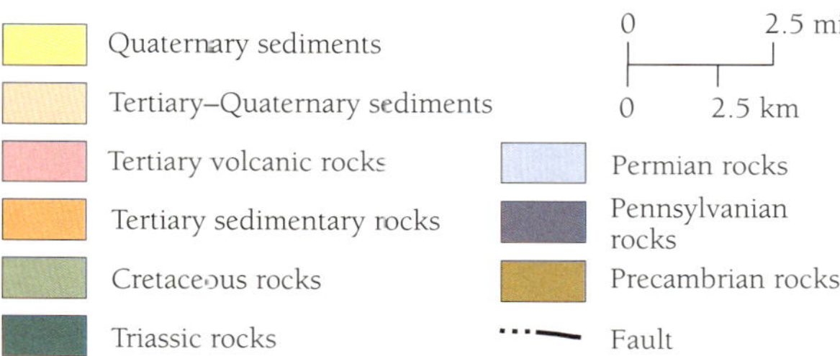

Great Unconformity, a surface that represents more than a billion years of missing rock record prior to deposition of Pennsylvanian strata.

The Pennsylvanian–early Permian Ancestral Rockies tectonic event, associated with the Pangean collision of North and South America, led to formation of numerous uplifted mountains and adjacent basins within southwestern North America. Contemporaneous southern hemisphere glaciations produced greatly fluctuating sea levels that caused an alternation between marine and nonmarine deposition in the Quebradas region. At the same time, northward continental drift caused New Mexico to experience a progressive change from a tropical to a highly arid climate. All of these processes added diversity to the character of Late Paleozoic sedimentary rocks that dominate most of the hills in the Quebradas region.

During the Mesozoic, deposition of fluvial and shallow-marine strata occurred, mainly during Triassic and Late Cretaceous time. This was followed by great geologic activity in the Tertiary, including Laramide crustal shortening, widespread volcanism, the creation of the Rio Grande rift by crustal stretching, and substantial filling of rift basins by young sediments. Geologic unrest continues to the present day with earthquakes, young lava flows to the east (near Carrizozo), and an active magma body that lies 12 miles deep beneath the Socorro region.

Geologic Features

Specific geologic features are exposed along the Byway at the ten roadside stops with pullouts and small numbered signs (some are hard to see, but a map and GPS coordinates are provided).

Generalized stratigraphy of the Quebradas region. Adapted from the International Commission on Stratigraphy Chart (2019).

AGE (million years)	GEOLOGIC AGE			STRATIGRAPHIC UNIT	THICKNESS IN REGION (feet)
0	Quaternary			unconsolidated sediment	0–100
2.6	Tertiary			Santa Fe Group	0–2,000
				Baca Formation and volcanics	300–3,200
66	Cretaceous	Upper		Mesaverde Group	~920
				Mancos Shale	820–980
				Dakota Sandstone	~70
100		Lower		absent	0
145	Jurassic			Morrison Formation	0–50
201	Triassic			Chinle Group	560–720
				Moenkopi Formation	
259	Permian	Lopingian		absent	0
		Guadalupian		Artesia Group	0–65
~273		Cisuralian		San Andres Formation	330–490
				Glorieta Sandstone	120–290
				Yeso Formation or Group	990–1,090
				Abo Formation	460–790
299	Pennsylvanian	Late		Bursum Formation	150–290
			Madera Group	Atrasado Formation	740–880
		Middle		Gray Mesa Formation	440–610
				Sandia Formation	280–580
323		Early		absent	0
359	Mississippian			absent	0
419	Devonian			absent	0
444	Silurian			absent	0
485	Ordovician			absent	0
541	Cambrian			absent	0
	Precambrian			granitic basement (1.6–1.7 billion years)	

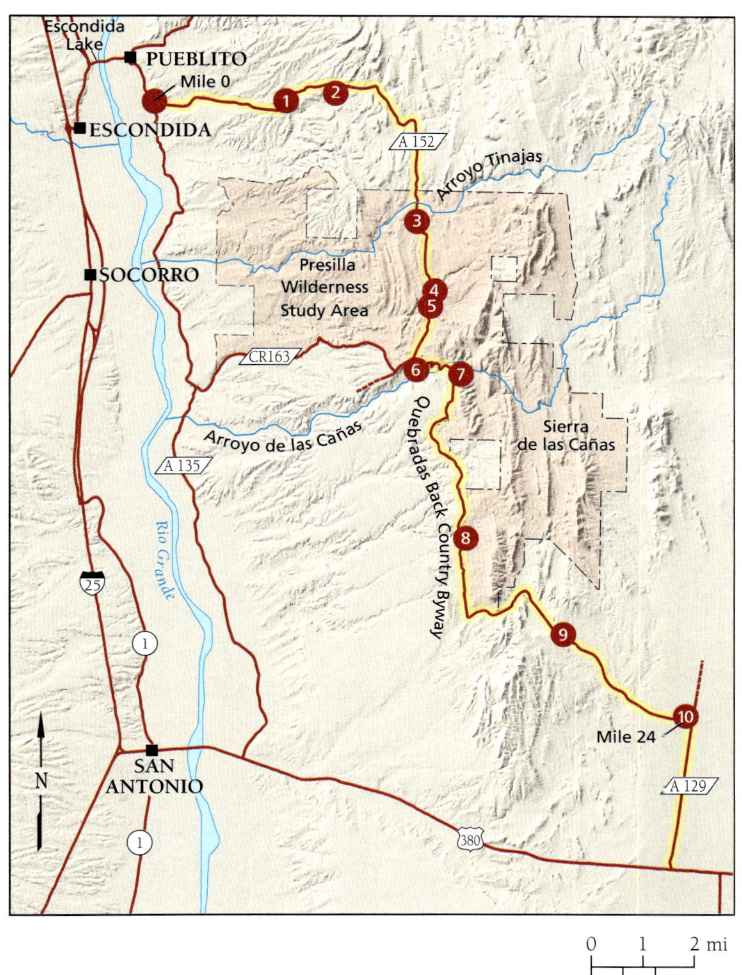

Route map for the Quebradas Back Country Byway. Distance is 24 miles from end to end. Motorized travel is limited to designated routes within Wilderness Study Areas.

STOP 1, MILE 2.7—Coordinates: N 34°06.592', W 106°49.495'; UTM Zone 13, 331860E, 3775840N—This overview site sits atop Late Tertiary alluvium and provides an excellent view of the Socorro Basin of the Rio Grande rift and the mountains on its western flank.

STOP 2, MILE 4—Coordinates: N 34°06.940', W 106°48.545'; UTM Zone 13, 333152E, 3776460N—Fluvial sandstones of the lower Permian Abo Formation, with visible stream channels and numerous plant fossils that are indicative of temperate to semiarid conditions are exposed here.

STOP 3, MILE 8.1—Coordinates: N 34°04.576', W 106°46.877'; UTM Zone 13, 335640E, 3772040N—The sandstones and conglomerates of the Upper Pennsylvanian Bursum Formation are best exposed

Laminated, irregular black crusts in Madera limestone. These have been interpreted as fossil plant root zones.

Folded siltstones and shales of the Yeso Formation northwest of Stop 4.

on the east side of the road. This unit contains mixed shallow-marine and nonmarine deposits with abundant feldspar (the pink mineral in the sandstones) shed from nearby granite-cored uplifts. A short walk down the creek takes you through limestone beds of the Upper Pennsylvanian Atrasado Formation that contain marine fossils, especially small and fragmented plates of phylloid algae as well as crinoids and brachiopods. At least three cycles of marine deposition capped by exposure surfaces (soils) are visible in this canyon.

STOP 4, MILE 9.7—Coordinates: N 34°03.325', W 106°46.483'; UTM Zone 13, 336205E, 3769720N—Sandstone and shale beds of the lower Permian Yeso Formation crop out at this locality. Careful examination of the white layers shows small cubic casts of former halite (salt) crystals. These casts are excellent evidence that these nonmarine to marginal-marine beds were formed under very arid conditions, comparable to the modern Arabian/Persian Gulf region. Also visible at this site are folds in the Yeso beds.

STOP 5, MILE 10—Coordinates: N 34°03.051', W 106°46.542'; UTM Zone 13, 336106E, 3769220N—Gray Atrasado Formation limestones of the upper Madera Group are visible in the distance to the west of the road at this site—the bedding shows the depositional effects of sea-level cycles. The rocks were subsequently folded and thrust faulted during the Laramide orogeny in the early Tertiary and are now standing nearly on end. In the foreground are the orange-red, gray, and white beds of the Yeso Formation, and the deeper-red siltstones and sandstones of the Abo Formation are in the middle ground.

STOP 6, MILE 11.3—Coordinates: N 34°02.048', W 106°46.736'; UTM Zone 13, 335775E, 3767370N—This is a panoramic overview of the west side of the Rio Grande rift that includes, from south to north, the Chupadera Mountains, the San Mateo Mountains, the Magdalena Mountains, Socorro Peak, Strawberry Peak and Sierra Ladrones. The rocks exposed in the canyon walls in the foreground consist entirely of Yeso Formation and mesa-capping Quaternary gravels.

STOP 7, MILE 12.2—Coordinates: N 34°01.938', W 106°45.957'; UTM Zone 13, 336106E, 3769220N—This is a complex exposure of the upper part of the Yeso Formation. The hillslope below the small parking spot has limestones that contain a restricted assemblage of molluscan fossils and large, carbonate microbially-coated grains, indicative of deposition in shallow, probably hypersaline marine waters. In the valley below, the contortion in the carbonate rocks is due to dissolution of underlying gypsum layers. Superb exposures of such evaporites, with nodular and laminated bedding and flow structures are found in low cliffs just a short distance upstream along the west-trending section of the creek.

STOP 8, MILE 16.1—Coordinates: N 33°59.033', W 106°45.721'; UTM Zone 13, 337148E, 3761770N—This stop features late-stage deformation of Permian rocks. Two generations of faulting are visible, including low-angle compressional faults (yellow arrows), probably of Early Tertiary (Laramide) age, cut by a high-angle normal fault (red arrows), probably of late Tertiary age.

STOP 9, MILE 20.4—Coordinates: N 33°57.340', W 106°43.720'; UTM Zone 13, 340269E, 3758590N—These small, isolated outcrops of the Triassic Chinle Formation only show a small portion of the roughly 700-feet maximum thickness of the Triassic section in this region. The pebbles, channel features, and nodular soil crusts in this section all indicate a fluvial (river) origin for these deposits.

STOP 10, MILE 23.5—Coordinates: N 33°55.812', W 106°41.280'; UTM Zone 13, 343981E, 3755700N—A panoramic view of the south Quebradas region: the broad plains of the northern part of the Jornada del Muerto are bordered by a series of mountain ranges along the flanks of the Rio Grande rift. This desolate region, now largely on the White Sands Missile Range, was the site of an early Spanish trail that extended northward from Mexico into central and northern New Mexico.

—Peter A. Scholle

Fault-propogation fold in the Yeso Formation at Stop 8. Several generations of faults are visible, including low-angle compressional faults (marked by yellow arrows), probably of Laramide age, cut by a high-angle normal fault (marked by red arrows), probably of late Tertiary age.

Additional Reading

A Geologic Guide to the Quebradas Back Country Byway, by Peter A. Scholle, New Mexico Bureau of Geology and Mineral Resources and the Bureau Land Management, 2010.

Printed brochure is available for free at the Bureau of Geology Bookstore in Socorro or as a free digital download at:
geoinfo.nmt.edu/publications/guides/quebradas

If You Plan to Visit

The Byway lies east of the Rio Grande between Escondida and San Antonio. The well-graded, 24-mile-long dirt road can be accessed from either the north or the south end.

From the north, take Exit 152 (the Escondida exit) off I–25, two miles north of Socorro. After exiting the freeway, turn east following the Back Country Byway signs, then north toward Escondida Lake. In just over a mile, turn right at the Escondida Lake sign and head east past the lake and across the Rio Grande. At the village of Pueblito (a T-intersection), turn right (south) and proceed about a mile to the junction of the Bosquecito Road with A 152. Turn left here; this is the beginning of the Byway.

From the south: Take Exit 139 off I–25 and head east on US 380. Eleven miles east of the village of San Antonio, turn north onto A 129, then drive 3 miles to the junction with A 152, where the Byway begins. Turn left; Stop 10 is just west of this junction. For more information:

Bureau of Land Management
Socorro Field Office
901 South Highway 85
Socorro, NM 87801-4168
(575) 835-0412
www.blm.gov/nm/st/en/prog/recreation/socorro.html

Magdalena and San Mateo Mountains
CIBOLA NATIONAL FOREST AND BUREAU OF LAND MANAGEMENT

The Magdalena and San Mateo mountains, adjacent mountain ranges southwest of Socorro, expose a wide spectrum of rock types, including Precambrian granite and metamorphic rocks, Paleozoic sedimentary strata, and huge volumes of younger volcanic rocks erupted from a series of ancient, now extinct supervolcanoes. Elevations of these two mountain ranges vary from about 6,000 feet to peaks approaching 11,000 feet. Vegetation zones range from piñon-juniper grasslands near the base to high-elevation Douglas fir forests and tundra-like meadows above tree line. These provide habitats for a diverse group of animals, including elk, deer, bear, and mountain lion. Lead, silver, and zinc mines in the area, including the world-famous Kelly Mine, were active from the 1870s through 1940s (see later section on the mining history). Scientific facilities along the summit of the Magdalena Mountains include the Langmuir Laboratory for Atmospheric Research and the Magdalena Ridge Observatory with its 7.8-foot-diameter optical telescope and interferometer (these are occasionally accessible to the public). A steep, graded road provides public access to within a 0.25 miles of these restricted research facilities high in the Magdalena Mountains. Travelers seeking seclusion may prefer the San Mateo Mountains, which host the Withington and Apache Kid wilderness areas.

View to the northeast from Timber Peak Trail across northern Magdalena Mountains. Barren ridge at left middle ground is on the north margin of the Socorro caldera. Light-colored sediments in La Jencia Basin form a backdrop for cliffs at the entrance to Water Canyon. Mountain peaks of the Rio Grande rift on the middle horizon, from left to right, are Ladron Peak, Polvadera Peak, Strawberry Peak, and Socorro Peak.

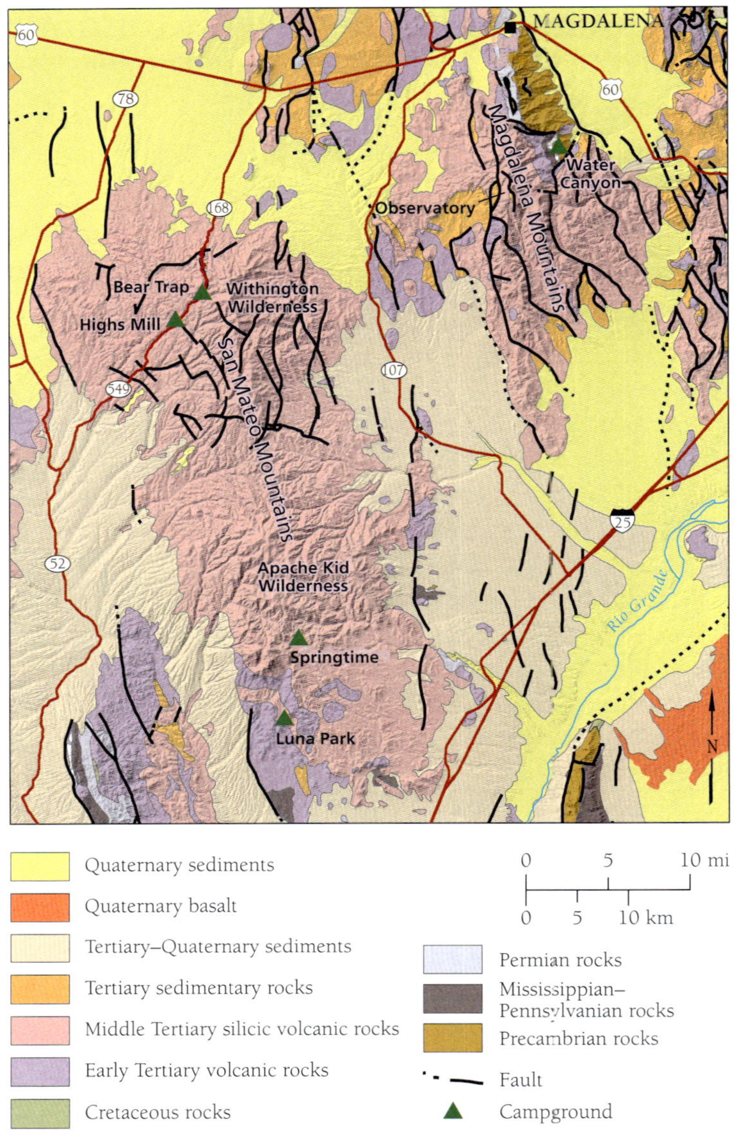

Generalized geologic map of basins and ranges of the Rio Grande rift in the Magdalena region. Alpine crests of the Magdalena and San Mateo mountains are underlain by buoyant granite intrusions that once fed large caldera-forming eruptions, the products of which are shown here as Middle Tertiary silicic volcanic rocks.

Regional Setting

The Magdalena and San Mateo mountains are part of the highly extended Basin and Range terrain on the western flank of the Rio Grande rift in south-central New Mexico. Rift-related faulting has stretched, broken, and tilted the Earth's crust to form tilted fault-block ranges and intervening basins filled with wedge-shaped volumes of sediment.

In the northern Magdalena Mountains, faulting and west-ward tilting has uplifted and exposed Precambrian and Paleozoic rocks in the "sunken" floor of the Socorro caldera (at Water Canyon campground) and on the adjacent rim of the caldera near the Kelly Mine. The southern Magdalena Mountains and the San Mateo Mountains are dominated by east-tilted fault blocks that expose caldera-related volcanic rocks of middle Tertiary age, many of which are related to supervolcano activity.

The Rock Record and Geologic History

In the northern Magdalena Mountains, granites and gneisses about 1.6 billion years old are overlain by a series of Paleozoic sedimentary units. Precambrian rocks exposed in the Magdalena Mountains formed deep in the crust during initial assembly of the North American continent. Rocks like these underlie all of New Mexico and are thus sometimes referred to as "basement rocks." These rocks were uplifted and eroded, then subsequently buried by younger Paleozoic sedimentary rocks deposited by rivers and seas. The boundary between Precambrian basement and Paleozoic sedimentary units is widespread throughout

the Southwestern U.S. and is termed the "Great Unconformity,"
because more than one billion years of the geologic record is missing.
Overlying Paleozoic sedimentary rocks include Mississippian to lower
Permian units near the ghost town of Kelly, just south of the village of
Magdalena. The Mississippian Kelly Limestone is a thin shallow-marine
deposit with basal sandstone beds overlain by fossiliferous (typically
crinoid-rich) limestones. The Kelly Limestone is overlain by about
1,200 feet of interbedded, dominantly marine limestones, sandstones
and shales of the Pennsylvanian Sandia Formation and Madera Group.
These units are overlain by nonmarine red shales and siltstones of the
Permian Abo Formation. Mesozoic rocks are absent in the Magdalena
and San Mateo mountains due to uplift and erosion during the
Laramide orogeny.

The majority of the rocks exposed in the Magdalena and San Mateo
mountains formed during an intense interval of volcanism from 40 to
25 million years ago. Andesitic stratovolcanoes,
similar to today's Cascade Mountains (Mount St.
Helens), formed first. These were succeeded by
a cluster of six Oligocene supervolcanoes (huge
explosive volcanoes similar to the Valles caldera
near Los Alamos) that extended toward the
southwest from Socorro to what is now the San
Mateo Mountains. These supervolcanoes are part
of the northeastern portion of the Mogollon–
Datil volcanic field, which extends south and
west to include other ancient calderas in the Gila
Wilderness and the Organ Mountains. Although
the original topography of these calderas has
been destroyed by younger faulting and erosion,
geologic studies have revealed a sequence of caldera eruptions, which progressed generally from
northeast to southwest between 32.5 and 24.6
million years ago. The southwestward march of
these calderas is thought to record progressive
westward sinking of oceanic crust previously subducted beneath
North America into the mantle. Each of the six calderas had its own life
cycle, starting with smaller andesite and rhyolite eruptions, followed
by catastrophic eruptions of rhyolite tuff, and ending with eruptions of
rhyolite lava. All these rock types are common in the San Mateo and
Magdalena mountains. Each of the large-volume rhyolite tuff eruptions
partially emptied the underlying magma chamber, causing collapse
of roughly circular calderas as large as 10 to 20 miles across. Erupted
rhyolite tuff pooled within these depressions, but some flowed outward

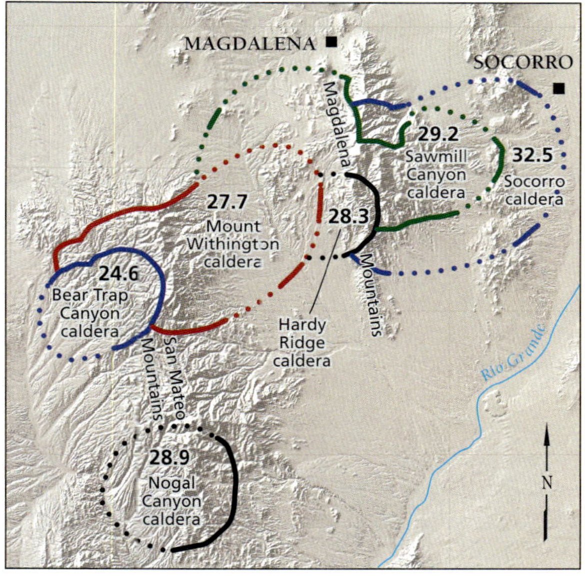

Relief map of the Socorro-San Mateo region showing exposed (solid) and concealed (dotted) margins of calderas (eruption ages in millions of years before present). Initially circular calderas are now stretched by later rifting. These calderas erupted a total of 1,200 cubic miles of magma, in six brief pulses.

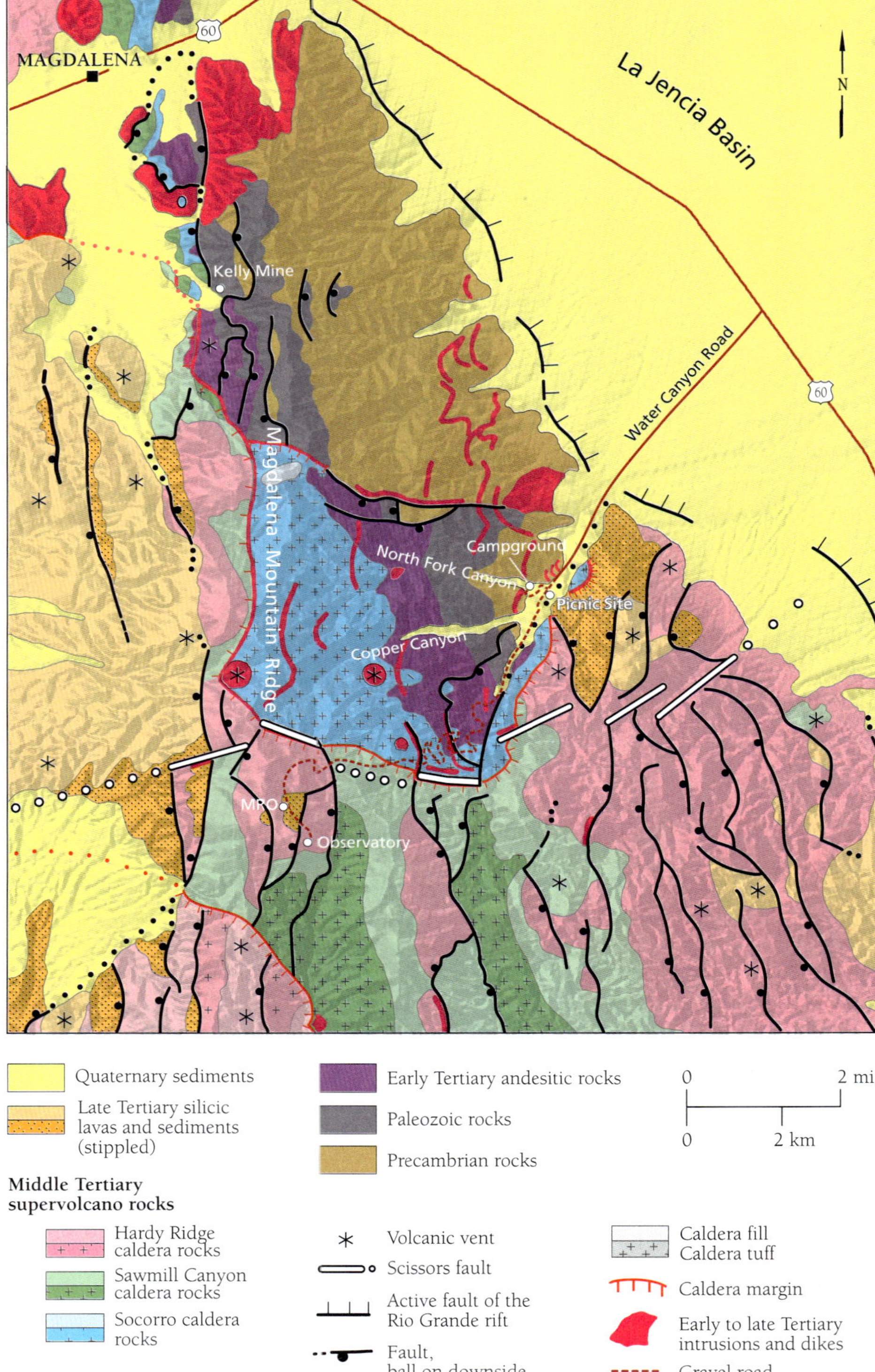

Geologic map of the northern Magdalena Mountains showing three cycles of caldera-forming eruptions and related intrusions. Calderas formed above a northeast–trending weakness in the crust that transported many pulses of basaltic magma from the mantle to the lower crust. Large volumes of dense basalt magma became the heat source for low density, caldera-forming magmas. Later rifting, uplift, and erosion exposed numerous dikes in the floor of the Socorro caldera near Water Canyon campground. The Hardy Ridge unit includes younger basalt lavas and tuffs.

in nearly all directions, forming sheets of welded tuff extending many tens of miles from each caldera. Numerous igneous dikes intruded both pre-volcanic and volcanic rocks and faults along the margins of calderas. As intrusions cooled, faults provided favorable pathways for geothermal fluids that formed deposits of lead, silver, and zinc, mainly in Mississippian limestones of the Kelly Mining District, as well as small gold deposits in the central Magdalena and San Mateo mountains.

Extension during and after caldera volcanism has broken the volcanic field into numerous tilted fault blocks that form the modern mountain ranges and intervening basins, causing erosion of uplifted rocks and local deposition of layered conglomerates and sandstones of the Santa Fe Group. Red, well-cemented sediments of the lower Santa Fe Group form scattered exposures near the base and at the top of the Magdalena Mountains, indicating multiple periods of uplift and tilting between 20 and 10 million years ago. Young fault scarps, including one exposed near the mouth of Water Canyon that continues northwestward to where it is crossed by US 60 about 5 miles east of Magdalena, demonstrate that faulting (rifting) is still active in this area. This fault last moved only about 10,000 years ago.

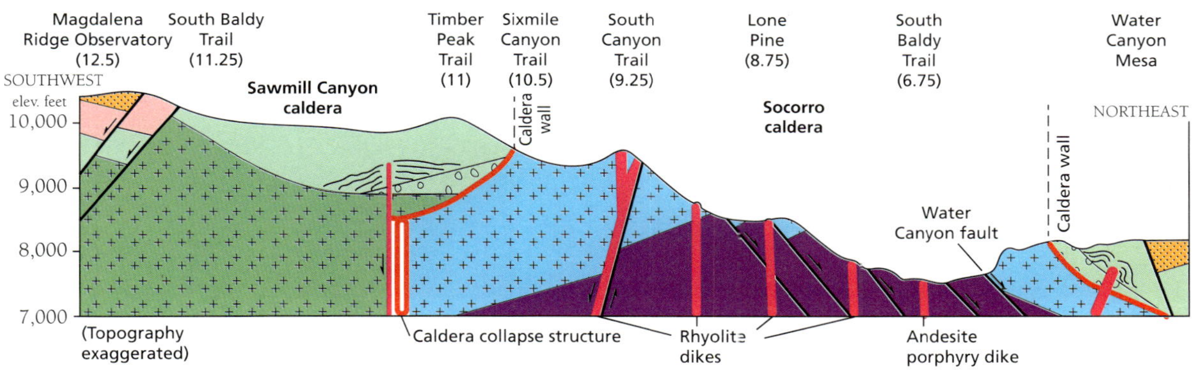

Profile illustrating subsurface geology of the Socorro and Sawmill Canyon calderas exposed along the Magdalena Ridge Observatory access road. Numbers in parentheses indicate mileage from US 60 (green mileposts). Colors match the geologic map shown on previous page.

Geologic Features

MAGDALENA RIDGE ACCESS ROAD—A variety of rock types are exposed along the road between Water Canyon and Magdalena Ridge. As the road gains elevation (see the diagram above), it exposes progressively younger rocks, although some rock formations reappear several times due to faulting. Roadcuts and natural outcrops near the Water Canyon picnic area (at 4.75 mi) consist of 1.6 billion-year-old gneiss and schist. Between the picnic area and mile post 6.75 are numerous outcrops of Pennsylvanian black shales and limestones of the Madera Formation that were "baked" and intruded by several dikes (approximately

0.9 to 1.3 miles from the campground). Starting near mile post 6.75 (the first switchback), dark-gray precaldera conglomerates, lavas, and many andesite dikes are well exposed for the next 0.75 miles. Between mile posts 7.50 and 9.00, white (altered) crystal-rich tuff and underlying andesite are repeated in fault blocks. One of several caldera-related rhyolite dikes forms a conspicuous white wall rising across the hillside below Lone Pine at mile 8.75. From Lone Pine, three hairpin switchbacks are needed to climb through 2,000 feet of caldera-forming tuff. At mile post 10.50, gray tuff terminates against dark andesite lava at the north wall of the Sawmill Canyon caldera. The road levels out across andesite lavas that backfilled the caldera. Down-faulted vesicular black basalt lava is visible in cuts between 11.25 miles and the gate at 12.25 miles. Red Santa Fe Group conglomerates cap the basalt at the observatory (12.50 miles).

Dome of the Magdalena Ridge Observatory houses a 7.8-foot-diameter mirror-based telescope primarily used to search the sky for celestial objects whose trajectory may pass close to the orbit of the Earth in the near future. This important effort is the modern high-tech version of looking for a needle in the proverbial haystack.

MAGDALENA MINING DISTRICT—From 1866 to 1940, four mining districts in the Magdalena Mountains produced more than 46 million dollars worth of zinc, lead, copper, silver, manganese, gold, and barite from carbonate-hosted, lead-zinc replacement deposits and volcanic-epithermal vein deposits. Nearly all production was from the Kelly, Juanita, and Graphic mines within the Magdalena District.

In 1867, J.S. Hutchason and P.H. Kelley discovered two separate deposits in the Magdalena District, but after the Mining Law of 1872, Hutchason restaked deposits that would host the Kelly, Juanita, and Graphic mines. Initially, the main mineral mined was silver-bearing cerussite. Ore was smelted in nearby adobe smelters until 1882. In 1896, the Graphic smelter was built, and the ore from the Kelly Mine was treated there. The recognition of zinc carbonate (smithsonite, a

A photograph (ca. 1905) captured the rare moment when the doubledrum hoist was delivered to the soon-to-be constructed Traylor Shaft head-frame at the Kelly Mine. Nearly three-dozen horses were called into service to pull the load upgrade.

vital ingredient in paint) within the limestones in 1903 rejuvenated the Kelly–Magdalena area and created a prosperous period of zinc production (and started a re-evaluation of other western mines).

The Kelly mine is best known for the remarkable and rare specimens of blue-green to apple-green smithsonite, many of which have a vibrant pearly luster and a rounded form. Large specimens from the mine are highly prized by collectors, and smaller specimens can still be found in the mine waste dumps. Unfortunately, the drab colorless variety of smithsonite is more common.

The town of Kelly grew near the mine in 1883, and nearly 3,000 people once lived there. Kelly had two churches, seven saloons, two dance halls and two hotels. By 1945, the town was abandoned due to the mines closing. Only a church, some foundations, and mine waste dumps remain.

This 5 x 6 inch gem-grade smithsonite originated in the famed Kelly Mine of the Magdalena Mining District. Produced in the heyday of zinc carbonate mining during the 20th century, the blue-green color and shimmering luster is highly prized and sought after by mineral enthusiasts around the world.

San Mateo Mountains

LUNA PARK CAMPGROUND—Picturesque rock towers at this delightful picnic area and campground were formed by erosion of the Luna Park Tuff, a thick welded tuff erupted from a nearby caldera. The cliff faces provide excellent examples of textures within this welded tuff, including flattened pumices and angular lithic fragments.

MT. WITHINGTON ROAD (FR 549) TO THE FIRE LOOKOUT (10,119 FEET)—The road leads south from US 60 and traverses part of the Plains of San Agustin, a branching segment of the Rio Grande rift, before ascending the range. Nearly all rocks from the mountain base to the peak are volcanic and are located within the Mt. Withington caldera, a supervolcano that erupted catastrophically about 27.7 million years ago, ejecting at least one thousand times as much magma as the 1981 Mount St. Helens eruption. At the fire-tower overlook, visitors are standing on a 2,000-feet-thick sequence of welded tuff that pooled in the giant depression of the collapsing caldera during the eruption. Subsequent faulting has elevated the San Mateo Mountains at least 10,000 feet skyward, uplifting the originally down-dropped caldera as part of the current mountain range. Views from the lookout are spectacular.

—*Richard M. Chamberlin, William C. McIntosh, and Matthew J. Zimmerer*

Aerial view of the southern San Mateo Mountains from Luna Park. The Luna Park Tuff forms east-tilted rock layer in foreground. Bold cliffs of Vicks Peak Tuff on the skyline are in the core of the 28.9 million-year-old Nogal Canyon caldera.

Additional Reading

Gustav Billing, the Kelly Mine, and the great smelter at Park City, Socorro County, New Mexico, R.W. Eveleth, in New Mexico Geological Society, 34th Annual Field Conference Guidebook, 1983.

If You Plan to Visit

The Magdalena and San Mateo mountains can be visited using unpaved U.S. Forest Service roads and hiking trails that extend from US 60, NM 107, and NM 52. The dirt roads throughout the San Mateo Mountains are rough and not well maintained; four-wheel drive vehicles are recommended. For more information:

Cibola National Forest Headquarters
Magdalena, NM 88310
(575) 434-7200
www.fs.usda.gov/attmain/lincoln/specialplaces

A world-class mineral museum, which includes the most comprehensive mineral collection in New Mexico, is located on the campus of
the New Mexico Institute of Mining and Technology in Socorro. The
museum collections focus on minerals from New Mexico but also
exhibit specimens from adjacent states and from famous localities
around the world. Since the museum's conception, the holdings
function as a "working collection," providing material for researchers,
displays for education, and preservation of minerals for posterity.

In 1889, the New Mexico Territorial Legislature established the
New Mexico School of Mines (now New Mexico Institute of Mining
and Technology). The charge to the school was to provide for higher
education in the Earth and related sciences. The school's board of
trustees (now "Regents") was assigned the responsibility of prescribing
courses of study, equipping laboratories and classrooms in a manner
appropriate for instruction, and assembling (for educational purposes)
a geological and mineralogical cabinet. Back in the 1880s, a "cabinet"
was a term for a permanent collection.

The collection was originally assembled by faculty, notably
Fayette A. Jones, the first president of the school. Other faculty,

Over 5,000 mineral specimens are
displayed in the main gallery
of the Mineral Museum at the
New Mexico Bureau of Geology
and Mineral Resources.

View of the original mineral museum in the Old Main building prior to the fire in 1928.

supporters, and graduates soon built the collection into one of the finest in the world, winning gold medals at the St. Louis World's Fair of 1904 and the Panama–California exhibition of 1915. Unfortunately, this early collection was lost in a fire in 1928. The museum was reestablished by donation and purchase, particularly based on the collection of School of Mines benefactor C.T. Brown.

The reestablished museum was housed in the main building on campus, Brown Hall, under the aegis of the geology department, where the minerals were displayed in cabinets in a room on the south end of the building. In 1939, the New Mexico Quarto-Centennial Commission officially named the collection "Coronado's Treasure Chest." After the fanfare of the 25th anniversary of New Mexico's Statehood, the collection was slowly boxed up and stored away with only a few pieces displayed as space requirements became critical. In 1960, during construction of a new facility, the collection was transferred to the New Mexico Bureau of Mines and Mineral Resources. The museum was located in Workman Center and was open 7 days a week, 24 hours a day! Under the direction of the Bureau, the first formal mineralogist/curator was hired to tend to the mineral collections.

In 1995, a new facility was remodeled for the museum in the "Gold Building." The building received its unofficial name (officially it was known as Workman Addition) from the fact that the money to build it came from recycling old government computers. The move to the new facility allowed for the expansion of displays and provided more efficient storage and curation opportunities.

The latest version of museum was completed in 2015 as part of a new Bureau of Geology building. An exhibit space specifically designed as a museum highlights top-quality minerals from New Mexico, the United States, and around the world. Over 5,000 mineral specimens are displayed in the main gallery. Spectacular mineral specimens from New Mexico mining districts are presented in thematic displays that illustrate the mineral wealth of each locality. These include copper minerals from the copper porphyry deposits near Silver City, spectacular quartz and feldspar crystals from the Organ Mountains, smithsonite and aurichalcite from Magdalena, and fluorite crystals from Bingham (near the Trinity Site). Other thematic displays include uranium mining, lapidary (rock art), gold and silver, New Mexico agates and geodes, meteorites, and petrified wood.

Minerals from the United States and around the world are also presented in a "new acquisitions" case that highlights recent additions to the collection from generous donors or purchased using the proceeds of the mineral museum shop. The most dynamic exhibit is the Guest Display, which is transformed by guest collectors on a yearly basis. Each year, three mineral enthusiasts showcase a portion of their collections for the public to admire. Mining memorabilia, a modest gemstone display, and a breathtaking ultraviolet-mineral exhibit are also found in the museum. A museum classroom includes a fossil display, educational demonstrations, and an "augmented reality topographic sandbox," all of which are part of the educational outreach mission of the museum.

In front of the Bureau of Geology's building is a display of large pieces of petrified wood, artistically polished, from the Petrified Forest area of Arizona. This outdoor exhibit is called the "Petting Forest" where visitors can climb on, touch, and observe the myriad of colors and patterns in these spectacular log sections. There are other "touchable" pieces in the atrium of the building, as well as a display that integrates minerals, commodities, artifacts, and Bureau of Geology publications.

Fluorite specimen from the Fishstick Prospect, Hansonburg District, Socorro County, New Mexico. This type of material is still being produced from the district by mineral specimen miners.

Turquoise from the Orogrande District, Otero County, New Mexico (gift of Rex Nelson). Orogrande was one of the great turquoise producers of southern New Mexico along with Santa Rita, Tyrone, and Hatchita.

Wulfenite specimen from the Stephenson-Bennett Mine, Organ District, Doña Ana County, New Mexico. This mine was one of the first great wulfenite-specimen-producing mines in the southwest, with specimens noted as early as the 1880s. Wulfenite was so abundant in the mine that it was mined as an ore for molybdenum.

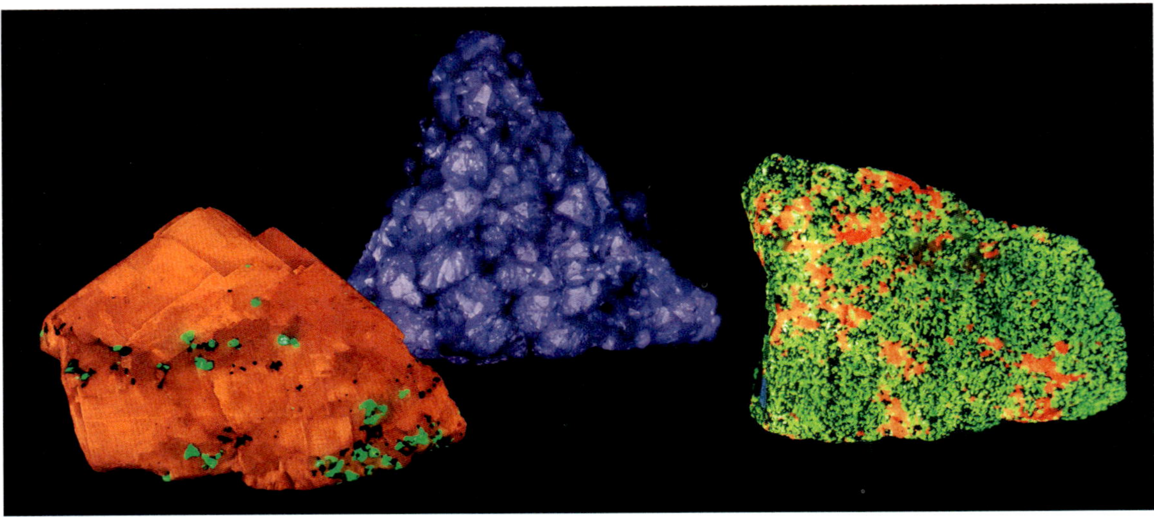

The Mineral Museum's fluoresent mineral exhibit includes examples of (from left to right); 1) Calcite, franklinite, and zincite; 2) calcite; and 3) franklinite, willemite and calcite.

The Mineral Museum is a public facility founded upon the need to understand the natural mineral resources of New Mexico and preserve materials of historical, aesthetic, and cultural value. The museum can be visited free of charge and is open most days of the year.

—Virgil W. Lueth

If You Plan to Visit

For teachers, students, and other groups, we offer free tours of the museum. Museum staff can also identify minerals or rocks for visitors. Please call ahead to ensure someone will be available.

New Mexico Bureau of Geology and Mineral Resources
801 Leroy Place
Socorro, NM 87801-4796
(575) 835-5490

Mineral Museum hours of operation:
9 a.m. to 5 p.m. Monday through Friday
10 a.m. to 3 p.m. Saturday and Sunday
(we are closed most university holidays).
For more information on the museum, please visit our website at:
geoinfo.nmt.edu/museum

Bosque del Apache National Wildlife Refuge

U.S. FISH & WILDLIFE SERVICE

West of the flocks of birds and birdwatchers along the wetlands of the Rio Grande, a story of cataclysmic volcanic eruptions, ancient dune fields, and long-gone towering mountains lie quietly awaiting visitors to the Bosque del Apache National Wildlife Refuge. The Chupadera Wilderness Trail and the Canyon Trail traverse this geologic record in the contemplative solitude of a wilderness setting, affording "rock-watchers" a quick trip through millions of years of Earth history.

Geography and Geologic Province

The refuge straddles the Socorro Basin, between the Chupadera Mountains on the west and a string of discontinuous low hills and Little San Pasqual Mountain on the east. This setting is a product of Rio Grande rift crustal stretching, which broke the crust beneath central New Mexico into fault-bounded blocks, each several miles wide, that very slowly tilted like falling dominoes during progressive crustal extension. Domino blocks near the Bosque have been tilted to the east. Up-tilted edges of blocks became mountains and hills, while down-tilted edges became basins. The rift here is superimposed on the eastern

Snow geese in flight over the Bosque del Apache with the Chupadera Mountains in the middleground and the Magdalena Mountains in the background.

Hackly cliffs of La Jencia Tuff along the Chupadera Wilderness Trail. The rough fabric was produced by open pockets in the rock formed by steam rising through the hot "taffy-like" tuff shortly after it was deposited.

part of the Mogollon–Datil volcanic field. The Canyon Trail and Chupadera Wilderness Trail traverse rocks related to both the rift and the volcanic field.

The Rock Record

Volcanic rocks, dune sands, river sands, and flood gravels line the trails of the refuge's western wilderness. The oldest rocks, 36- to 27 million-year-old lavas and welded tuffs, lie along the Chupadera Wilderness Trail at the base of Chupadera Peak. The tuffs, which are typically gray, brown, or reddish layers of fused ash and crystals, erupted from Mogollon–Datil calderas, supervolcanoes characterized by violent eruptions of large volumes of material and collapse of the land surface into the underlying magma chamber as they emptied. Features of the Mogollon–Datil field, both welded tuffs and calderas, are discussed in more detail in chapters of that section. The lavas, which are typically rubbly and black to dark reddish brown, erupted from a now-buried Mount St. Helens-type volcano in the vicinity of the refuge.

The lavas and tuffs are no more than a few hundred feet thick at the base of Chupadera Peak, but as the trail climbs westward, it ascends through 1,000 feet of a single tuff, the 29.2 million-year-old La Jencia Tuff. This unusual thickness is the result of the molten La Jencia ash filling the colossal depression left by the prior eruption of one of the largest of the Mogollon–Datil calderas, the 32.5 million-year-old Socorro caldera. The caldera depression once extended 16 miles from Chupadera Peak northward to Socorro Peak, but subsequent filling by the La Jencia Tuff and disruption by Rio Grande rift faults now obscure most of its original extent.

Rocks of an entirely different character straddle the Canyon Trail, where 15 to 8 million-year-old, pale-brown sandstones interfinger with discontinuous, darker-brown conglomerates. The sandstones are characterized by large, curved crossbeds indicating the sands accumulated in eolian dunes that traversed eastward across the area in an arid or semiarid climate. In contrast, conglomerates were deposited by high-energy flood events. The nature of the gravels and various other features of the conglomerate beds indicate that the floods originated from the east. At the time, there was no Rio Grande here, and the low hills that today lie to the east must have once been formidable mountains to have shed such coarse material this far westward.

The youngest rocks occur along the lower reaches of the Chupadera Wilderness Trail, where uncemented gravel beds interfinger with loose, light-gray sands. The gravels clearly originated from the Chupadera Mountains to the west and not the ancient eastern mountains. The sands are deposits of the ancestral Rio Grande, which arrived in this area only about 5 million years ago.

Geologic History

Lavas and tuffs blanketed the area from 36 to 27 million years ago, constructing a high volcanic plateau extending westward into the Mogollon–Datil volcanic field. Rio Grande rift crustal stretching began around 25 million years ago, dissecting the plateau and uplifting mountain blocks while down-dropping the refuge area. From 15 to 8 million years ago, a large dune field occupied the southern Socorro Basin and lapped against an eastern mountain block, which shed gravels westward into the dunes during flood events. Continued stretching, fault-block motions, and erosion subsequently rearranged the topography, uplifting the Chupadera Mountains while eroding the eastern mountains. The rising western mountains blocked westerly winds, effectively killing the dune field. The Rio Grande spilled into the area from the north around 5 million years ago. Beginning about 800,000 years ago, the Rio Grande began to cut its valley, and the modern setting was established.

Geologic Features

All that remains of the volcanoes, dunes, and ancient mountains are recorded in the east-tilted rock layers along the trails in

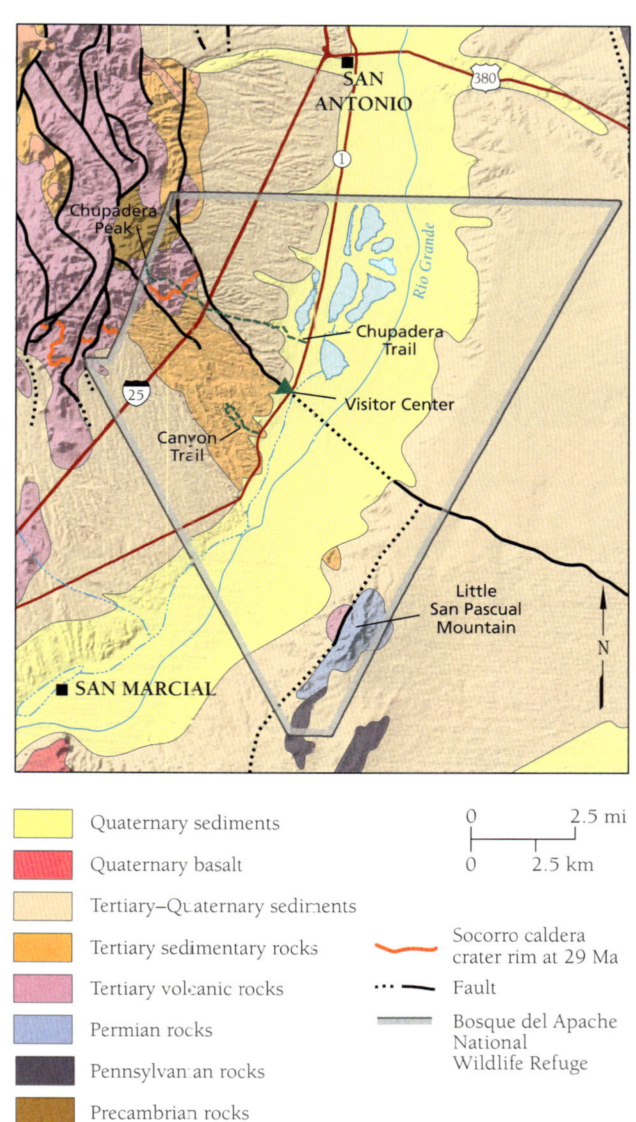

the western refuge wilderness areas. Along the upper reaches of the Chupadera Wilderness Trail (9.5 miles round trip), examine the tilted lavas and tuffs, and while climbing to the top, notice the abundant elliptical pockets in the red cliffs of La Jencia Tuff; these pockets were produced by steam rising through the tuff shortly after eruption and before it solidified. Along the 2.2 miles Canyon Trail, notice the great thickness and continuity of the sandstone beds. The desert dune field in which these sands accumulated must have been quite extensive.

—Colin Cikoski, Richard M. Chamberlin, and David W. Love

Pale-brown eolian sandstones along the Canyon Trail, capped by a thin gray conglomerate ledge. The steeply-dipping crossbeds in the sandstones, particularly apparent in the middle of the photo, are characteristic deposits that accumulated in a dune field.

If You Plan to Visit

From the north, take I–25 south to the San Antonio/US 380 exit (Exit 139). Go east on US 380 for about 0.6 miles to the junction with NM 1 (Old Highway 1). Turn right and take NM 1 south for approximately 3.3 miles to the refuge boundary. The Visitor Center is another 4.5 miles on the same road. From the south, one can follow the same directions or take Exit 124 (San Marcial). Go east on the dirt road 1.5 miles, then north on NM 1 to the southern boundary of the refuge.

Many pullouts are available along NM 1 for birdwatching, and larger, signed parking areas are provided for the two trails. The refuge also has about a 12 mile driving tour loop with birdwatching platforms and trails that is accessible from the Visitor Center area. For more information:

Bosque del Apache National Wildlife Refuge
P.O. Box 280
San Antonio, NM 37832
(575) 835-1828
www.fws.gov/refuge/bosque_del_apache

Fort Craig and El Camino Real Trail
BUREAU OF LAND MANAGEMENT AND NEW MEXICO HISTORIC SITE

Fort Craig and El Camino Real Historic Trail are in south-central New Mexico, about 35 miles south of Socorro, between I–25 and the Rio Grande. The sites overlook the Rio Grande and route of the El Camino Real de Tierra Adentro, the Royal Road to the Interior Lands. Established by Spain in 1598 along an older Native American trail, the road passed through the arid lands east of the Rio Grande (the Jornada del Muerto) and connected Mexico City to the south with Ohkay Owingeh Pueblo, north of Santa Fe, a distance of more than 1,600 miles. It was traversed for nearly three centuries by Spanish, Mexican, and early Anglo settlers and traders. Fort Craig was established in 1854 to help control raids by Apache and Navajo tribes and remained a major military supply base (one of the largest in the West) until it was abandoned in 1885. During the Civil War, it played a key role in the Battle of Valverde, fought near the fort in 1862.

An interesting historical note relates to the Pope siding on the Burlington Northern Santa Fe Railway, which lies across the river from Fort Craig. The siding was the drop-off point for supplies for the Manhattan Project test of the first atomic bomb. "Government Road" connected the siding to the Trinity Site, 28 miles to the east, where the bomb

Route of El Camino Real de Tierra Adentro between Mexico City, Mexico, and Ohkay Owingeh Pueblo, New Mexico.

Adobe walls of a large, semisubterranean storehouse within the walls of Fort Craig. The basalt-capped Mesa del Contadero is in the middle distance. To the north-northeast along the skyline is the jagged profile of the Chupadera Mountains on the western edge of Bosque del Apache Wildlife Refuge.

Camino de Sueños (Road of Dreams) sculpture by Greg Reiche near I–25 with the southern San Mateo Mountains in the background.

was detonated on July 16, 1945. "Jumbo," the containment vessel, designed to catch parts of the bomb if it did not detonate, was the largest and heaviest item ever shipped by rail (217 tons, 25 feet long, 10 feet in diameter). The bomb, the equivalent of 20 kilotons of TNT, initiated the atomic age, and introduced new radiogenic time markers for future geologists.

Geographic Setting

Fort Craig is situated on terraces along sizeable tributaries to the Rio Grande. These ephemeral streams drain parts of the San Mateo and Magdalena mountains to the west and northwest. The Rio Grande and its tributaries have incised through older deposits in response to ice-age climate changes during the past 800,000 years. Basaltic lavas that erupted east of the Rio Grande mark past levels of the river.

Geologic History

Fort Craig is within the little-studied San Marcial Basin of the Rio Grande rift. Rift-flank uplifts form the Magdalena and San Mateo mountains west and northwest of the basin. The deepest part of the San Marcial Basin lies beneath the river. The basin was initially filled by extensive alluvial fans fed by streams that drained mountains to the west. About 5 million years ago, the Rio Grande spilled southward into the basin. Sediments delivered by the Rio Grande and its tributaries aggraded the basin floor to about 4,700 feet elevation about 800,000 years ago, when the river began to cut downward to form the modern Rio Grande valley. Basalts that cap Mesa del Contadero, about 300 feet in elevation above the modern Rio Grande, were erupted 820,000 years ago, prior to downcutting.

Geologic Features

Features of note include: 1) large, deeply incised tributaries west of the Rio Grande such as Milligan Gulch and Sheep Canyon; 2) terraces along the Rio Grande and its tributaries; 3) volcanoes and lava flows of Mesa del Contadero and the Jornada del Muerto volcano; and 4) the high, planar surface deposited by the ancestral Rio Grande east of the valley.

The incised tributaries and badlands west of the Rio Grande made north-south travel there difficult. The lack of deep canyons and the relatively gentle slopes east of the Rio Grande made travel there somewhat easier. Despite the scarcity of water on the Jornada del Muerto, El Camino Real became established as a better alternative for north-south travel than west of the river, particularly for carts and wagons.

Two ages of lava are present east of the Rio Grande. About 820,000 years ago and

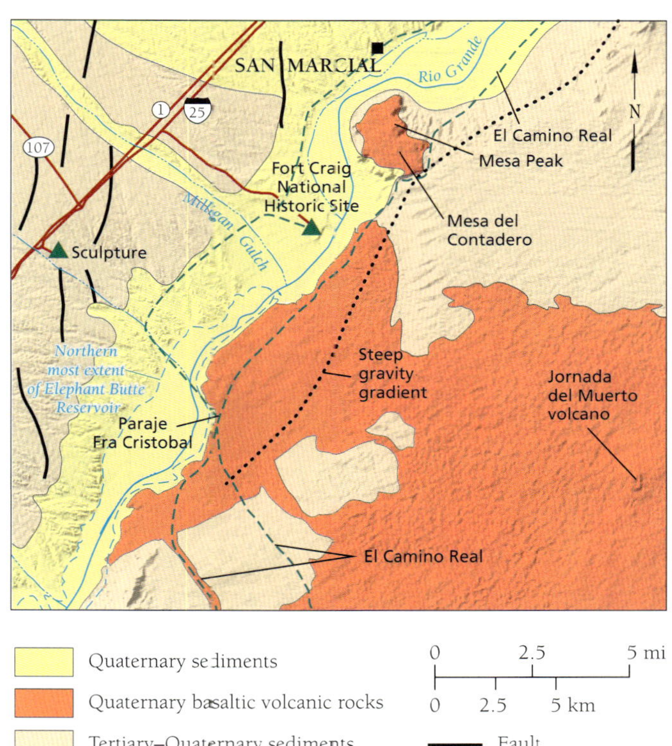

Generalized geologic map of the Fort Craig area with routes of El Camino Real on both sides of the Rio Grande. Paraje Fra Cristobal was one of the important campgrounds along the Camino Real.

3 miles northeast of Fort Craig, an explosive eruption created a crater (called a tuff ring) on what is now Mesa del Contadero. The crater was partly filled by an eruptive mixture of volcanic blocks and ash, as well as pebbles and sand derived from the underlying Rio Grande deposits. The tuff ring then filled with basaltic lava that eventually flowed beyond the crater and now forms the cap of Mesa del Contadero. The lava flows were followed by the final event of this volcanic episode, the construction of the cinder cone now called Mesa Peak. The eroded cinder cone stands about 100 feet higher in elevation than the basalt cap on Mesa del Contadero. Some of that basalt was used in construction of the fort.

About 78,000 years ago, following much erosion in the Rio Grande valley, lava issued from Jornada del Muerto volcano, a complex shield volcano with collapse craters, lava tubes, and cinder cones, located about 10 miles southeast of Fort Craig. At least three flows of distinct ages were erupted during this episode. Lava flowed westward into the Rio Grande valley, where it now forms a low, continuous bench about 50 feet above the modern floodplain south of Mesa del Contadero.

The Rio Grande cut its present valley in a series of erosive episodes during glacial periods of high water discharge. These alternated with episodes of partial backfilling during drier, interglacial periods to form terraces and, most recently, the modern floodplain. The western tributaries followed suit, forming as many as seven terraces along their courses; each were graded to the Rio Grande at the time. The low

Northeastward panorama of the Fort Craig area, with fall colors of the Rio Grande bosque and Mesa del Contadero, topped by Mesa Peak cinder cone in the middle distance and Little San Pasqual Mountain beyond on the skyline.

terrace of Milligan Gulch underlying Fort Craig is at the same elevation as the top of the 78,000 year-old lava flow on the opposite side of the Rio Grande. Sparse river gravel on the lava flow suggests that the flow briefly dammed the river, which then shifted farther west.

The present-day character of this stretch of the Rio Grande has changed with the downstream construction of the Elephant Butte Reservoir, which filled gradually after completion of the dam in 1916 to reach full capacity in 1942. The gradient of the Rio Grande at the north end of the reservoir has decreased, and it now functions as a delta, with additional sediment buildup occurring with each new flood.

—*David W. Love*

If You Plan to Visit

The Fort Craig Historic Site is about 35 miles south of Socorro. From the north, take I–25 to Exit 124 (San Marcial), then east over the I–25, and south on Old Highway 1 (about 11 miles). Then turn left at the signs to Fort Craig, which is 4.7 miles to the east. From the south, take I–25 to Exit 115. Turn east and then north to take NM 1 about 2.9 miles; then turn right (east) on the signposted road to the fort.

BLM Socorro Field Office
901 S. Highway 85
Socorro, NM 87801-4168
(575) 835-0412
www.blm.gov/visit/fort-craig-historic-site

El Camino Real Historic Trail Site is about 40 miles south of Socorro. Take I–25 to Exit 115 and south on Old Highway 1 to a signposted road to the site, which is 3 miles to the east. At the time of this writing, the site and museum are closed, but the Camino de Sueños (Road of Dreams) sculpture by Greg E. Reiche (visible from I–25) and the geologic scenery can still be viewed along the site access road.

New Mexico Historic Sites
New Mexico Department of Cultural Affairs
PO Box 2087
Santa Fe, NM 87504
(505) 476-1130
www.nmhistoricsites.org/el-camino-real

Websites of interest:
www.nps.gov/elca/index.htm
www.blm.gov/visit/el-camino-real-nht
www.caminorealcarta.org
socorro-history.org/CAMINOREAL/map_general.jpg

The Valley of Fires Recreation Area is an ideal place to examine very young Hawaiian-style basaltic volcanism. The term "Malpais" is Spanish for "bad country" and relates to the fact that the roughness and fracturing of the flows created difficult conditions for hoofed animals or wagons and presented a serious barrier to early travel in the region. The area has long been recognized as a feature of geological interest, first becoming Valley of Fires State Park in 1986, and then changing status to a national recreation area in 1991.

The lower flow of the malpais, with a total length of 46 miles, is one of the longest young lava flows in the United States and is a classic example of a tube-fed pahoehoe (ropy lava) system. Only about 5,200 years old, it is one of the younger (but not the youngest) lava flows in the United States. The flows of Carrizozo Malpais are beautifully preserved because of New Mexico's dry climate. Visitors are encouraged to hike on the lava flows and observe the well-preserved geologic features first hand.

OPPOSITE: Well-preserved pahoehoe, or ropy, lava of the Carrizozo Malpais.

Geography and Geologic Province

The Carrizozo Malpais is located at the north end of the broad, sediment-filled Tularosa Basin. The reason the lava erupted here, however, is not related to the Tularosa Basin itself but rather to an east-west zone of crustal weakness (the Capitan Lineament) along which magma was able to rise. As it was erupted, the highly fluid lava flowed far to the south through the topographically lowest part of the basin. The middle part of the malpais is narrow because the valley is restricted by high-standing bedrock. Farther south, the lava was able to spread out over a broad area where the valley widens again.

Basaltic lava flowing in a tube system on Hawaii, similar to tubes that would have existed during eruption of the Carrizozo Malpais.

The Rock Record

Two main rock types are present in the Valley of Fires Recreation Area. The young basaltic lava of the Carrizozo Malpais is the dominant rock. However, on the margins of the flows, and in some areas protruding through the lava flow, there are much older rocks that range in age from Permian to Cretaceous. These rocks are dominantly shales, limestones, and sandstones of marine and riverine origins. Directly adjacent

Satellite image of the young, black, lava of the Carrizozo Malpais. The flow varies in width from 0.6 miles at the narrowest point to 3 miles at its widest.

to the lava flows in some areas, and probably underlying much of the lava, are young alluvial sediments and wind-blown sand. Given the very young age of the lava flows, the appearance of the valley when the flows were erupted is probably much as it is today.

Geological History

The Carrizozo Malpais is a beautiful example of a very young basaltic lava. However, the exact age of eruption was difficult to determine because the composition of the lava is not well suited to conventional dating techniques. An early estimate of the age of the lava was 1,500 years, based on its youthful appearance and lack of weathering. More recently, a geological dating technique that measures how long the surface of the lava flow has been exposed to cosmic radiation has yielded an age of around 5,200 years.

The volcano from which the Carrizozo Malpais lava flows erupted is a small cinder cone less than 100 feet high, called Little Black Peak, located at the north (uphill) end of the lava field. The early stages of eruption of the lava flows were likely from a fissure vent, which, as the eruption progressed, focused down to a single vent, producing the small cinder cone that can be seen today.

The lower flow attained its great length by developing a series of lava tubes through which lava flowed. A lava tube provides insulation to the flow, which in turn allows the lava to travel further before solidifying. In Hawaii, when similar lavas flow through tubes, they cool only slightly over tens of miles of travel. Lava within the tube may flow as fast as 50 feet per second, faster than an Olympic sprinter can run.

The upper of the two Carrizozo Malpais lava flows was also tube fed. It flowed over and partially buried the lower flow but traveled less than half as far. In the rare places where both flows can be seen, no soils have developed between the two features. Furthermore, the chemical composition of the two flows is indistinguishable, and ages of the upper and lower flows are the same. It thus appears that the two flows were erupted during a single, nearly continuous volcanic event. This event produced around one cubic mile of lava during a probable eruption interval of about three decades. The total thickness of lava is typically between 30 to 50 feet, and the flows cover an area of about 130 square miles.

Geologic Features

The Carrizozo Malpais lava flows exhibit excellent examples of features produced by Hawaiian-style basaltic volcanism. One of the most visible and interesting are the ropey, or pahoehoe, flow tops characteristic of basaltic volcanism. Pahoehoe ropes develop in very fluid lava, when the top of the flow starts to cool and solidify, but hotter, more fluid lava continues flowing beneath the surface. Dragging of the cooler top by flowing lava produces wrinkles, similar to a rug being pushed across a hard floor. Other features include deep fissures that develop when the surface of a partially cooled lava flow is pushed up and cracked by lava pressure from

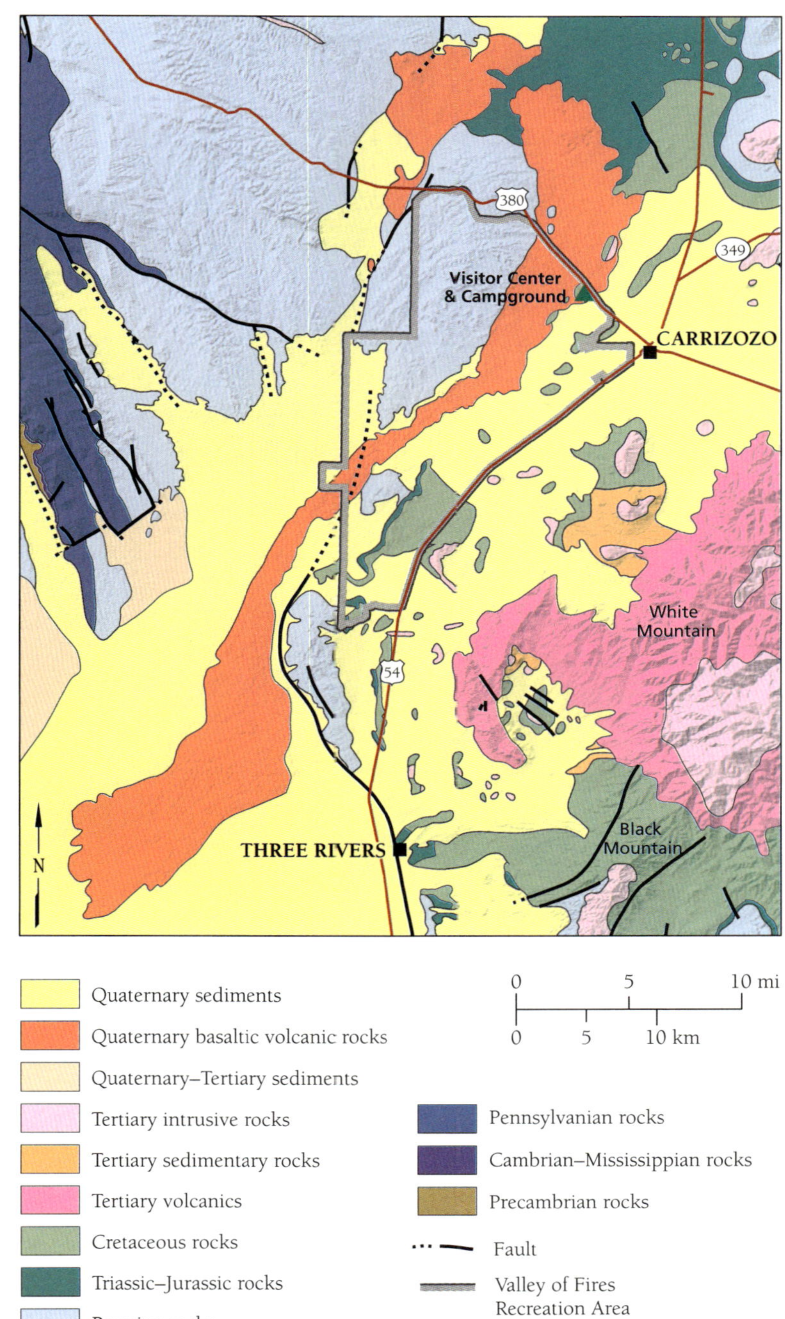

below. The flow interior can be observed deep in these cracks; it contains abundant vesicles (formed by volcanic gas cavities). Finally, nice examples of collapsed lava tubes can be seen at numerous locations on the malpais.

—*Nelia W. Dunbar*

If You Plan to Visit

The Valley of Fires Recreation Area is accessed from US 380, 4 miles west of Carrizozo. This fee area features a bookstore, a paved nature trail with interpretive signs, picnic areas, and campsites. Visitors are encouraged to explore the lava field off the trail but should use caution due to the rough and sharp lava. For more information:

Valley of Fires Recreation Area
P.O. Box 871
Carrizozo, NM 88301
(575) 648-2241
www.blm.gov/visit/valley-of-fires

Capitan Mountains Wilderness and Smokey Bear Historical Park

LINCOLN NATIONAL FOREST AND NEW MEXICO STATE FORESTRY DIVISION

The Capitan Mountains are a prominent landmark in south-central New Mexico, towering a mile or more above the Pecos Valley and the High Plains region to the east. The mountains are 20 miles long and two to six miles wide, with elevations varying from about 5,500 feet near the eastern boundary to 10,083 feet at Capitan Peak (sometimes called El Capitan Mountain).

The Capitan Mountains Wilderness, the birthplace of Smokey Bear, is located in the eastern part of the Capitan Mountains, northeast of the town of Capitan in Lincoln County. It is part of the Smokey Bear Ranger District of the Lincoln National Forest and was formally designated as a wilderness area in 1980. It now comprises a total of 35,067 acres.

Rock-glacier deposits in the Capitan Mountains.

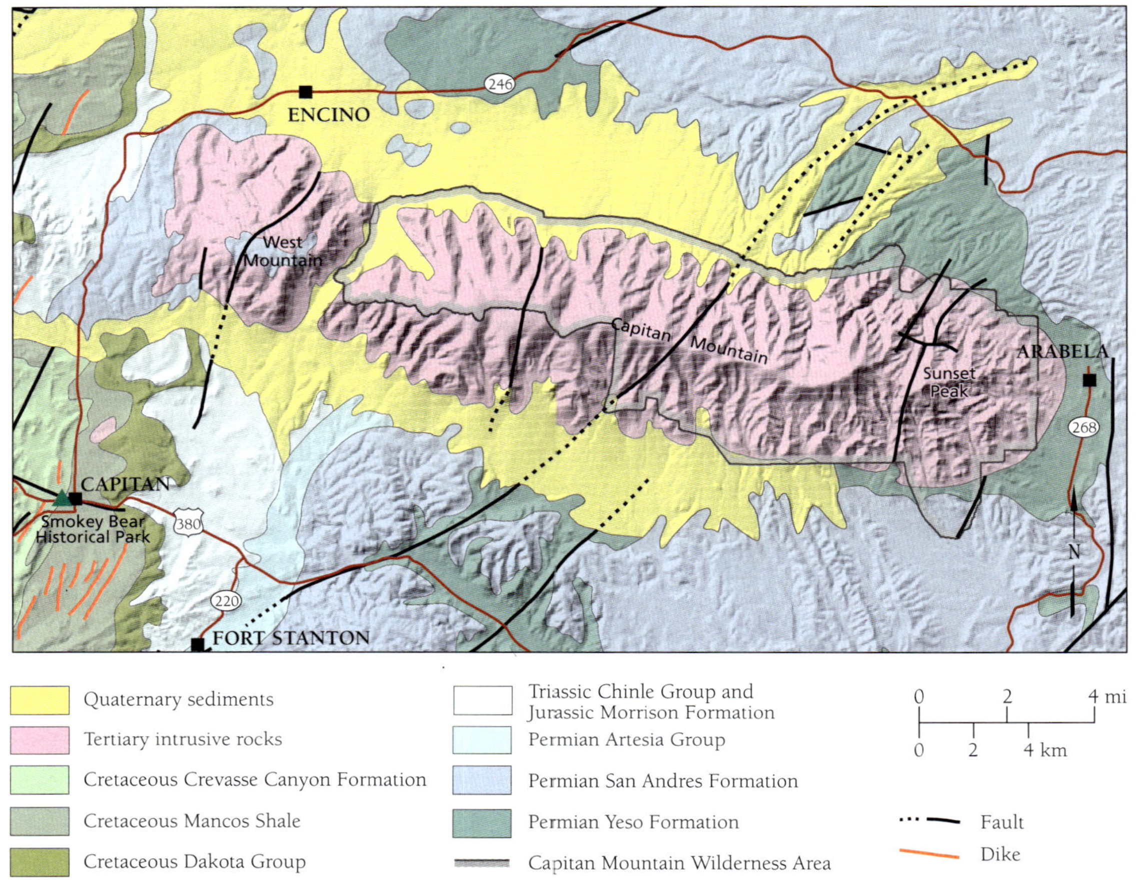

Quaternary sediments	Triassic Chinle Group and Jurassic Morrison Formation
Tertiary intrusive rocks	Permian Artesia Group
Cretaceous Crevasse Canyon Formation	Permian San Andres Formation
Cretaceous Mancos Shale	Permian Yeso Formation
Cretaceous Dakota Group	Capitan Mountain Wilderness Area

0   2   4 mi
0   2   4 km

······—— Fault
—— Dike

Regional Setting

The Capitan Mountains are located in the easternmost part of the Rio Grande rift physiographic province and include the largest exposed Cenozoic intrusive pluton in New Mexico and the only one with an east-west trend.

The Rock Record

The Capitan Mountains are formed by the Capitan pluton, a large granitic intrusion. The roof zone of the pluton, about 600 feet thick and exposed in the western part of the mountains, consists of granite that contains many small cavities that are commonly filled with very tiny crystals (typically less than 1 millimeter) of fluorite, feldspar, quartz, and other minerals. The middle zone is predominantly fine-grained granite (termed aplite). The core of the pluton, consists of porphyritic granite. The term "porphyritic" refers to a texture consisting of large crystals of feldspar surrounded by a fine-grained matrix. The core-zone granite is at least 3,000 feet thick.

The steep upper slopes of the Capitan Mountains locally are covered by barren talus. Some of these formed as rock glaciers during the Pleistocene and consist of cobbles and boulders of the Capitan granite. The areas between the boulders were formerly filled with ice, which allowed the rock glacier to flow downhill under the influence of gravity.

Geologic History

The oldest exposed rocks in the area are limestone, dolomite, siltstone, and gypsum beds of the Permian Yeso and San Andres formations. Depositional settings included shallow-marine seas, coastal dunes and hypersaline lakes.

One mile east of Capitan along US 380 (just off the edge of the map), purplish to greenish sandstones and shales of the Jurassic Morrison Formation are visible in a roadcut. These rocks were deposited by rivers that flowed near the southern edge of the Morrison depositional basin.

In the hills east of Capitan, Upper Cretaceous rocks are visible. The tan beds of the Dakota Sandstone represent coastal beach deposits along the Cretaceous Interior Seaway. Within the Capitan Valley, the younger Upper Cretaceous Mancos Shale contains ammonites and bivalves indicating that the Mancos was deposited in marine to brackish-water environments about 88 million years ago. Fossils can be found in the shale along Salado Creek east of Capitan.

Overlying the Mancos Shale, northwest of Capitan on US 380, are faulted exposures of the Gallup Sandstone and the Crevasse Canyon Formation. These consist of sandstones, shales, and coals deposited in nearshore and coastal-plain environments as the sea retreated to the northeast. In 1884, coal was discovered west of Capitan and became known as the Capitan or Salado coal field; the town of Coalora grew around the mines. At least 20 mines were opened in the Crevasse Canyon Formation with total production from 1895 to 1939 amounting to more than 700,000 short tons.

Prior to the intrusion of the Capitan pluton, the Sierra Blanca Basin formed. Capitan lies on the eastern margin of the basin, which formed in response to Eocene Laramide crustal shortening.

Close-up of the Capitan granite core.

The Capitan pluton was intruded about 28 million years ago. The trend of the Capitan pluton reflects the influence of a deep linear flaw in the crust, termed the Capitan Lineament, which can be traced beyond the Capitan Mountains to the east and west. The textural zones within the granite suggest that crystallization of the magma occurred rapidly and at shallow depths. About 25 million years ago, after emplacement of the pluton, the Capitan Mountains region began to be uplifted. The region surrounding the pluton has since been deeply eroded to more than a mile in depth, leaving the resistant pluton standing high. Unlike many mountain ranges in the Rio Grande rift to the west, the Capitan Mountains were not raised by faulting along the mountain flanks.

Extensive alluvial fans surround the eastern part of the pluton. These fan deposits contain clasts of granite, limestone, sandstone, and magnetite. The presence of eroded magnetite demonstrates that Permian sedimentary rocks containing local iron deposits capped and surrounded the pluton prior to erosion. The rock glaciers, noted earlier, formed during the Pleistocene ice ages.

Smokey Bear Historical Park

Smokey Bear Historical Park lies in the Capitan Valley in the eastern part of the Laramide Sierra Blanca Basin, in the transition zone between the Rio Grande rift and the Great Plains province. It commemorates a famous fire prevention icon—Smokey Bear. Since 1944, a bear was used as a symbol for many posters about fire prevention by the U.S. Forest Service. In 1950, the Capitan Gap wildfire blackened part of the Capitan Mountains in the Lincoln National Forest, north of Capitan. As firefighters brought the blaze under control, a black bear cub was found clinging to the remains of a charred tree. Although the firefighters didn't realize it then, that cub would become the living symbol for fire prevention. The cub, named Smokey Bear, eventually went to the National Zoo in Washington, D.C.

After Smokey died in 1976, his body was returned to his "hometown" of Capitan, New Mexico, where he was buried at the Smokey Bear Historical Park. A new bear cub, also a victim of a forest fire in the Capitan Mountains, was sent to the National Zoo to carry on as Little Smokey. Little Smokey passed away in 1990, but he was not replaced in the National Zoo.

—Virginia T. McLemore

The Smokey Bear icon was introduced in 1950.

Capitan Mountains looking north from US 380.

If You Plan to Visit

The rugged dirt roads of the Capitan Mountains and the Capitan Wilderness Area are accessible from US 380 but require four-wheel drive, high clearance vehicles. A road follows the summit ridgeline of the mountain range. Other roads allow access to the flanks of the mountain ranges. Numerous hiking, and pack-and-saddle trails cross the rugged mountains, and caution is advised as water is generally not available.

The three-acre Smokey Bear Historical Park sits in the center of Capitan on US 380 about 70 miles west of Roswell and about 20 miles east of Carrizozo. A Visitor Center houses extensive exhibits commemorating Smokey Bear and the Forest Service's fire prevention program and a short trail featuring native plants. The city-owned log museum, next to the state museum, was built in the early 1950s. For more information:

Lincoln National Forest
Smokey Bear District
901 Mechem Dr.
Ruidoso, NM 88345
(575) 257-4095
www.fs.usda.gov/attmain/lincoln/specialplaces

Smokey Bear Historical Park
118 Smokey Bear Blvd.
Capitan, NM 88316
Phone: (575) 354-2748
www.emnrd.state.nm.us/SFD/SmokeyBear/SmokeyBearPark.html

Fort Stanton-Snowy River Cave National Conservation Area

Fort Stanton, located approximately 5 miles south of Capitan, was established in 1855 and named for Captain Henry W. Stanton, who was killed in a skirmish with Mescalero Apaches. The fort was built to protect settlements along the Rio Bonito. In 1861, the Union Army abandoned the fort to Confederate soldiers, who were attempting to occupy New Mexico. The Confederate occupation was shortlived. In 1862, Kit Carson and Union troops recaptured the fort and began rounding up the Mescalero Apaches, who were subsequently confined to the Bosque Redondo Reservation, established at Fort Sumner. Later, Fort Stanton became the headquarters for the Indian Agency for the original Mescalero Apache Reservation, which was established in 1873 and relocated further south in 1883. Fort Stanton was decommissioned in August, 1896.

Civil war re-enactors firing a canon at Fort Stanton during Memorial Day celebrations in 2016.

But that was not the end of Fort Stanton. From 1899 to 1953, the fort was used as a Merchant Marine Hospital, a Civilian Conservation Corps (CCC) camp during the great depression, and an internee camp for German sailors during World War II. In 1953, the fort was transferred to the state of New Mexico and became a branch of Los Lunas Hospital until 1995. In 1996, the fort was used as a drug rehabilitation center for state prisoners. Later, the state attempted to close the fort, but local residents convinced the state legislature to preserve the facility. In 2007, Fort Stanton became a State Monument. This long and diverse history is, in part, because Fort Stanton's buildings were constructed of durable local stone that helped many of the original structures survive a century and a half of hard use.

In 2009, Fort Stanton State Monument partnered with the Bureau of Land Management (BLM) to create the Fort Stanton-Snowy River Cave National Conservation Area, a 25,080-acre area established to protect, conserve, and enhance the unique historic, cultural, and scientific resources of the Fort Stanton-Snowy River cave system. There are many hiking, biking, and horse trails, as well as dirt roads that provide access to some of the best rock exposures of Permian, Triassic, and Cretaceous rocks in central New Mexico.

Regional Setting

Fort Stanton-Snowy River Cave National Conservation Area is located on the eastern margin of the Rio Grande rift physiographic province, near the northern end of the Sacramento Mountains. It lies on the south side of the Rio Bonito valley, which separates the Sacramento Mountains from the Capitan Mountains to the north.

The Rock Record

Most of the rocks exposed in the eastern portion the Fort Stanton-Snowy River Cave National Conservation Area are sedimentary rocks of the Permian San Andres Formation. In the western portion of the conservation area, Triassic Chinle Formation sandstones are exposed, along with outcrops of Cretaceous marine and riverine sedimentary rocks. Small Tertiary dikes, sills, and plugs are found intruding the sedimentary rocks in the southern portion of the area.

Geologic History

The San Andres Formation predominantly consists of limestones and dolomites that were deposited in shallow seas that covered most of the area about 270 million years ago. Triassic red beds were then deposited by rivers within a large continental basin. Cretaceous sediments were deposited in streams and beaches along the margin of an interior seaway that extended from the Gulf of Mexico to the Arctic Ocean. Ammonites and bivalves (clams) are found in shales along Salado Creek, east of Capitan. These fossils indicate marine to brackish-water environments existed about 90 million years ago. The Capitan Mountains form the skyline north of Fort Stanton-Snowy River Cave National Conservation Area and

In 2001, the Snowy River passage of Fort Stanton Cave was discovered. The floor is coated with white calcite that precipitated from intermittent waterflow through the cave. The deposit, over 10 miles long, is thought to be the longest cave formation in the world.

represent the largest exposed Tertiary pluton in New Mexico. Other smaller intrusions cut the sedimentary rocks in the southern portion of the Fort Stanton-Snowy River Cave National Conservation Area. The caves are relatively young features (late Cenozoic), and dating of gypsum needles from Fort Stanton Cave suggests that crystals grew in the caves as recently as 500 to 7,000 years ago.

Geologic Features

Fort Stanton Cave, currently estimated to be 31.4 miles long, is the second-longest cave in New Mexico, exceeding the length of the better-known Carlsbad Cavern, and is currently the 62nd longest known cave in the world. Fort Stanton Cave is formed in limestones of the Permian San Andres Formation. The cave was first recorded in about 1855 when the nearby Fort Stanton was established, and in 1877 it became the second cave west of the Great Plains to be surveyed. Fort Stanton Cave is administered by the Bureau of Land Management, which has gated the cave and controls access via a permit system.

In 2001, a team working on the Priority Seven Breakdown Dig in Fort Stanton Cave broke through into a new passage. This newly discovered section of the cave was named Snowy River because of the presence of a snow-white pool deposit that occupies an old stream channel in the passage. The Snowy River pool deposit consists of opaque white calcite with a corrugated texture that was deposited on top of red-brown mud that originally made up the bed of the stream. The mapped length of the Snowy River deposit is now more than 11 miles, and the passage continues to extend southward for an unknown distance. It has been suggested that the calcified bed of Snowy River may be the longest continuous cave formation on Earth.

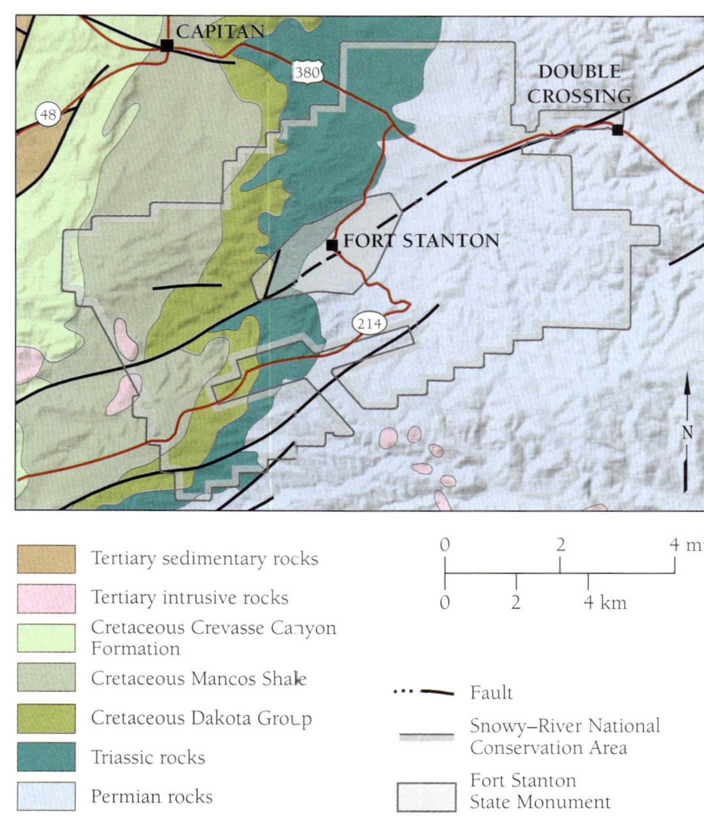

—Virginia T. McLemore and Lewis Land

195

If You Plan to Visit

Most of the Fort Stanton-Snowy River Cave National Conservation Area, including the Fort Stanton State Monument, is open year round. However, a permit is required to access Fort Stanton Cave.

From Roswell, travel west for about 59 miles on US 380 and then turn south on NM 220 for 2.6 miles. From Carrizozo, travel east on US 380 for 24.2 miles and then turn south on NM 220 for 2.6 miles. For more information:

Fort Stanton Historic Site
PO Box 36
Fort Stanton, NM 88323
(575) 354-0341 (museum)
www.nmhistoricsites.org/fort-stanton

Bureau of Land Management
Roswell Field Office
2909 West 2nd Street
Roswell, NM 88201
(575) 627-0272
www.blm.gov/nlcs_web/sites/nm/st/en/prog/NLCS/FSSRC_NCA.html

View to the north across the
Rio Hondo of the Lincoln folds,
showing tightly folded beds of
the Yeso Formation overlain by
gently tilted beds of the San Andres
Formation along the skyline.

Lincoln State Monument was established in 1937 in the town of
Lincoln, west of Capitan. It is most famous for Billy the Kid and the
Lincoln County War of 1878. The town was established about
1849 by Hispanic farmers and was originally called Las Placitas del
Rio Bonito, "Little Settlements on the Pretty River." The town was
renamed Lincoln in 1869 when Lincoln County was created from the
eastern part of Socorro County. In 1959, the area surrounding the
town of Lincoln was designated as the Lincoln National Historic
Landmark to preserve the area's rich history; Lincoln State Monument
was established in 1979.

Regional Setting

Visitors to Lincoln are surrounded by historical structures and can
easily miss the geology surrounding the town. Lincoln is in the central
mountains portion of the Rio Grande rift province, between the
Capitan Mountains to the north and Sierra Blanca to the southwest.
Rocks in the Lincoln area are Permian, Tertiary, and Quaternary in age.

The Rock Record

The oldest rocks exposed in the area are the Permian Yeso Formation. Regionally, it is more than 1,000 feet thick, but only the upper part is exposed in the Lincoln area where it forms most of the surrounding hillsides. Yeso is Spanish for gypsum. Most of the roadcuts near Lincoln are in the Yeso Formation, which is commonly folded and distorted. The lower Permian Yeso strata were deposited in coastal salt flat (sabkha) and eolian (wind-blown) environments about 275 million years ago. The younger (early to middle Permian) San Andres Formation overlies the Yeso Formation and caps the low ridges and hills surrounding Lincoln. The San Andres predominantly consists of limestones and dolomites that were deposited in shallow inland seas that covered the area about 270 million years ago. Pleistocene and recent sediments fill the valley, partially covering the Yeso Formation.

A hiking trail runs along the Rio Bonito, which borders the town of Lincoln. The stream gravels and terraces attest to cycles of erosion and deposition by Rio Bonito throughout Pleistocene and recent times.

Hispanic settlers first used local geologic materials for construction of their houses and other buildings. Original Hispanic homes from the 1840s, termed *jacales*, were constructed from vertical posts of cedar or juniper that were caulked with mud and gypsum. The interiors were whitewashed with a gypsum solution, called jaspé. A jacal is on

Folds in the Yeso Formation.

display at the Lincoln Heritage Trust Center. Many of the more recent houses are constructed of adobe, made from local sand, silt, and clay. Plaster was made from gypsum from the Yeso Formation. The Torreon, a structure built in the 1850s as a defense against Apache raids, is made of sandstone and dolomite boulders from the Yeso Formation.

Geologic History

The town of Lincoln lies on the Pedernal uplift, a major north-trending uplift, which developed during the Ancestral Rocky Mountain deformation about 340–300 million years ago. During the Permian period, the eroded remnants of the uplift were buried by shallow-marine deposits including the Yeso and San Andres formations. The area continued to accumulate sediments episodically during the Mesozoic. During the Laramide orogeny (75–45 million years ago), the region was again deformed, creating the Sierra Blanca basin to the west in which Eocene deposits accumulated. Major intrusion and volcanism began about 38 million years ago near Sierra Blanca and culminated about 28 million years ago with the emplacement of the granite pluton that

The Torreon at Lincoln, New Mexico, which is made of sandstone and dolomite boulders from the Yeso Formation.

now forms the Capitan Mountains to the north of Lincoln. Since that time, the region has experienced mostly uplift and erosion.

Geologic Features

The most spectacular geologic features in the area are the Lincoln folds, which can be seen on both sides of the valley in and around Lincoln. The folds are a series of anticlines and synclines that resemble a rumpled rug. The Lincoln folds are restricted stratigraphically to the weak, gypsum-bearing Yeso Formation. The overlying San Andres and younger formations are only broadly arched over the Lincoln folds. Similar folds in the Yeso Formation are locally present in regions west of Lincoln. The origin of such folds is uncertain. Hypotheses include deformation by igneous intrusions, dissolution of evaporite minerals and local sag of overlying rocks, and folding related to compression or gravitational sliding.

Landslides and slump blocks of San Andres limestone and dolomite are common in canyons and along ridges surrounding Lincoln. The San Andres Formation is more resistant to erosion than the underlying, less competent Yeso Formation. Erosion, freeze-thaw action, and truck traffic occasionally dislodge the limestone blocks, causing them to slide downslope on the less resistant Yeso beds. This process is aided when the Yeso beds become saturated with water or ice, resulting in a slick surface on which the overlying limestone blocks can easily move. The contact between the San Andres and Yeso formations is partly obscured by these landslides and slump blocks. It is wise to check slopes for such features before building a house or other structures in this area.

—Virginia T. McLemore and Steven M. Cather

If You Plan to Visit

The Lincoln State Monument encompasses most of the town of Lincoln and is located directly on US 380 about 57 miles west of Roswell and 30 miles east of Carrizozo. For more information:

Lincoln State Monument
Highway 380
Lincoln, NM 88338
(575) 653-4372
www.nmhistoricsites.org/lincoln

The Lincoln Highlands of South-Central New Mexico

The Lincoln National Forest was established in 1902 and covers more than a million acres of the highlands of south-central New Mexico. This is the land of Smokey Bear and Billy the Kid, and the geology is as fascinating and diverse as the history and landscape, which ranges from the Chihuahuan Desert in the Tularosa Basin to the southernmost alpine terrain in the United States at the crest of Sierra Blanca. This National Forest has a northern section centered around Capitan and a seperate southern section that includes Cloudcroft.

OPPOSITE: Travertine mounds and terraces at Monument Spring.

Regional Setting

The forest extends from the heavily faulted Basin and Range/Rio Grande rift geologic provinces on the west to the margin of the geologically quiescent Pecos valley in the east. The Jicarilla Mountains and the Capitan Range are composed of intrusive igneous rocks, part of a string of such mountains along the east side of the Rio Grande rift. The Capitan Mountains, intruded along an east-west structural trend about 28 million years ago, are notable for being one of the few ranges in New Mexico that do not trend north-south. Sierra Blanca is the deeply eroded remnant of a huge volcanic field, and the summit is the southernmost glaciated peak in the United States. In contrast, the central and southern Sacramento Mountains are composed of layered sedimentary rocks, which are uplifted along faults of the Rio Grande rift at the base of their steep and rugged west flanks. These rocks dip shallowly to the east under the Pecos River valley and extend thousands of feet beneath the surface of southeastern New Mexico.

The Rock Record

The west face of the Sacramento Mountains exposes layered sedimentary strata of the Paleozoic era, from Cambrian at the foot to Permian at the crest. Limestones are more abundant than sandstones and shales. Large limestone mounds, easily seen from Tularosa to south of Alamogordo on US 54 and east on US 82, stand out amidst the otherwise flat-bedded Mississippian- to early Permian-age strata. These ancient organic buildups were bioherms (reef-like structures) that formed lens-shaped topographic highs on the seafloor.

Faults and folds of the Ancestral Rocky Mountains orogeny tilted and severed the sedimentary layers north and south of US 82. The prominent "step" halfway up the west side of the Sacramento

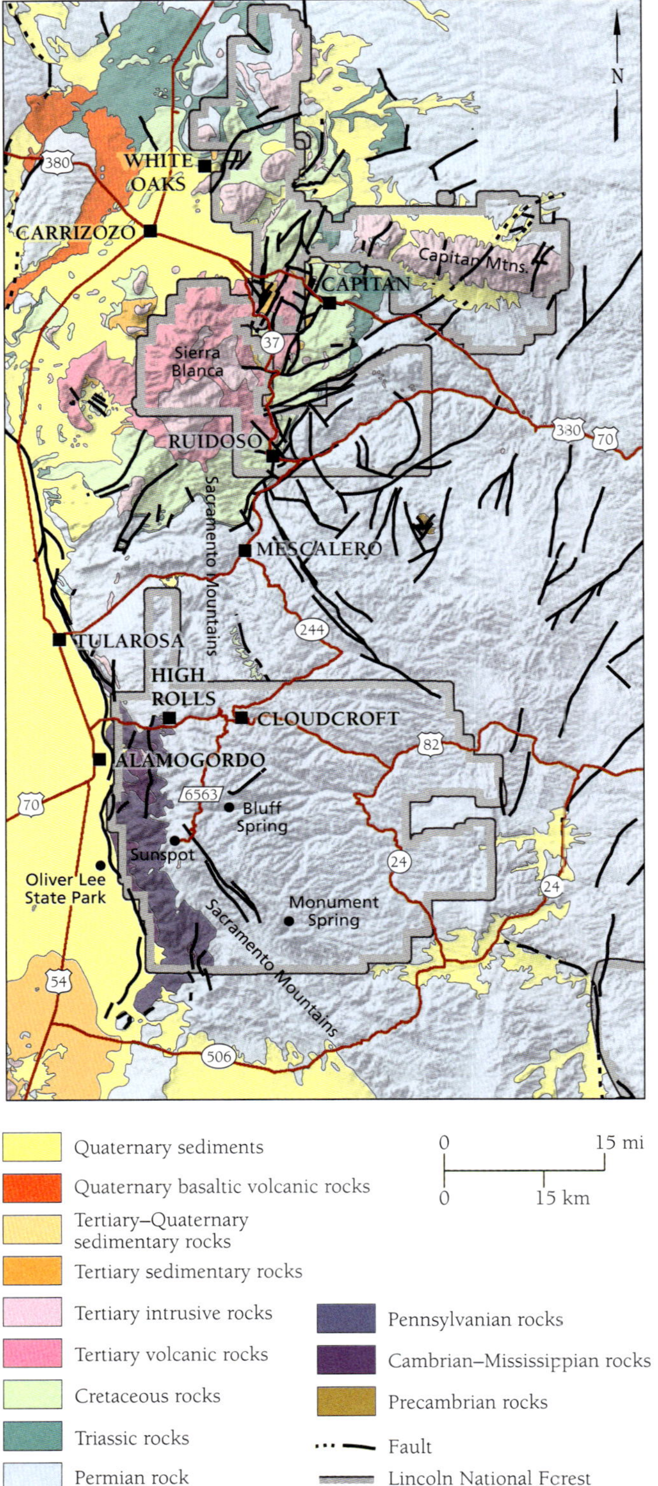

Mountains escarpment is eroded into gypsum and red, orange, and yellow shales and sandstones of the Permian Abo and Yeso formations. The forested crest is capped by the resistant Permian San Andres Limestone. This formation is exposed over a huge area, including most of the eastern portion of the forest, north and south of the Capitan Mountains. The colorful Yeso Formation rocks are exposed in the eastern canyon bottoms. The Tertiary igneous intrusions and volcanic rocks that make up the northern peaks and Sierra Blanca intruded and lie on Mesozoic and Cenozoic sandstones and shales from Ruidoso to Capitan and White Oaks.

Geologic History

For most of Paleozoic time, the area was periodically covered by warm, shallow seas, which deposited layers of limestones and shales, followed by periods of exposure when the seas withdrew. This exposure resulted in erosion or deposition of nonmarine shales, sandstones, and conglomerates. Especially during the Pennsylvanian, glacially controlled sea-level changes deposited numerous cycles of marine carbonates during highstands, followed by fluvial sediments deposited during periods of exposure.

Uplift of the Ancestral Rocky Mountains during Pennsylvanian time resulted in highlands to the east and northeast of the present-day Sacramento Mountains. Permian erosion of this ancient range buried the older rocks and structures with red and orange sandstones and shales. Pajarito Mountain and Pedernal Peak in central New Mexico are remnants of these mountains. Warm seas returned to deposit limestones of the San Andres Formation, which was followed by terrestrial and marine deposition of sandstones and shales derived from distant mountains during the Mesozoic. Onset of the Laramide orogeny deformed the crust into a bowl-shaped basin northwest of Ruidoso, which filled with sediments eroded from nearby, growing mountain ranges.

This was followed by massive volcanic eruptions and igneous intrusions about 40 to 25 million years ago. Hot fluids associated with the magmatism deposited gold, silver, iron, and other metallic ores in the adjacent rocks. As the Rio Grande rift opened, beginning about 25 million years ago, fault movement occurred along the west side of the Sacramento Mountains. Erosion dominates the landscape now, but seismic activity is still possible along the rift faults.

Geologic Features

The crest trails north of Sierra Blanca and in the southern Sacramento Mountains offer stunning views of the topography and geology of the region. FR 90 south of High Rolls (the West Side road), offers access to the Abo and Yeso formations and great views of faults and folds. Be prepared for driving in a remote, rugged area if you take this road—high-clearance and four-wheel drive are recommended.

One can hike from Oliver Lee State Park, south of Alamogordo, up the steep Dog Canyon Trail to Joplin Ridge and the West Side road, through sheer cliffs of Paleozoic rocks (see the chapter on Oliver Lee State Park).

NM 6563 (Sunspot Scenic Byway) winds from Cloudcroft to the National Solar Observatory at Sunspot along the crest of the southern Sacramentos. It displays many exposures of the Permian San Andres and Yeso formations, and offers spectacular views of the Tularosa Basin, White Sands National Park, and the San Andres and Organ mountains to the west.

Southwest of Capitan hundreds of dikes emanate from beneath the Sierra Blanca volcanic field. Dikes are vertical sheets of magma that filled faults and cracks in the host rock. Faults, folds, and intrusions exposed in roadcuts along US 380 near Indian Divide (8–10 miles east of Carrizozo) show how complex the geology of the Sierra Blanca area really is!

Volcanic breccia exposed along Trail 26 to Nogal Peak.

The blue minerals corundum (sapphire) and lazulite in hydrothermally-altered lavas at the junction of the south fork of the Rio Bonito and Bluefront Canyon in the White Mountain Wilderness.

206

Salado Canyon, Bridal Veil Falls, and Grandview trails near High Rolls lie along the route of the abandoned Alamogordo and Sacramento Mountains Railroad within Pennsylvanian limestones and shales. The rocks dip steeply where they were deformed by the Ancestral Rocky Mountains orogeny. Several folds and faults are well exposed along the trail. Lower Permian conglomerates near the Salado Canyon Trailhead are typical of sedimentary rocks deposited near steep, rising mountains.

Bluff and Monument springs are two of the most spectacular springs of the dozens within the mountains. Bluff Spring is easily reached on Otero CR 17. Here, a waterfall flows over a cliff formed of travertine and tufa deposited from the spring water as it emerges from the ground. This water is saturated with calcium carbonate from the dissolution of limestones by percolating groundwater. Monument Spring is more of an adventure, as it requires a canyon hike from the end of steep and rugged FR 5600. The reward is a spring-fed stream emerging from a cliff and flowing over beautiful travertine terraces and pools brimming with life. See also the chapter on Sitting Bull Falls Recreation Area, which also features travertine waterfalls and pools in the areas southwest of Carlsbad.

—Geoffrey C. Rawling

If You Plan to Visit

The Lincoln National Forest surrounds the towns of Capitan, Alto, Ruidoso, and Cloudcroft, among other areas, and is crossed by highways US 380, US 70, and US 82. It's headquarters are in Alamogordo, and it has district offices in Carlsbad, Cloudcroft, and Ruidoso. For more information:

Lincoln National Forest
3463 Las Palomas Road
Alamogordo, NM 88310
(575) 434-7200
www.fs.usda.gov/lincoln

The Sunspot Solar Observatory was established in 1947 and is a facility for solar research.

The approach to Nogal Peak along Trail 26. The peak is not a volcanic cone but rather an erosional remnant composed of layered lava flows.

The White Mountain Wilderness is the heart of Lincoln National Forest and encompasses rugged and beautiful terrain formed by the eroded Sierra Blanca volcanic field. The sharp ridges and deep canyons expose volcanic and intrusive igneous rocks, along with small ore deposits, abandoned mines, and prospects. The Little Bear Fire of June, 2012 radically changed much of this landscape, and visitors can see the dramatic effects of catastrophic wildfire on the vegetation and geomorphology.

Regional Setting

The ancient volcanic complex comprising the northern Sacramento Mountains and Sierra Blanca Peak overlies the Sierra Blanca basin, a downwarp in the Earth's crust formed during the Laramide Orogeny, about 75–45 million years ago. To the west of Sierra Blanca are faults

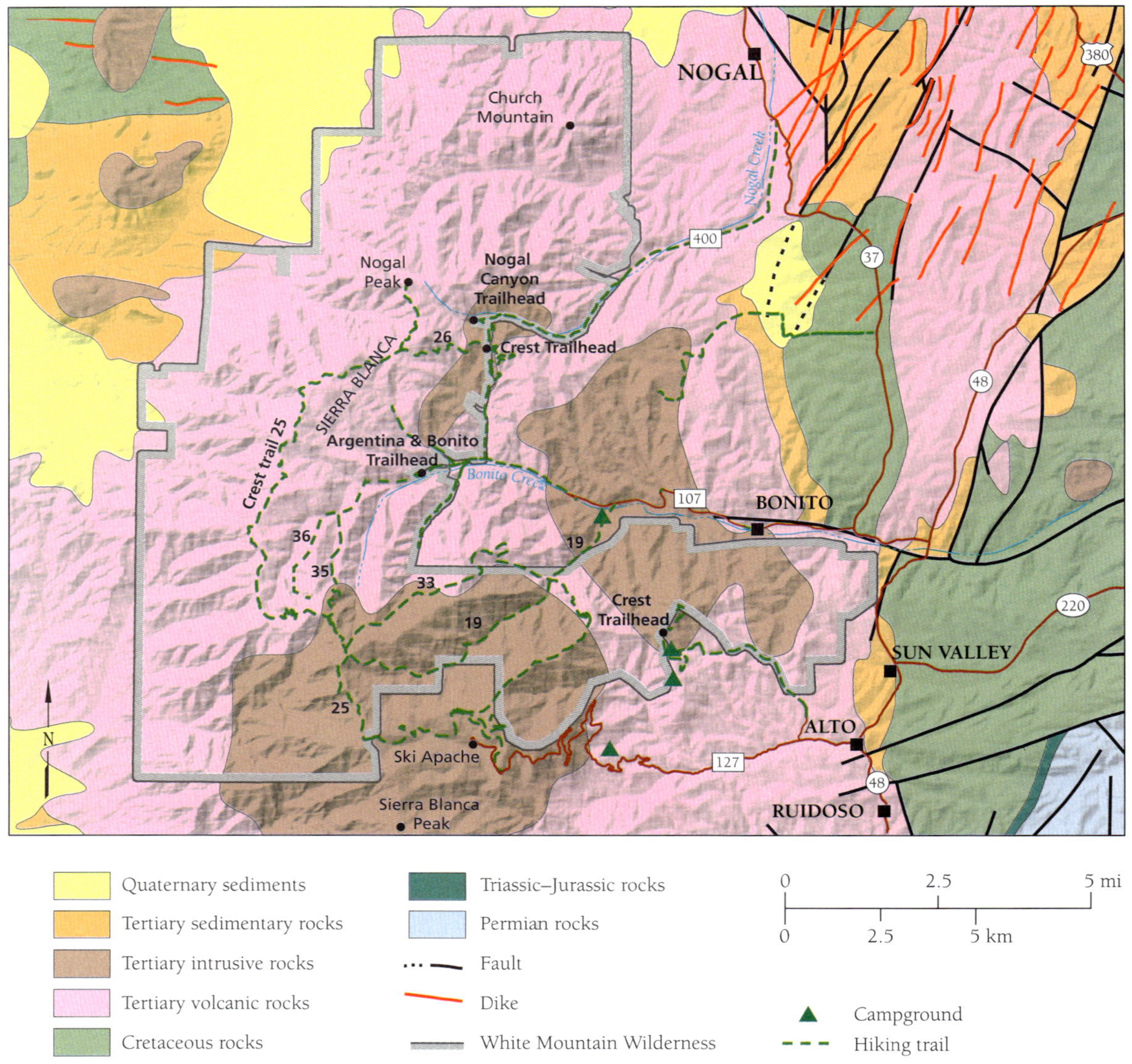

Quaternary sediments

Tertiary sedimentary rocks

Tertiary intrusive rocks

Tertiary volcanic rocks

Cretaceous rocks

Triassic–Jurassic rocks

Permian rocks

Fault

Dike

White Mountain Wilderness

Campground

Hiking trail

0   2.5   5 mi
0   2.5   5 km

**Geologic map of the
White Mountain area.**

of the Rio Grande rift that bound the Tularosa Basin. There is nearly 7,900 feet of local topographic relief from the floor of the basin to the summit of Sierra Blanca at 11,981 feet, the most anywhere in New Mexico. The views from the high peaks are spectacular. One can see various mountain ranges bordering the Rio Grande rift, including the Franklin Mountains near El Paso to the southwest, the Black Range and the San Andres mountains to the west, and the Manzano and Sandia mountains to the northwest. Igneous rocks of the Jicarilla and Capitan mountains are close to the north and northeast, and the great expanse of the Pecos River valley and the southern High Plains forms the eastern horizon.

The Rock Record

The Sierra Blanca volcanic field consists of rocks of Eocene to Oligocene age (approximately 38 to 27 million years ago) more than 6,500 feet thick. Lava flows erupted, volcanic breccias formed from debris flows and slope collapses, and sedimentary layers were deposited by streams eroding the volcanic centers. All these rocks form a complex, roughly layered pile. The Rialto, Bonito, and Three Rivers stocks are magma bodies that intruded, and are younger than, the lower part of the volcanic pile. Intrusion of these magmas and release of hot, mineral-rich fluids during cooling formed many small ore bodies throughout the mountains and extensively altered many of the host rocks. The ore bodies were mined for gold, silver, lead, zinc, molybdenum, and copper. Vertical dikes and horizontal sills, formed by injection of magma into cracks and between layers of volcanic rocks, are common. Erosion has removed several thousand feet from the top of the mountain—thus, the summit of Sierra Blanca is composed of coarse-grained intrusive rocks of the once-buried Three Rivers stock.

Geologic Features

TRAILS 25 AND 26—NOGAL PEAK—This hike is a 2-mile round trip from the junction of FR 400 and FR 108 at Nogal Pass to the isolated summit of Nogal Peak. Along the way are outcrops of pale lavender, coarse-grained intrusive rocks of the Rialto Stock and light- and dark-gray lavas, and multicolored volcanic breccias. The steep slopes of Nogal Peak are composed of ledgy and angular exposures of light-gray lavas with vesicles (cavities created by gas bubbles in the originally hot magma).

TRAILS 19 AND 33—SOUTH FORK OF THE RIO BONITO AND BLUEFRONT CANYON MINE—This hike is a 2.5-mile round trip from the abandoned South Fork Campground gate through an area that was severely burned in 2012. Hikers should be in good shape and comfortable with navigating off trail in rugged terrain. The South Fork Canyon along Trail 19 alternates between beautiful streambed outcrops of white to gray intrusive rocks of the Bonito Stock and gravels eroded from the steep, burned hillslopes. After about 1 mile, head west towards the north side of the mouth of Bluefront Canyon. Look for a mine dump of white boulders several feet across. Hot, acidic waters along the margin of the Bonito Stock altered lavas to alunite and a white, clay-rich rock with lovely bands of the blue corundum and lazulite.

TRAILS 36, 25, 35—BONITO CREEK AND ASPEN CANYON LOOP—This beautiful 6- to 7-mile hike loops through wooded, unburned canyons up to grassy meadows at the range crest. Rocks encountered mostly are gray, purple, and brick-red lavas and breccias. Blocky, fine-grained, dark-gray rocks were probably intruded as dikes and sills. The canyon bottoms have little gravel and debris in contrast to the severely burned South Fork of the Rio Bonito. Lavas exposed before the junction with Crest Trail 25 have abundant apple-green epidote on fracture surfaces and in vesicles, another product of hydrothermal alteration. For a spectacular view, hike the extra 250 feet up to the wide-open summit of White Horse Hill.

—Geoffrey C. Rawling

If You Plan to Visit

The White Mountain Wilderness can be accessed from forest roads and trails heading west from NM 37, NM 48, and NM 32 in Lincoln County. For more information:

Lincoln National Forest
3463 Las Palomas Rd.
Alamogordo, NM 88310
(575) 434-7200
www.fs.usda.gov/attmain/lincoln/specialplaces

The Three Rivers Petroglyph Site is located in the Tularosa Basin in Otero County, north of Tularosa and south of Carrizozo. The site features hiking trails, a picnic/camping area, and restrooms. The confluence of three small streams that merge just northeast of the site provide the origin of the name of the park. Native Americans first occupied the region about 10,000 years ago, and between 900 and 1200 AD, the Jornada Mogollon people scratched and pecked more than 21,000 petroglyphs into the desert varnish that coats the weathered surfaces of andesitic rocks. Requiring thousands of years to form, desert varnish is a relatively thin, orange, purple, or black layer of wind-deposited clay coated with manganese and iron oxides. These oxides were deposited by microbes living on the surface of the rock.

Regional Geography and Geology

The Three Rivers site lies on the southwestern margin of the Sierra Blanca Basin, a Laramide basin that contains sedimentary rocks as young as Eocene in age, and near the northeastern edge of the Tularosa Basin, a younger, rift-related basin. Plutonic and volcanic rocks of the Sierra Blanca volcanic field intruded and buried the Sierra Blanca Basin about 38 to 27 million years ago. The Sierra Blanca volcanic field is located in the transition between the tectonically stable Great Plains and the tectonically active Rio Grande rift, and it formed at about the same

One of many petroglyphs at the Three Rivers site. Note that the dark coat of desert varnish is broken on the lower left corner of the block, revealing the underlying crystal-rich andesitic basaltic rock.

AGE		UNIT
Cenozoic	Oligocene	Godfrey Group
Cenozoic	Eocene	Walker Group
Cenozoic	Eocene	Sanders Canyon Formation
Cenozoic	Eocene	Cub Mountain Formation
Mesozoic	Cretaceous	Crevasse Canyon Formation
Mesozoic	Cretaceous	Gallup Sandstone
Mesozoic	Cretaceous	Upper Mancos Shale
Mesozoic	Cretaceous	Tres Hermanos Sandstone
Mesozoic	Cretaceous	Lower Mancos Shale
Mesozoic	Cretaceous	Dakota Sandstone
Mesozoic	Triassic	Moenkopi Formation

Triassic to Oligocene rock units exposed near the Three Rivers site. The height of the boxes is scaled to the relative thickness of the units in this region.

time as the Mogollon–Datil volcanic field to the west. The Sierra Blanca volcanic rocks were erupted and emplaced during the transition from Laramide crustal shortening to Rio Grande rift crustal stretching. Faulting within the rift has since disrupted parts of the Sierra Blanca volcanic field.

The Rock Record and Geologic Features

Rocks near the site record the retreat of the Cretaceous Western Interior Seaway, followed by the formation of the Sierra Blanca Basin, eruption of the Sierra Blanca volcanic field, and development of the Tularosa Basin during Rio Grande rift crustal stretching. Outcrops south of the county road between US 54 and the site are tilted red beds of the Triassic sedimentary rocks (Moenkopi Formation) and overlying Upper Cretaceous formations, including Dakota Sandstone, Mancos Shale, Tres Hermanos Sandstone, Gallup Sandstone, and Crevasse Canyon Formation. These beds generally become younger to the east, toward the axis of the Sierra Blanca Basin. The petroglyph site itself is on a northwest-trending ridge of dark-gray andesitic rocks, that intrude along the bedding of the Cretaceous rocks (sill). Beneath the andesitic rocks are yellowish-gray Cretaceous beds exposed on the western slope of the ridge. The Gallup Sandstone lies at the base of the slope and contains fossil oysters that lived in the Western Interior Seaway approximately 80 million

years ago. The Gallup Sandstone grades upward into the nearshore swamp and river deposits of the Crevasse Canyon Formation, also exposed on the western slope.

Eocene sedimentary units deposited within the developing Sierra Blanca Basin include the Eocene Cub Mountain and Sanders Canyon formations. The Cub Mountain Formation consists of red sandstones, mudstones, and conglomerates derived from the Laramide highlands to the southwest. The Sanders Canyon Formation overlies the Cub Mountain Formation and is made of mudstones and sandstones derived from volcanic source areas near Las Cruces. The Sanders Canyon Formation is exposed in small arroyos about 1 mile east of the campground.

The Sierra Blanca volcanic field formed between about 38 and 27 million years ago. The earliest eruptions originated from now-eroded stratovolcanoes in the northern and central parts of the field, producing lava flows and debris flows that traveled long distances over a subdued topography. Younger volcanism, about 29 million years ago, is best preserved in the Godfrey Hills just northeast of the petroglyph site. The main ridge at the petroglyph site is capped by an andesitic sill, a sheet-like intrusion. The sill was formed about 36 million years ago, and it contains hornblende crystals that are as large as 2 inches across. These andesitic rocks provide the canvas for the many artistic petroglyphs at the site.

Another example of rock art at Three Rivers.

Godfrey Hills to the left and Sierra Blanca to the right. The prominent black cliff in the Godfrey Hills is the Palisades Tuff. The petroglyph site is just to the right of the photo.

During the last 25 million years, the intrusive plumbing system of the Sierra Blanca volcanic field was uplifted by faults of the Rio Grande rift and exhumed by water and wind erosion. The Tularosa Basin also formed during this time as blocks were down-faulted into the rift. As a result the western escarpment of Sierra Blanca is the tallest in New Mexico, with a difference in elevation of nearly 7,900 feet between the peak and the floor of the northern Tularosa Basin. During the last few million years, debris shed from Sierra Blanca and the Godfrey Hills was deposited as alluvial fans and stream terraces that surround the main ridge at the petroglyph site.

—*Shari A. Kelley*

If You Plan To Visit

The Three Rivers site is located 17 miles north of Tularosa and 28 miles south of Carrizozo on US 54. Turn east from US 54 at Three Rivers (site of a beautiful, redbrick 1904 schoolhouse) onto CR B30 and travel 5 miles on paved road to the parking/picnic area and campsites. For more information:

Bureau of Land Management
Las Cruces District Office
1800 Marquess Street
Las Cruces, NM 88005
(575) 525-4300
www.blm.gov/visit/three-rivers-petroglyph-site

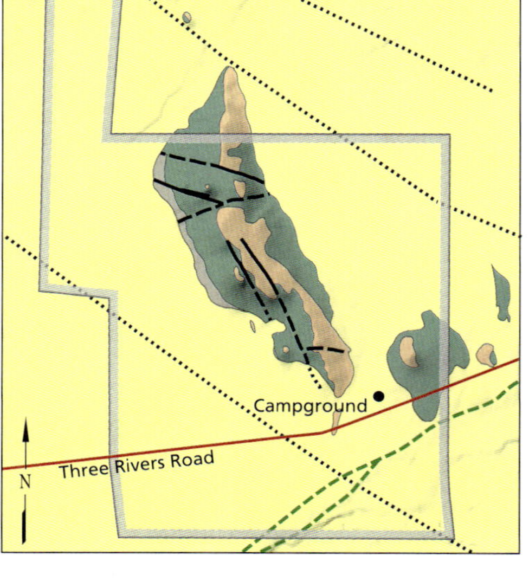

Tunnel Vista Observation Site and Fresnal Canyon

The Tunnel Vista Observation Site is on the western slope of the Sacramento Mountains along US 82, just west of the entrance to the High Rolls tunnel. The only tunnel on a US Highway in New Mexico, it was built between 1947–1949. The Observation Site displays cliff-forming Pennsylvanian limestones that rise 600 feet along the north side of Fresnal Canyon. These rocks form the ridge through which the tunnel was excavated and are a popular site for rock climbing and rappelling. From this site, one can also view a panorama to the west that includes the western slope of the Sacramento Mountains, the Tularosa Valley beyond, and the northern part of the White Sands gypsum dune field. On a clear day, one can see the San Andres Mountains forming the distant skyline.

West end of High Rolls tunnel, from near Tunnel Vista parking lot.

Regional Setting

The Sacramento Mountains, a north-trending range more than 50 miles long, attain a maximum elevation of 9,700 feet. The range is an east-tilted fault block that was uplifted along its western boundary fault during mid- to late Cenozoic crustal stretching associated with the Rio Grande rift. Vertical movement along the western fault, which uplifted the Sacramento Mountains and lowered the adjacent sediment-filled Tularosa Basin, was approximately 7,000 feet.

The Bug Scuffle Limestone visible from the Tunnel Vista Observation Site.

The Rock Record

The Sacramento Mountains expose a thick sequence of Cambrian to Permian beds, but near Tunnel Vista, the sedimentary section is composed entirely of middle Pennsylvanian to lower Permian strata. The oldest stratigraphic unit in this sequence is exposed opposite the Tunnel Vista site—the Middle Pennsylvanian Bug Scuffle Limestone Member of the Gobbler Formation. This unit primarily consists of cliff-forming limestones separated by thin shale intervals. The limestones formed on a stable, shallow-marine shelf in a tropical environment by accumulation of fine-grained carbonate sediments composed of shells of invertebrate fossils and algae.

The base of the limestone cliff shelters a seasonal campsite used by Archaic hunter-gatherers. Archaeological excavations at this Fresnal Shelter site yielded sandals, fragments of baskets, hide, bones, twine, feathers, wooden and stone tools, and other artifacts, but no pottery. Radiocarbon dates suggest occupation from about 7,000 to 1,650 years ago. Remains of corncobs and beans indicate that these were being cultivated in the region as much as 3,000 years ago.

Units that overlie the Bug Scuffle Limestone—the Beeman, Holder, and Laborcita formations (Upper Pennsylvanian to lower Permian) and the lower Permian Abo Formation—are visible at lower elevations west of Tunnel Vista. The older Bug Scuffle Limestone crops out above the younger formations because it was uplifted along the north-trending Fresnal fault, which lies 0.2 miles to the west (marked by steeply dipping strata along the highway), whereas the younger formations west of the fault have been down-dropped about 1,600 feet.

Illustration of a sandal which was found during an excavation between 1996 to 2000, in the High Rolls Cave.

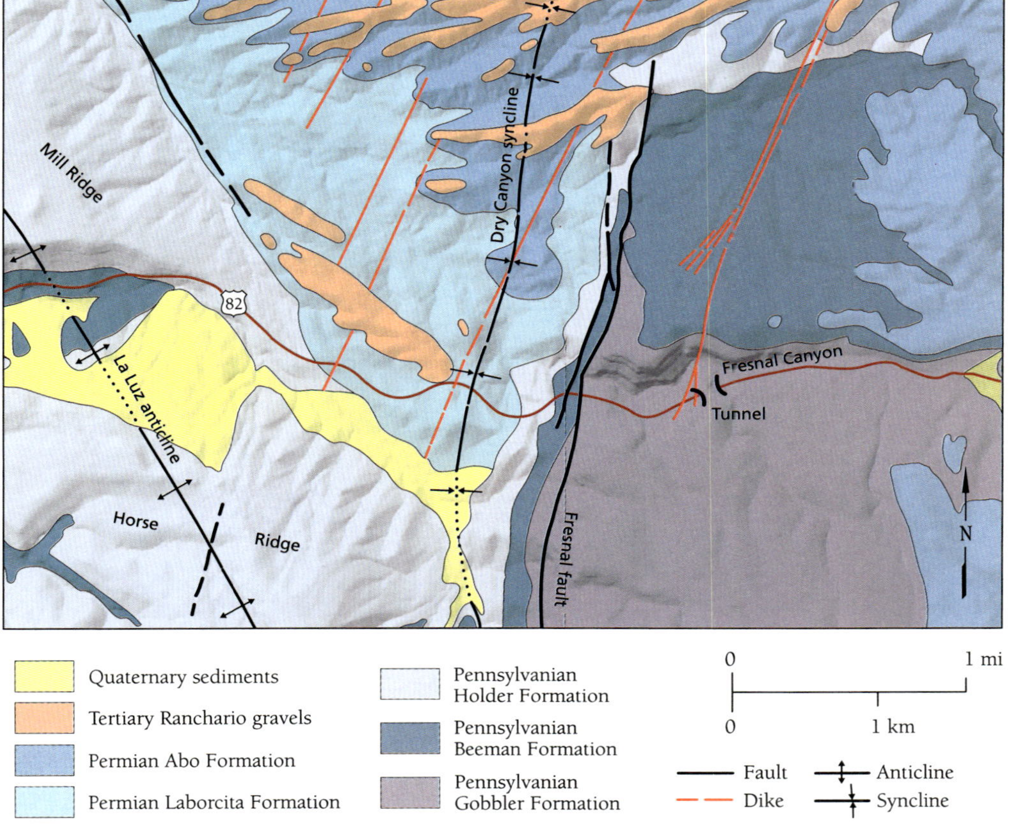

Geologic map of the Tunnel Vista site.

To the west, in the near distance, low reddish hills just west of the Fresnal fault consist of coastal plain and stream-channel sediments of the lower Permian Laborcita Formation, a unit that transitions into marine sediments to the north and is overlain northwest of Tunnel Vista by red beds of the Abo Formation. The high broad ridge (Mill Ridge) that can be seen farther down-slope north of US 82, and continues south of the highway as Horse Ridge, is the axis of the La Luz anticline, an arch-like fold. The ridge is composed of the Holder Formation, a predominantly marine unit of mudstone with interbedded limestones and local, thick algal mounds. The end of one such mound can be seen from Tunnel Vista, along the edge of Mill Ridge north of the road.

Younger Permian strata (the Yeso and San Andres formations), chiefly limestones and gypsum, are widely exposed east of the tunnel near Cloudcroft. Cenozoic dikes—thin, nearly vertical sheets of dark andesite—have intruded the limestones near the west entrance to the tunnel and along the highway about 1 mile downhill (west) of the tunnel.

View to west, showing reddish Laborcita exposures in foreground, the Holder Formation on Mill Ridge, with the end of a phylloid algal mound on lower slope, and across the Tularosa Valley (with White Sands) to the San Andres Mountains.

Geologic History

During Late Paleozoic time, New Mexico was strongly affected by Ancestral Rocky Mountains tectonism that produced numerous uplifts and marine basins across the state. Pennsylvanian sediments in the Tunnel Vista area were deposited on a shallow-marine shelf between the large, north-trending Pedernal uplift to the east and the rapidly subsiding Orogrande Basin to the west. Variations in uplift activity, as well as glacially regulated sea-level changes, produced cycles of sedimentation now preserved in the Pennsylvanian rocks.

Near the beginning of the Permian, increased erosion on the Pedernal uplift resulted in rapid deposition of conglomerates, sandstones and shale along its western margin, ultimately causing a transition from marine deposition to nonmarine deposition.

Starting at the end of the Permian, and continuing through the Mesozoic and early Cenozoic, the Tunnel Vista area became increasingly buried. Uplift of the Sacramento Mountains, beginning around 25 million years ago, allowed erosion to expose the sedimentary sequence we see today.

—Barry S. Kues

If You Plan to Visit

From Alamogordo, go about 3 miles north on US 54, then turn east on US 82 for about 7 miles. The Tunnel Vista parking area is on the left side of the road just west of the entrance to the High Rolls tunnel. From the east, take US 82 west from Artesia, descend the steep grade from Cloudcroft, pass through the tunnel, and turn right into the parking area. For more information:

Lincoln National Forest
3463 Las Palomas road
Alamogordo, NM 88310
(575) 434-7200
www.fs.usda.gov/recarea/lincoln/recarea/?recid=34234

White Sands National Park
NATIONAL PARK SERVICE

America's newest national park (as of December 2019), White Sands is famous for its extensive sea of white gypsum dunes—indeed, it is the largest gypsum dune field in the world. Located in the southern Tularosa Basin, the park was established as a national monument in 1933 and encompasses nearly 176,000 acres (275 square miles, including 115 square miles of gypsum sand dunes). The park not only contains the large dune field but also a saline mudflat called Alkali Flat, a smaller ephemeral salt lake (or playa) named Lake Lucero, parts of the gypsum-dust plains east of the dune field, and alluvial fans from the surrounding mountains. The dune field and Alkali Flat extend more than 12 miles to the north of the park onto the White Sands Missile Range.

Six major factors account for the exceptional accumulation of gypsum sand in this area:

1. The area has **internal drainage** (there is no surface exit from the basin for water). In this respect, the basin is similar to the evaporitic Great Salt Lake of Utah.

2. The area has a **prolific source of calcium sulfate** (gypsum) from Permian rocks found in nearby mountains, especially to the north, as well as from equivalent rocks present beneath the Tularosa Basin. Those older gypsum deposits (mainly in the Yeso and San Andres formations) are easily dissolved and transported by surface and ground-water into the basin.

3. The **arid climate** of the region led to reprecipitation of gypsum following the last ice age (about 15,000 years ago) as Lake Otero dried up.

4. **Persistent southwesterly winds** erode gypsum from the exposed lake sediment and deposit it in dunes that migrate progressively to the northeast.

5. A very **shallow water table** (typically just 1 to 3 feet below the surface in interdune areas) maintains high soil moisture and helps to stabilize the wind-blown gypsum.

6. Dunes are further **stabilized by vegetation and soil crusts**, especially along the eastern margin of the dune field.

OPPOSITE: **Gypsum dunes and grasses in interdune flats with the San Andres and Organ mountains in the background.**

NASA EO-1 satellite image of the White Sands gypsum deposits in the Tularosa Basin. The large X (arrowed) visible on the Alkali Flat west of the dunes is a 10-mile-long backup landing strip for Space Shuttle missions. It was used only once for Columbia in 1982.

222

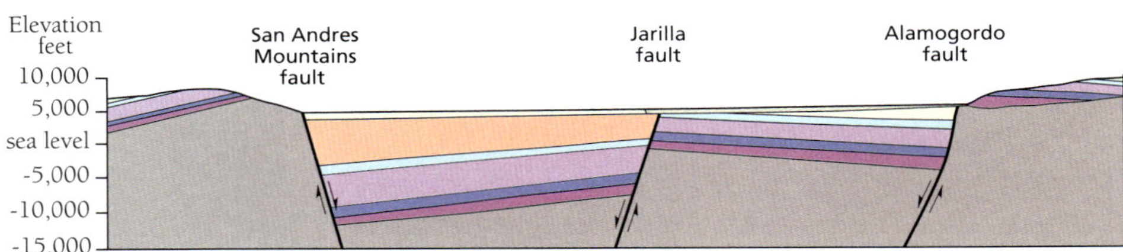

West-east cross section through the Tularosa Basin showing fault blocks that underlie the basin. Down-dropped Precambrian, Paleozoic, and Cenozoic rocks are overlain by younger basin-fill deposits.

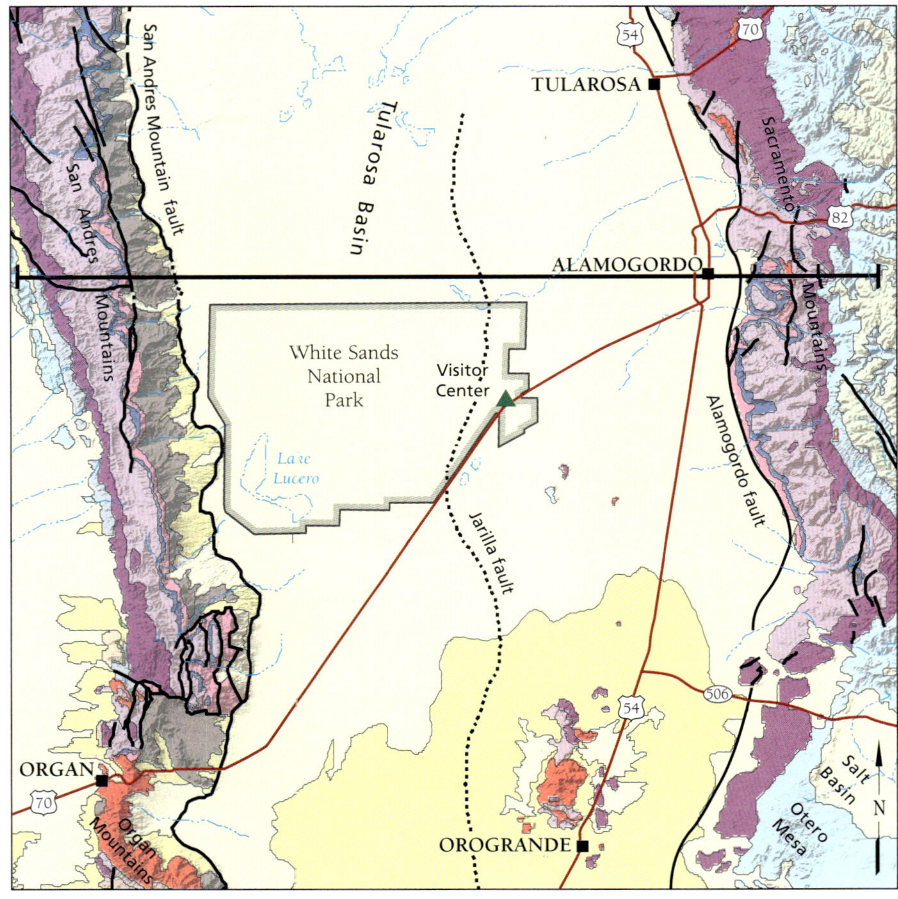

Geologic map of the White Sands National Park region. The line through Alamogordo marks the location of the cross section above.

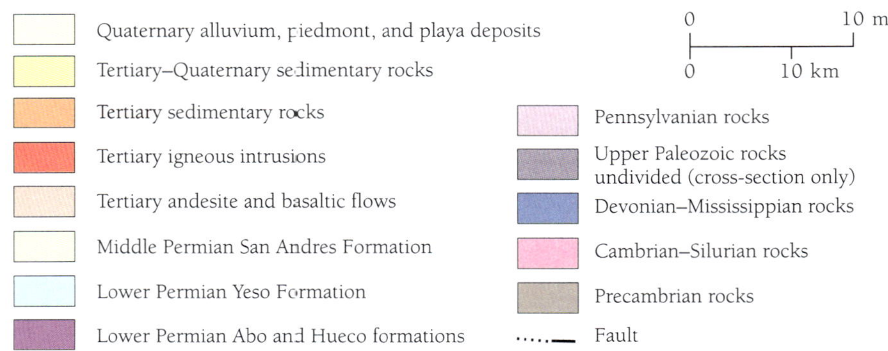

Geologic Province and Geography

The White Sands dune field is located in the southern part of the Tularosa Basin, a major down-dropped part of the Rio Grande rift. The basin was formed by sunken and tilted fault blocks that were subsequently buried by sediments eroded mostly from the nearby San Andres and Sacramento mountains.

Today, the basin has no surface water discharge, and inflow either evaporates or infiltrates and ultimately leaves the basin via groundwater transport. Lake Lucero (elevation 3,890 feet) is an ephemeral lake (commonly termed a playa) near the southern end of Alkali Flat. Accumulations of gypsum sand increase downwind (northeast) from Alkali Flat and Lake Lucero, with the main dune field rising about 100 feet in elevation above the playa. Individual dunes typically are 15 to 35 feet high and advance downwind as much as 40 feet per year near Alkali Flat. This dune movement requires continual plowing of the roads and parking areas.

White Sands National Park lies in the semiarid northern Chihuahuan Desert, with high temperatures in the summer reaching 110°F and temperatures in the winter as low as -25°F. Although average annual precipitation is only about 10 inches, monsoonal summer rains and decadal rainfall events can produce short-lived flooding in the region. Prevailing winds are from the southwest, and gusts exceeding 55 miles per hour are relatively common.

White Sands provides a natural laboratory for studies of the adaptation and rapid evolution of various gypsophile plants, animals (reptiles, amphibians, fish, insects, spiders, and mammals), and microorganisms. They all have undergone natural selection to survive in harsh conditions in terms of the semiarid climate, bright, hot sunlight with barely filtered ultraviolet rays, an uncommon gypsum substrate, and limited nutrients. Most noticeable are pale-colored insects and lizards, but other arthropods, reptiles, amphibians, and small mammals also are adapted to the blindingly white landscape. Some of the more uncommon plants that

Composite of aerial photographs, acquired on October 10, 1996, of White Sands National Park. The main gypsum source areas of Alkali Flat and Lake Lucero are labeled, as are three areas with distinctive dune types. The arrows indicate the direction of the predominant winds that transport gypsum from the source areas to the dunes.

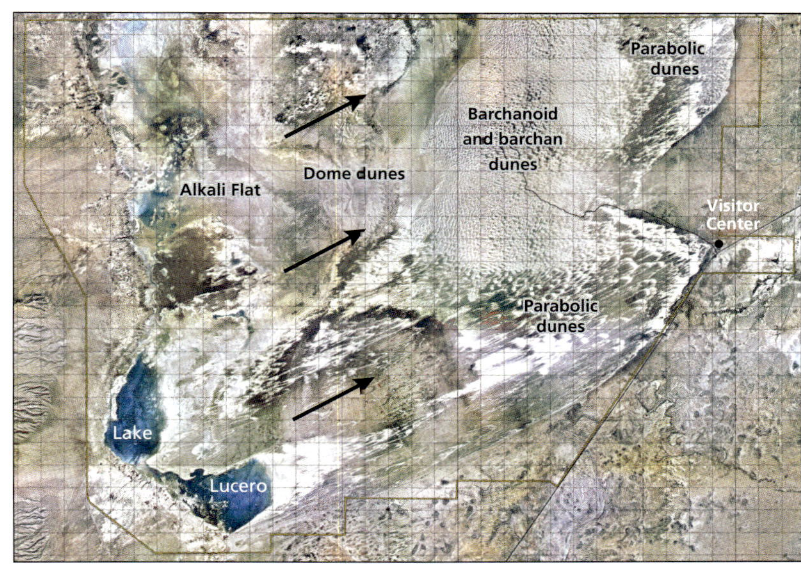

Photograph taken from the International Space Station in late Feburary, 2012, shows that dust from White Sands can be carried as a visible plume for hundreds of miles, well into west Texas.

are found in the gypsum are small (6 to 12 inches high) bushes with odd names such as "hairy crinklemat" (*Tiquilia hispidissima*) and "gyp moonpod" (*Acleisanthes acutifolia*). Although surface water is rare, scattered Rio Grande cottonwoods are found throughout the dunes. The gypsum crusts developed on some soils contain many active microorganisms as well. Visitors are referred to the White Sands National Park Visitor Center and webpages for more information about plants and animals.

The Rock Record and Geologic History

Deposits exposed within the park are geologically young (less than 45,000 years old to only hours old). The only exception is a small outcrop of Permian limestone on the hill with the water tower east of the Visitor Center. Despite the recent formation of most of the features within the park, an important part of the geologic evolution of White Sands began over 270 million years ago, during Permian time. Southern New Mexico was then within 5 degrees of the equator and submerged beneath a shallow sea in which limestones, mudstones, sandstones, and, importantly, gypsum were deposited. These deposits were subsequently buried and preserved in the subsurface for hundreds of millions of years.

Development of the Rio Grande rift, including the Tularosa Basin and its bordering uplifts, began about 25 million years ago. As the crust was slowly stretched in a east-west direction, subsiding blocks became basins that eventually filled with thousands of feet of sediment eroded from the surrounding uplifts. In addition, the Rio Grande at one point

A windstorm carrying gypsum-rich dust northeast from White Sands National Park as seen from the crest of the Sacramento Mountains near Sunspot.

flowed into the southern Tularosa Basin through Filmore Pass, 40 miles southwest of the park. Between 2 and 3 million years ago, the river rapidly dumped nearly 1,500 feet of sands and gravels into the basin as a broad fan. The uplifted mountains surrounding the Tularosa Basin, and the buried rocks under the basin, contain Permian-age gypsum. The highly soluble gypsum is dissolved and transported by surface and subsurface water. Because the basin has no outlet for surface water, the dissolved materials accumulate in basinal waters and sediments.

During the ice age, a large lake (Lake Otero) occupied much of the southwestern part of the Tularosa Basin. The climate at that time is estimated to have been several degrees cooler with more precipitation. Locally, deposits of Lake Otero contain fossils of small and large mammals, fish, amphibians, gastropods, and microfossils of plants and other aquatic organisms. Tularosa Basin began to dry out about 15,000 years ago, eventually producing Alkali Flat and the modern playa of Lake Lucero. As the water of Lake Otero evaporated, gypsum was precipitated. Alkali Flat developed by wind erosion of dry lake muds on the floor of the basin. Sand-size gypsum crystals began to accumulate in dunes, downwind to the northeast of the playa. Formation of a large field of gypsum sand dunes was well underway several thousand years ago. The gypsum sand derives from older lake sediments as well as from modern gypsum crystals precipitated at the surface of Alkali Flat. A few factors limit the size of the gypsum dune fields of this area. Gypsum is both a very soft and a very soluble mineral. Dune migration leads to rapid

Large gypsum (selenite) crystals exposed in the Alkali Flat area are weathering out of Pleistocene Lake Otero deposits. These crystals are one of several sources for the White Sands dunes. Crystals shown are about 2 to 3 inches long.

abrasion of gypsum sands and eventual reduction of those grains to dust-size particles. The ultra-fine gypsum either leaves the basin during dust storms or settles as fine dust on the east side of the basin, where much of it is dissolved by rain and adds to the dissolved materials in regional groundwater.

Geologic Features

THE LOOP DRIVE AND ASSOCIATED HIKING AREAS—Large white gypsum dunes and associated interdune areas are the main geologic features visitors see along the loop drive in the park. They are part of the larger system of wind erosion, transport, and deposition across the southern Tularosa Basin. Eolian erosion mobilizes gypsum crystal fragments from Alkali Flat and the Lake Lucero playa and moves them to the northeast. Sand grains gather into small sand ripples and low dunes under vegetation. Larger dome dunes form farther downwind and, as they advance, they may form barchan dunes. The margins of barchan dunes form "horns" pointing downwind. Individual barchan dunes coalesce to form barchanoid ridges with sinuous crests. These are the most common and extensive features found along the loop drive in the "heart of the dunes." Straight-crested linear dunes are rare. In areas partially stabilized by vegetation, sand sheets and older dunes may remobilize to form parabolic dunes with long arms trailing from upwind margins. The dunes along the eastern margin of the dune field move very slowly, in large part because they have been stabilized by vegetation. Archaeological sites along the margins show that the past inhabitants took advantage of the diverse ecological environments that developed there.

Characteristic shapes of some dune types found in the White Sands area.

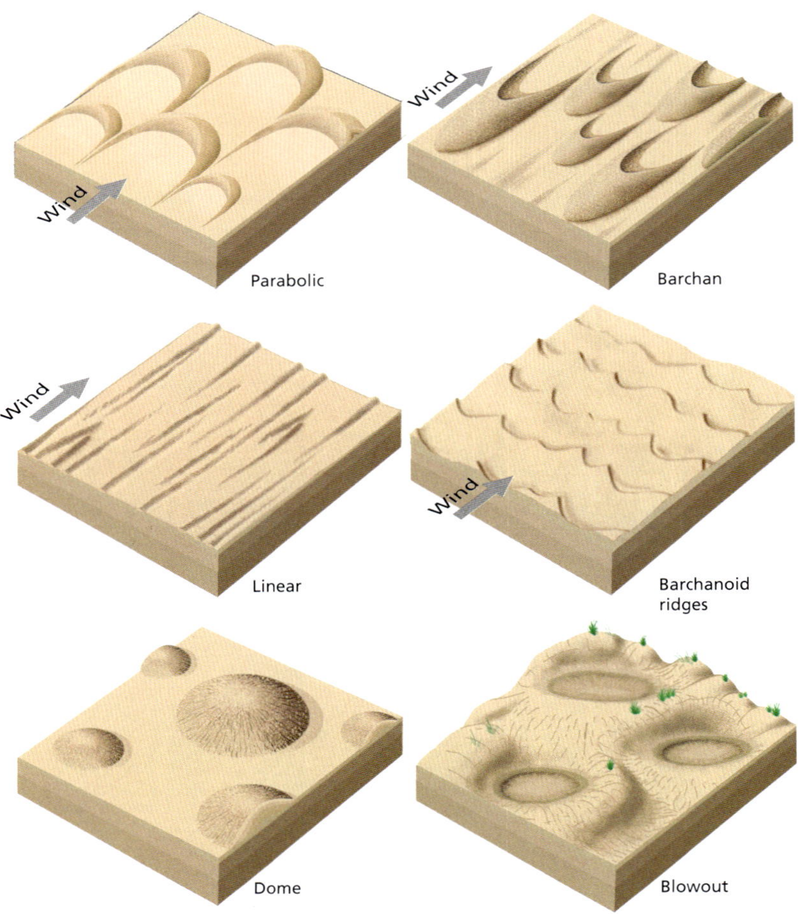

A dune field has two major zones—one of material transport represented by the dunes themselves and one of material storage in the interdune areas. Roadcuts through dunes, exposures beneath vegetation-protected pedestals, and eroded interdune areas reveal that dune interiors have complex cross-laminated sands with cross-cutting features related to constant erosion and redeposition of dune sands.

The dunes migrate by wind blowing grains up the gentle, hard-packed backsides of the dune that then cascade as small avalanches down the sheltered lee side of the dune. That forms slopes of loosely packed sediment, typically with slope angles of about 30 degrees (termed the "angle of repose"). Differences in slope angle and packing account for the variations in effort needed to walk across the dune landscape.

Dune interiors commonly are partially saturated with water, derived from precipitation, that is held by capillary forces and slowly percolates downward to the shallow water table. The remaining water makes dune interiors cool to the touch, although the dry surfaces may be quite hot.

In interdune areas, zones of sediment accumulation, or storage, are largely stabilized by a zone of dampness above the very shallow (1–3 feet deep) water table. Variations in rainfall lead to fluctuations in the elevation of the water table that can trap sediment, accounting for the sometimes spectacular "scrollwork ridges" in eroded interdune flats. These ridges reflect small differences in grain size of sediment layers formed by avalanching of sediment down the lee slopes of dunes. The curvature of the ridges results from the cusp shapes of the common barchan, barchanoid, and parabolic dunes that deposited the sand.

On rare occasions, interdune areas may receive enough precipitation to make ephemeral ponds that can be extensive enough to flood the main loop drive. Water in both dunes and interdune flats or ponds seeps downward and then moves laterally, evaporating near the surface and undergoing evapotranspiration by plants. Much of the shallow groundwater is perched above impermeable clay layers. Regional groundwater slowly flows southward toward the Rio Grande. Because

Crest of a typical barchan dune (at sunrise) with its steep downwind face showing "scars" of many small avalanches of gypsum sand.

227

Aerial view of closely spaced barchanoid sand dunes with narrow, brown, sparsely vegetated interdune areas.

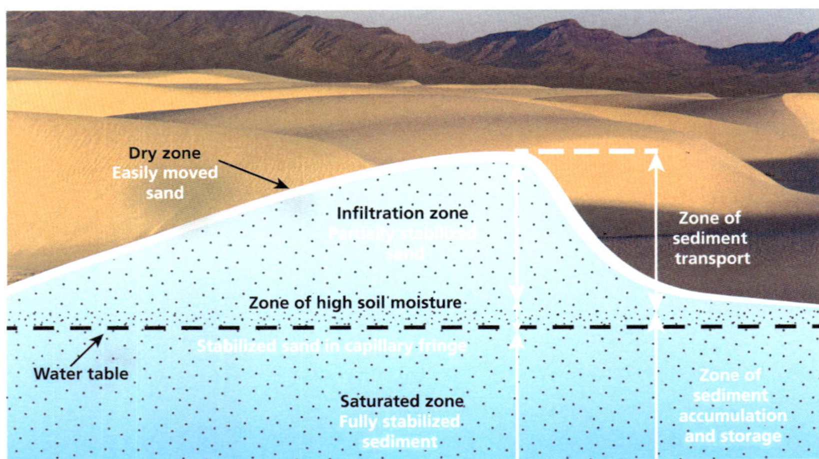

228

Diagram showing the distribution of interstitial water in dunes and its effect on sediment mobility versus sediment stability. Background photo shows sunrise on the dunes with the San Andres range in the distance.

Flooded interdune areas following an exceptionally intense late summer rainstorm in 2006. The flooding forced closure of the main road into the park for nearly six months.

the exceptionally shallow groundwater plays a vital role in the formation and stabilization of dunes in this area, there is great concern that increased use (including desalination) of groundwater in the region will affect the future of the dune field.

Because groundwater beneath much of the park is saturated with dissolved gypsum and other minerals, it is not potable. Fresh water for human use is piped into the park area from wells near the Sacramento Mountains.

OTHER PARK AREAS—Although most visitors will only visit the main dune area, infrequent (once a month in January, February, March, April,

November, and December) National Park Service-led tours to Lake
Lucero and Alkali Flat reveal a very interesting and different landscape.
During times of wet weather, Lake Lucero is flooded and large areas
of Alkali Flat contain standing water. Under more common, drier
conditions, the wind-eroded sediments of ice-age Lake Otero
are exposed, revealing large gypsum crystals (up to 4 feet in length)
that grew in the lake muds, as well as many fragments of the mineral
as it weathers. Accumulations of gypsum fragments have produced
odd mounds and pinnacles on Alkali Flat. Additional evaporite
minerals precipitated in the area include halite, thenardite, mirabilite,
hexahydrite, dolomite, and others. As the wind erodes Alkali Flat,
12- to 20-feet high, prow-like "mesitas" are carved from the resistant
beds of Lake Otero to form features called "yardangs" that point into
the predominant wind direction.

Along the margins of Lake Lucero and on Alkali Flat, numerous
fossil tracks of mammoths, camels, sloths, and rare predators (dire
wolves and saber-tooth cats) are briefly exposed before being eroded.
One east-west camel trackway, with individual tracks filled with resistant dolomite, was traced across Alkali Flat for more than 1.5 miles.
In addition, footprints have been found of humans interacting with
megafauna, and throughout the park, there are thousands of fossilized
human prints from adults and children—the greatest concentration of
Pleistocene fossil footprints in the Americas.

The Lost River flows into the park along incised drainages coming
from La Luz Creek and the foothills of the Sacramento Mountains
to the northeast. The path of Lost River within the park is westward
into the dune field, reaching a small playa before being overrun by
dune sand. It deposits red mud on the playa that contrasts with the

Artist Karen Carr's rendering of the Late Pleistocene environment and now extinct megafauna at White Sands National Park.

Vegetation-stabilized dunes at the eastern edge of the White Sands dune field.

A vegetation-stabilized erosional remnant of a dune (called a pedestal). The steeply inclined layering is typical of the internal structure of dunes. It is formed by sand avalanches on the lee (downwind) side of dunes. The non-native tamerisk tree seen at the top of the pedestal can transpire over 100 gallons of water per day.

white gypsum sand in surrounding dunes. A rare and endangered fish, the White Sands pupfish (*Cyprinodon tularosa*, up to 2.5 inches long), was apparently introduced into the Lost River in the late 20th century and has established itself in the perennial reach of the creek.

—*David W. Love, Bruce D. Allen, Peter A. Scholle, and David Bustos*

Additional Reading

White Sands National Monument, Rose Houk, Michael Collier, and Sandra Scott, Southwest Parks and Monuments Association, 1994.

White Sands National Monument Geologic Resources Inventory Report, National Park Service, 2012.

If You Plan to Visit

The entrance to White Sands National Park is on the north side of US 70, 15 miles southwest of Alamogordo and 52 miles northeast of Las Cruces, New Mexico. The park may be closed for unusual conditions, such as extreme weather, flooding, or missile tests on the adjacent White Sands Missile Range. US 70 can also be closed for an hour or more during missile tests. Check the park web pages for scheduled closures.

The main access to the park is a 16-mile loop drive into the heart of the dunes. Along the drive, several nature trails start from parking lots. Because the park is open during the daylight hours, entrance and exit times vary during the year. In addition, moonlight hikes are scheduled monthly and special escorted visits to Lake Lucero require prior registration. Check the park website for hours and to sign up for special tours.

White Sands can be a harsh environment, so please follow all National Park Service safety rules concerning appropriate clothing and sunscreen. Take plenty of water and food. In an environment where each dune looks similar to the previous one, it is easy to become disoriented—so a charged cellphone, GPS device, and compass are all essential for longer hikes. For more information:

White Sands National Park
PO Box 1086
Holloman Air Force Base, NM 88330
(575) 479-6124
www.nps.gov/whsa

For Lake Lucero tour information:
www.nps.gov/whsa/planyourvisit/lake-lucero-tour.htm

An eroded interdune area with scrollwork patterns that reflect the preserved "toes" of former barchan dunes, now removed by the wind. The dune toes are retained by capillary water associated with the shallow, underlying water table.

Oliver Lee Memorial State Park
NEW MEXICO STATE PARKS

Oliver Lee Memorial State Park lies in the foothills of the Sacramento Mountains, approximately 15 miles southeast of Alamogordo. The main attractions in the 180-acre park are the hiking trails near the Visitor Center and along Dog Canyon and a restored historic ranch house that was owned by Oliver Lee (1865–1941). He was a well-known rancher who helped bring the railroad to Alamogordo in 1898 and later became a State Senator. Mr. Lee and his predecessor, Francois Jean "Frenchy" Rochas, settled in Dog Canyon because of the perennial springs that flow from fractures and bedding planes in the Ordovician Montcya Formation in the lower reaches of the canyon and from Mississippian limestones in the upper part of the canyon.

OPPOSITE: **View of the north wall of Dog Canyon looking north from the Dog Canyon Trail. The greenish-black band that cuts at an angle across the layering in the early Paleozoic sedimentary rocks is one of the Cenozoic sills.**

Regional Setting

The park is on the east side of the Tularosa Basin, a basin of the Rio Grande rift, and on the western edge of the fault-bounded Sacramento Mountains escarpment. The escarpment provides magnificent exposures of Ordovician to Pennsylvanian sedimentary rocks that are accessible in the park and on the trail extension to adjacent Lincoln National Forest lands. In addition, gypsum dunes of White Sands National Park and the San Andres Mountains to the west are easily visible from the hiking trail in Dog Canyon.

Geologic History

Sedimentary rocks exposed along the Sacramento Mountains escarpment record a roughly 200 million-year history of sea-level rises and falls separated by intervals of erosion along the margin of a shallow sea. The oldest rocks in the park belong to the Ordovician El Paso Group. The El Paso Group was deposited in a shallow sea and consists of about 430 feet of predominantly dolomite with a few quartz sandstone beds and chert lenses. A sea-level fall then resulted in exposure and erosion. This was followed by a rise in sea level and deposition of about 350 feet of Ordovician Montoya Formation. This unit includes a basal quartz sandstone overlain by dark-gray and light-gray dolomite beds. Shallow-marine conditions continued during the accumulation cf the Valmont Dolomite in late Ordovician time. The Valmont Dolomite

Early spring blooms of a White Sands giant claret cup cactus.

is characterized by abundant chert nodules and thin shales interbedded with dolomites.

After a period of sea-level fall and erosion, the Silurian Fusselman Dolomite reflects another influx of shallow-marine waters. The brownish-gray, ledgy dolomitic beds of this unit contain abundant chert nodules and, although the Fusselman is less than 85 feet thick, it forms a distinctive erosion-resistant cliff. The Fusselman Dolomite is overlain by slope-forming Devonian siltstone, sandstone, carbonate, and black shale beds of the Oñate and Sly Gap formations, that record the continuation of shallow-marine conditions in this region. These rocks are overlain by the Mississippian Caballero Formation,

View of the Sacramento Mountains escarpment south of the Dog Canyon Trail.

Lake Valley Limestone, and Rancheria Formation, which are fossiliferous limestones with minor shales and sandstones deposited on a south-sloping marine shelf along the fringes of a mid-continent arch about 345 to 320 million years ago. Fossils include crinoids, brachiopods, and corals.

A feature called Little Sugarloaf is visible in the cliffs north of the mouth of Dog Canyon—although not visible from within the park, it can be seen from the Dog Canyon entrance road to the park. This reeflike mound (often termed a bioherm) is composed of carbonate mud, bryozoans, crinoids, and brachiopods that formed on the continental shelf during accumulation of the Lake Valley Limestone. Little Sugarloaf is the southernmost of a series of such mounds exposed in the Sacramento Mountains escarpment from Alamogordo southward.

Sea level dropped again following Mississippian deposition, and karst features (sinkholes and caves) developed on the top of the Mississippian strata. The Pennsylvanian rocks, the youngest units exposed along the trail, include the Gobbler, Beeman, and Holder formations. Erosion of the nearby Pedernal uplift that formed during Ancestral Rocky Mountain deformation provided the detritus for the sandstones, siltstones, and shales of the lower Gobbler Formation; the Gobbler Formation was deposited in a marginal-marine environment. Deposition of sandstones, shales, and carbonates in the Beeman and Holder formations was also influenced by Ancestral Rocky Mountain tectonic activity, as well as periodic sea-level rises and falls associated with extensive glaciation of the southern hemisphere continents during Pennsylvanian time.

The two sills that intruded the Devonian section, which stand out as distinctive thin, black horizons, were likely emplaced during a time of significant volcanism between about 38 and 27 million years ago in the Sierra Blanca volcanic field 35 miles north of the park. The sills are composed of green porphyry, with large crystals of hornblende and feldspar generally about 0.25 inches long. The largest of the hornblende crystals, however, are 5 inches long.

Crustal stretching associated with the Rio Grande rift began in the Sierra Blanca volcanic field about 25 million years ago, which caused the Tularosa Basin to drop and the Sacramento Mountains to rise relative to each other along the Alamogordo fault near the base of the mountain front. Development of significant topographic relief between the basin and the mountains allowed streams to carve canyons into the rising escarpment. These streams carried sand, silt, and clay westward from the mountains to form alluvial fans on the margin of the basin, a process that continues to this day. The youngest parts of these alluvial fans are cut by the fault, indicating that the fault is still active.

Geologic Features

The Dog Canyon Recreational Trail starts at the Visitor Center and ends at FR 90B on Joplin Ridge. The trail, which is 5.5 miles long and has a vertical elevation gain of 3,144 feet, climbs through Ordovician, Silurian, Devonian, Mississippian, and Pennsylvanian marine carbonates, sandstones, and shales and passes by two Cenozoic sills intruded into the Devonian section.

Rock units exposed along the Dog Canyon Trail.

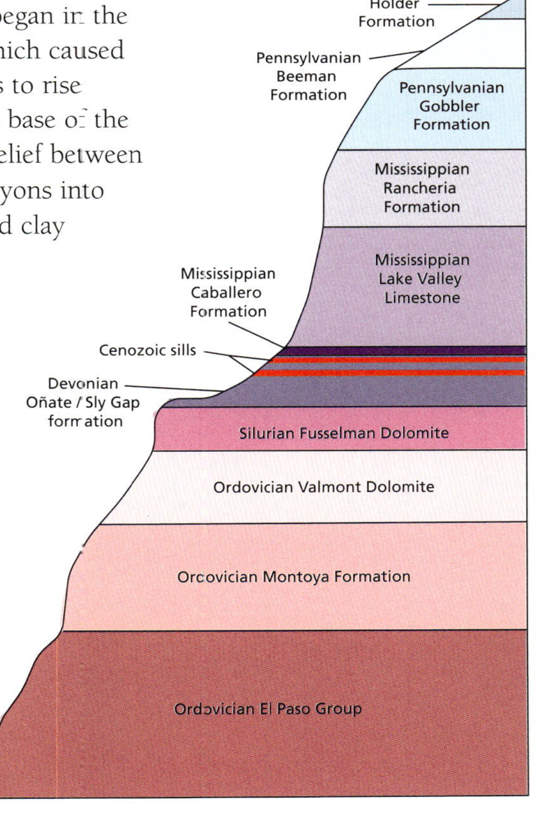

The Riparian Nature Trail is a shorter hike that also leaves from the Visitor Center. It visits one of the lower springs that flow out of the Ordovician Montoya Formation. This oasis on the floor of the canyon hosts large cottonwood trees, maidenhair ferns, mosses, and even stream orchids, in contrast to the surrounding Chihuahuan Desert vegetation that is more typical of the area.

—*Shari A. Kelley*

Close-up of large hornblende crystals surrounded by a finer-grained groundmass in one of the sills exposed along the Dog Canyon Trail. The large crystals are about 0.25 inch across.

If You Plan to Visit:

Oliver Lee Memorial State Park is 12 miles south of Alamogordo via US 54, and then drive east for 4 miles on Dog Canyon Road. A Visitor Center and camping facilities are available. For more information:

Oliver Lee Memorial State Park
409 Dog Canyon Rd.
Alamogordo, NM 88310
(575) 437-8284
www.emnrd.state.nm.us/spd/oliverleestatepark.html

Water shortages of the 1880s necessitated the planning of a large reservoir on the Rio Grande in southern New Mexico. The deep and narrow bedrock walls of Elephant Butte Canyon were chosen as the ideal foundation for the dam. When completed in 1916, Elephant Butte Dam, originally called Engle Dam, was the largest irrigation dam in the nation—a 301-foot tall, 1,674-feet wide behemoth that today impounds New Mexico's largest body of water and is the state's largest hydroelectric power generating facility. A century later, Elephant Butte Lake is New Mexico's most popular state park, an oasis in a picturesque desert setting that is perfect for boating and escaping the desert heat.

Geography and Geologic Province

Elephant Butte Lake lies in the Engle Basin of the Rio Grande rift along the west side of the Fra Cristobal Mountains and Cutter Sag which is a region of low bedrock hills that separate the Fra Cristobal range from the Caballo Mountains. To the southwest are the Mud Springs Mountains, and to the northwest and west the San Mateo

Elephant Butte is all that remains of a 2–3 million-year-old volcano that erupted through Cretaceous rocks of the Cutter Sag. Purplish rocks at the base of the butte are Cretaceous mudstones. Note the "bathtub" rings around the butte reflecting various times when the lake levels were higher during wetter years.

Mountains and the Sierra Cuchillo form an imposing skyline. These mountains and intervening basins primarily are the result of crustal stretching along the Rio Grande rift, which broke the crust from Leadville, Colorado, to northern Mexico, into tilted, fault-bounded blocks several miles wide. The uplifted parts of the blocks became mountain ranges, whereas the down-dropped parts became basins. In New Mexico, these basins collected sediments shed from the surrounding mountains as well as sands, gravel, silt and clay delivered by the ancestral Rio Grande and its tributaries.

Elephant Butte Dam consists of almost 620,000 cubic yards of concrete that spans Elephant Butte Canyon. At full reservoir capacity, the dam restricts about 2 million acre-feet (over 600 billion gallons!) of water for boating in the lake and to support downstream agriculture.

The Rock Record

Two distinct packages of rocks and sediments are present along the shores of Elephant Butte Lake. The soft gray sands of the western lakeshore were carried into the basin by the Rio Grande between about 5 million and 800,000 years ago. These sands accumulated in a mountain-bounded river valley similar to that of today. In contrast, rocks of the southeastern shore and around the dam site are much older. They were deposited about 90 to 66 million years ago during the Cretaceous time period, when the area was a broad depositional plain formed by northeastward-flowing rivers that originated in mountains far to the southwest. Brown sandstones and conglomerates, common around Winding Roads Park just east of the dam, were laid down in river channels, while the purplish-gray mudstones around Elephant Butte and the greenish-gray mudstones that flank the Paseo del Rio Campground, just downstream of the dam, accumulated on floodplains. Petrified wood and dinosaur fossils, including a *Tyrannosaurus rex* jaw, have been found in these rocks. Cretaceous rocks and younger Rio Grande sands are juxtaposed across the Hot Springs fault, which separates the Cutter Sag from the Engle Basin and extends northeastward from just west of the dam site. In addition, Paleozoic and Precambrian rocks are well exposed in the Fra Cristobal Range (see back cover photograph), along the northeastern edge of the lake and extensive stretch of the park along the Rio Grande. Although exposures are excellent, access is difficult.

More than a dozen extinct volcanoes, each about 2 to 3 million years old, are found around the lake. Some volcanoes erupted broad, sheet-like flows of black basalt that now cap the mesas to the east of the lake, including Kettle Top Butte. Other volcanoes, termed maar volcanoes (described in more detail in the Kilbourne Hole chapter), formed as volcanic explosion craters in the Rio Grande valley floor. The underlying "throat" of a maar volcano, called a diatreme (a volcanic pipe formed by a gaseous explosion), contains fragmented volcanic material and clasts of the host rocks that record the explosive eruption. Volcanic plugs and diatremes are commonly more resistant to erosion than the surrounding rocks, causing them to stand high in the landscape. Elephant Butte itself is an exhumed volcanic plug, while diatremes are found at Three Sisters Point and west of Rock Canyon Marina. Shiprock, in northwestern New Mexico, is a more famous and dramatic example of a diatreme.

Geologic History

About 25 million years ago, crustal stretching in the Rio Grande rift caused the Engle Basin to subside and begin to collect sediments from adjacent, rising mountains. Beginning less than 5 million years ago, the Rio Grande spilled into the area from the north, and the subsiding basin started to accumulate sands delivered by the Rio Grande in addition to gravels from the nearby mountains. Sedimentation continued until about 800,000 years ago, when the Rio Grande began to cut down into the basin sediments and locally into the Cretaceous rocks of the Cutter Sag, carving the broad valley containing Elephant Butte Lake and the narrow, bedrock-walled

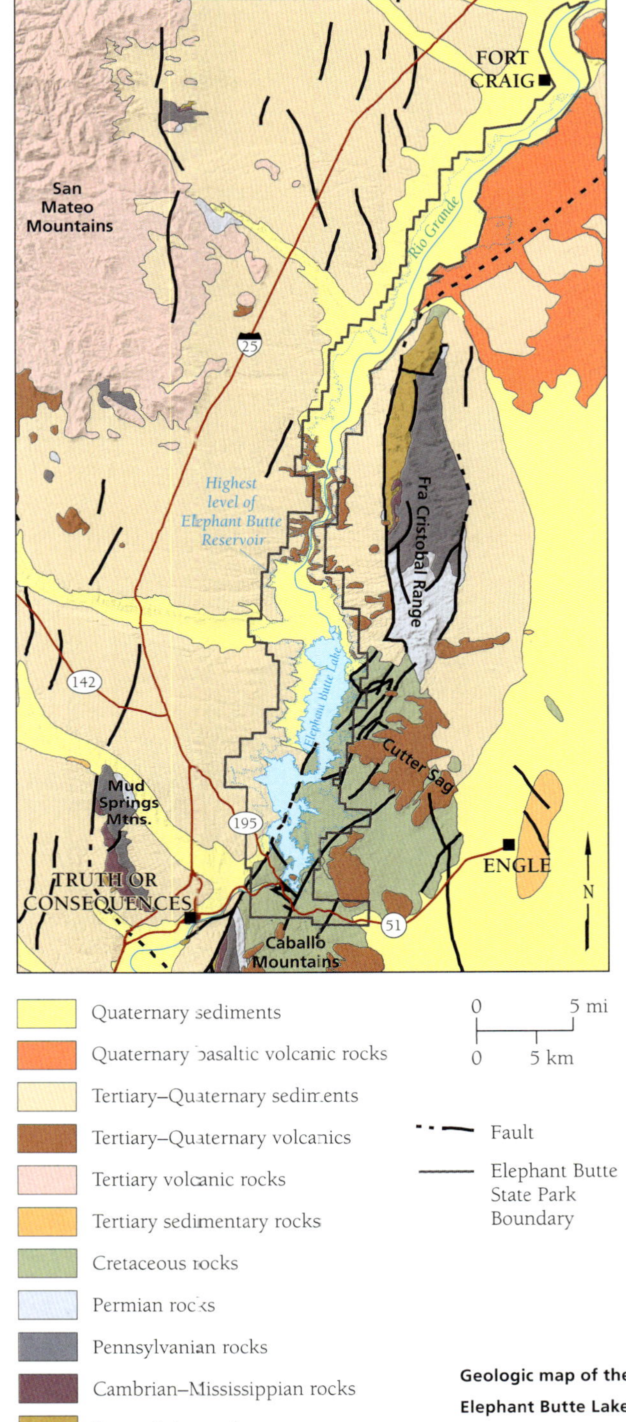

Quaternary sediments

Quaternary basaltic volcanic rocks

Tertiary–Quaternary sediments

Tertiary–Quaternary volcanics

Tertiary volcanic rocks

Tertiary sedimentary rocks

Cretaceous rocks

Permian rocks

Pennsylvanian rocks

Cambrian–Mississippian rocks

Precambrian rocks

0   5 mi

0   5 km

- - - Fault

——— Elephant Butte State Park Boundary

Geologic map of the Elephant Butte Lake State Park area.

Elephant Butte Canyon that provides the foundation for the dam. The downcutting by the Rio Grande in this area coincided with the river finally reaching the Gulf of Mexico (see river history in the introduction to this part of the book).

—Colin Cikoski

Additional Reading

Elephant Butte–Eastern Black Range Region: Journeys from Desert Lakes to Mountain Ghost Towns, Richard P. Lozinsky, Richard W. Harrison, and Stephen H. Lekson, New Mexico Bureau of Geology and Mineral Resources, Scenic Trips to the Geologic Past, no. 16, 1995.

If You Plan to Visit

The Elephant Butte Lake State Park Visitor Center can be accessed from the north by taking Exit 83 off I–25. Take NM 181, make an immediate turn on to NM 195 and follow that road about 4 miles southeast to the town of Elephant Butte and the park. Access from the south is most easily made by taking Exit 79 off I–25. Follow Business I–25 (North Date Street) to NM 51 (East Third Avenue). Take that route east/northeast for about 2 miles. Then, a left turn on to NM 179 will take you to the main west-side Elephant Butte State Park areas. Alternatively, continuing straight on NM 51 will lead to the Damsite Historic Fish Hatchery Park, several overview sites for Elephant Butte dam, Winding Roads Park (with its old Civilian Conservation Corps-built structures), the Damsite Resort and Marina, and other west-side locations. For more information:

Elephant Butte Lake State Park
101 Highway 195
Elephant Butte, NM 87935
(575) 744-5923
www.emnrd.state.nm.us/SPD/elephantbuttelakestatepark.html

Caballo Lake and Percha Dam State Parks
NEW MEXICO STATE PARKS

Caballo Lake and Percha Dam State Parks, located south of the town of Truth or Consequences, offer striking views of the Caballo Mountains and Red Hills, prominent uplifts in the southern Rio Grande rift. The Percha and Caballo dams were built for the Bureau of Reclamation's Rio Grande Project in 1918 and 1936–38, respectively. Percha Dam, although only 18.5 feet tall, diverts water into the Rincon Valley, irrigating farmland where much of New Mexico's famous green chile is grown, along with many other crops. Two miles upstream, the 96-foot-tall and 4,590-feet-wide earth-fill Caballo Dam stores water released from hydro-electric-power generation at Elephant Butte Dam and regulates delivery of that water to downstream users during irrigation season.

Percha Dam during spring waterflow of the Rio Grande. The dam provides water for agricultural irrigation in the lower Rio Grande valley.

Regional Setting

These parks are located in the Rio Grande rift, a series of basins extending from Colorado to northern Mexico. The crustal stretching that created the rift resulted in broad basins flanked by mountain ranges. Over time, sediments shed from the surrounding mountains partly filled the basins, forming valleys that contain precious groundwater aquifers throughout New Mexico.

The Rock Record

Although the rocks within the boundaries of the parks are geologically young, much older rocks are exposed in the nearby Caballo Mountains and Red Hills. The massive, pinkish rocks at the bases of these uplifts are composed of granite and metamorphic rocks formed

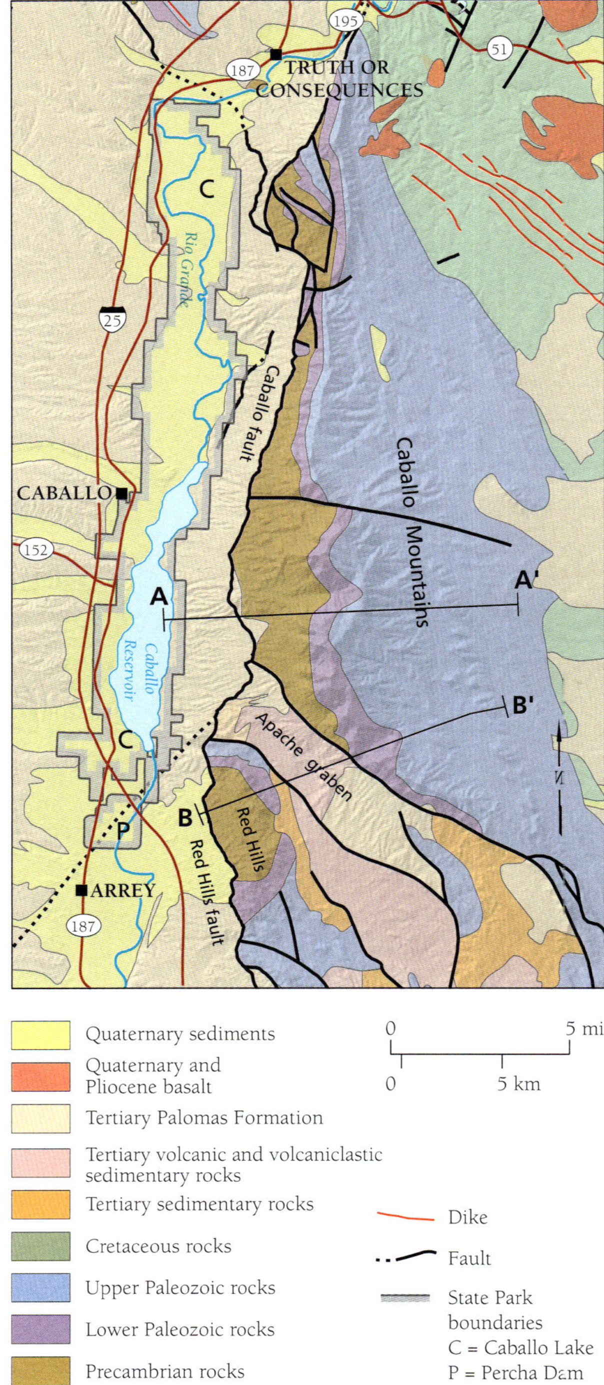

Generalized geologic map of the Caballo Mountains and vicinity.

during the Precambrian, (Proterozoic) 1.7–1.4 billion years ago. These rocks were subjected to extensive erosion before a shallow sea covered this part of New Mexico about 500 million years ago, at which time the state was located south of the equator. The dark band above the granite along the western escarpment of the Caballo Mountains is the Cambrian–Ordovician Bliss Sandstone, which was deposited along an ancient shoreline. A great thickness of rock was eroded prior to deposition of these shoreline deposits, and the widespread gap left in the rock record is termed the Great Unconformity. Above the shoreline sandstones are cliffs of Paleozoic limestones, dolomites, and shales that record the repeated rise and fall of sea level over hundreds of millions of years. Laramide deformation affected this area 75-45 million years ago, creating reverse faults and folds.

Rocks exposed along the modern shores of the Caballo Reservoir and in side drainages were deposited in basins along the margins of the ancestral Caballo Mountains. These deposits are known as the Palomas Formation and initially accumulated on alluvial fans that surrounded playa lakes in the lowest parts of the basins. Later, the ancestral Rio Grande arrived and deposited sands and muds that interfingered with locally derived alluvial-fan gravels. Fossils recovered from local Palomas Formation outcrops include tortoises, camels, and elephant-like gomphotheres.

Geologic History

Formation of the Rio Grande rift began as early as 25 million years ago as the crust started to extend. The Caballo Mountains block rose haltingly by movement on the Caballo fault, which also helped to create

a series of basins to the south and west that evolved into the modern
Hatch–Rincon and Palomas basins. These basins filled with sediments
both eroded from surrounding mountain ranges and deposited by
the ancestral Rio Grande, which entered the region around 5 million
years ago.

The Rio Grande began cutting down about 800,000 years ago,
sculpting the modern valley through alternating episodes of incision
and backfilling of its valley with sediment. Local tributary streams
underwent a similar pattern of events, ultimately carving deep arroyos
along the Caballo Mountains escarpment as well as for many miles
across the high plain to the west.

The Caballo Mountains and Red Hills are bounded on the west by
the Caballo and Red Hills faults. These faults form prominent scarps
best viewed in morning light. The Red Hills fault is a southern extension
of the Caballo fault that domed up
and exposed the granite that gives
the Red Hills their name. This fault
has likely been active in the last
100,000 years, whereas the Caballo
fault near Truth or Consequences
moved in the last 5,000 years.
These scarps only document the
most recent uplift; cumulative vertical displacement on the Caballo
fault over the last 25 million years
may exceed 5 miles.

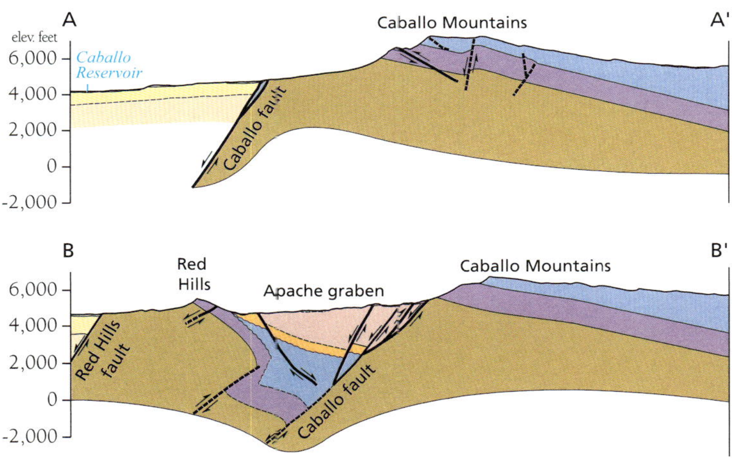

Cross sections showing east-tilted structure of the central Caballo Mountains (A–A') and Red Hills and southern Caballo Mountains (B–B'). See geologic map for symbol key.

Geologic Features

The Palomas Formation forms extensive outcrops along Caballo
Reservoir and the Rio Grande below Caballo Dam. Exposures southeast
of, and across the river from Riverside Campground show interfingering
between ancestral Rio Grande sands and alluvial-fan gravels shed from
nearby uplifts. The ancient Rio Grande deposits contain a mix of pebbles
and cobbles originating from as far away as northern New Mexico and
as near as the central and northern parts of the Caballo Mountains.

Mineral Resources

Over 3,300 ounces of gold were extracted from sand and gravel
deposits exposed in the canyons just east and north of Caballo Dam.
These stream deposits, known as the Shandon placers, were mined-out
almost entirely during the early 1900s, although modern weekend

prospectors still recover a few flakes or nuggets from the sands. Sources of gold are probably quartz veins in the outcrops of Precambrian granite above the placer mines.

—Andrew P. Jochems and William R. Seager

If You Plan to Visit

Much of the geology described above lies outside state park boundaries and can be accessed by dirt roads on the east side of Caballo Reservoir. Before entering these areas, please consult land ownership maps, available from the Bureau of Land Management.

Caballo Lake State Park is located about 16 miles south of Truth or Consequences along NM 187 and can be accessed from I–25 at Exit 59 (then north on NM 187) or Exit 63 (then south on NM 187). It features numerous campsites on the west side of the reservoir and along the Rio Grande below the dam.

Panoramic view of the Caballo Mountains from near Caballo Lake State Park headquarters. Note red to pinkish Precambrian rocks at the base of the escarpment underlying dark band of lower Paleozoic sandstones.

Percha Dam State Park is about 21 miles south of Truth or Consequences via I–25. Take Exit 59, then a short jog south on NM 187, followed by a turn to the east on B 040. In about a quarter of a mile, stay right on Riverside Park Road, following a turn south to the park entrance.

Caballo Lake/Percha Dam State Park
Highway 187
Caballo, NM 87931
(575) 743-3942
www.emnrd.state.nm.us/SPD/caballolakestatepark.html
www.emnrd.state.nm.us/SPD/perchadamstatepark.html

Lake Valley Back Country Byway
BUREAU OF LAND MANAGEMENT

The Lake Valley Back Country Byway (LVBB) follows NM 152 and NM 27 for 47 miles through a scenic mountainous region of south-central New Mexico and passes through old mining sites as well as modern mining and energy development areas.

These directions start at the north end of the byway, at the intersection of NM 152 and I–25. Zero your odometer there (If you began your trip at the village of Nutt and travel in the opposite direction, the mileages from the south end are given in parentheses). Traveling westward from I–25 on NM 152, the Lake Valley Back Country Byway climbs onto the Cuchillo surface, a gently east-sloping plain dissected by arroyos that drain to the Rio Grande. To the east, the Rio Grande and Caballo Reservoir are visible, with the Caballo Mountains in the background. To the west, the Animas Mountains lie beneath the towering Black Range on the western skyline. At mile 9.9 (36.7), just past the entrance to the Copper Flat Mine, a byway display is on the left. The highway then leaves the Cuchillo surface and winds through Paleozoic rocks before descending into the Percha Creek valley and the town of Hillsboro. Hillsboro, first known as Hillsborough, was established in 1877 and served as the Sierra County seat from 1884 to 1938.

After passing the Black Range Museum in Hillsboro, the byway climbs out of the Percha Valley and turns south on NM 27 at mile 16.6 (30.4). It crosses several streams that drain the Mimbres Mountains

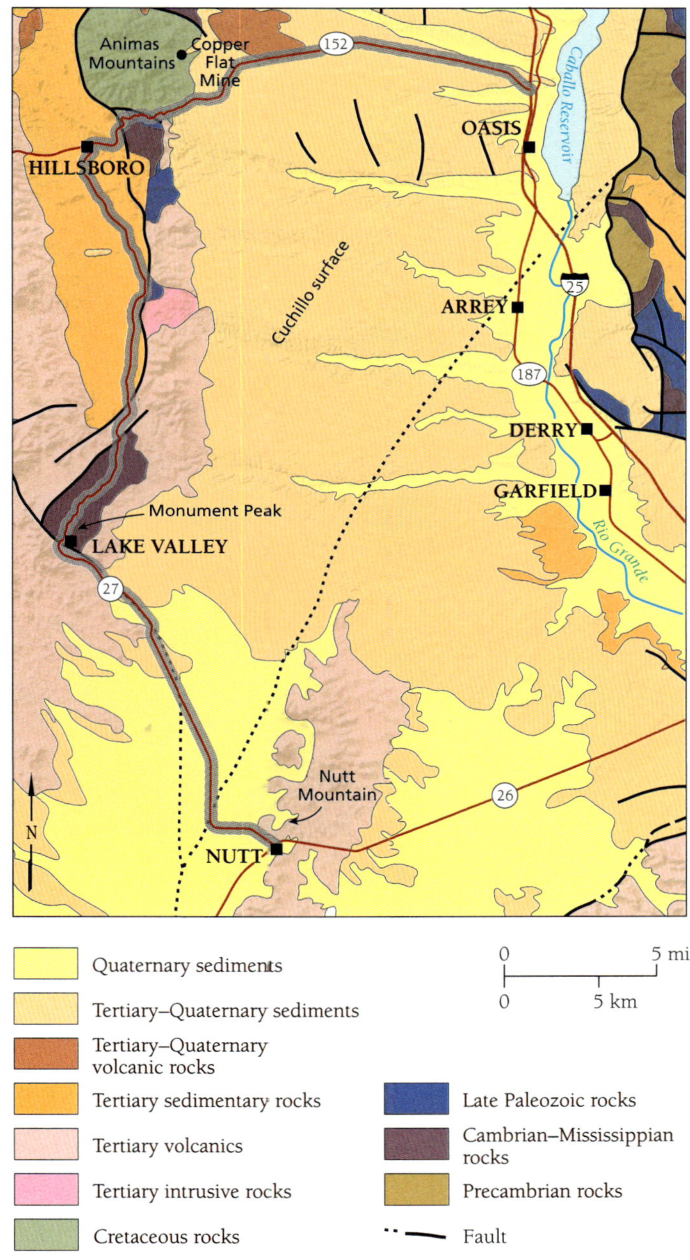

Geologic map of the Lake Valley Back Country Byway.

to the west and passes Sibley Mountain, composed largely of dark basaltic andesite and the Kneeling Nun Tuff, on the east. The small valley at mile 27.4 (19.6), where Jaralosa Creek joins Berenda Creek, is the site of the former Lake Valley reservoir, which supplied the 1880s mining town of Lake Valley with potable water via a pipeline. The ruins of the Lake Valley townsite are at 33.5 miles (13.5). South of Lake Valley, the Cookes Range forms the western skyline, and Nutt Mountain and the Good Sight Mountains can be seen to the south. The byway follows the remains of an Atchison, Topeka, and Santa Fe Railway spur built to Nutt in 1884 to transport rich silver ore from Lake Valley to market. Mile 47 (0), the southern end of the byway at NM 26 near Nutt, provides views of the Macho Springs wind and solar facilities that supply electricity to southern New Mexico.

Panorama of the Animas Mountains west of Copper Flat with hydrothermally altered Pennsylvanian and Mesozoic rocks beneath.

Geologic History

The Lake Valley Back Country Byway mainly traverses volcanic rocks of the eastern Mogollon–Datil field and younger sedimentary rocks of the Rio Grande rift. Most of the mountains in the region were uplifted during the formation of the Rio Grande rift beginning about 25 million years ago. These mountains locally expose Paleozoic marine strata that lie beneath the volcanic rocks, such as along NM 152 north of Hillsboro. Apache Hill, approximately 1.6 miles north of Lake Valley on the east side of NM 27 at mile 31.9 (15.1), has exposures of the Mississippian Lake Valley Limestone that are among the best fossil collecting sites in southern New Mexico. Some of the best outcrops are located just below the top of the hill where brachiopods, bryozoans, crinoids, horn corals, gastropods, pelecypods, cephalopods, and trilobites weather out of the shaly limestone.

Volcanic rocks, formed during three main episodes, dominate exposures in the region. Late Cretaceous flows erupted about 75 million years ago from the Copper Flat volcano in the Animas Mountains near the north end of the byway. The Copper Flat stock intruded beneath this volcano and now hosts copper porphyry mineralization at the Copper Flat Mine. The Hillsboro Mining District owes its origin to this deeply eroded volcano.

Following a nearly 40 million-year break in eruptive activity, volcanism resumed, and volcanic rocks of the Mogollon–Datil field buried the Copper Flat volcanic/intrusive complex. This volcanic episode included development of the Emory caldera, a supervolcano, which erupted the Kneeling Nun Tuff about 35 million years ago The most recent volcanic activity is recorded by Pliocene basalt flows that

were extruded about 4 million years ago and now cap many of the peaks and mesas of the Animas Mountains.

The byway begins and ends on sedimentary fill of the Rio Grande rift, which ranges in age from about 25 million years to 800,000 years. The Cuchillo surface represents a constructional surface built by streams that drained to the Rio Grande, prior to the onset of drainage incision about 800,000 years ago. The river and its tributaries have since cut down about 425 feet.

Mining History

Mining in the Hillsboro District began in 1877 along veins that radiate outward from the Copper Flat volcano. Placer gold was found in Snake and Wicks gulches in 1877, and Gold Dust, a tent city, was founded in 1881. Placer dumps are visible along the creeks as one travels south from the Copper Flat road. Currently, THEMAC Resources Group, Ltd. is applying for mining permits to resume operations at the Copper Flat Mine to produce copper ore with some gold, silver, and molybdenum. The Hillsboro Mining District lies on the eastern edge of the Arizona–Sonora–New Mexico copper porphyry belt, and the late Cretaceous Copper Flat copper porphyry deposit is the oldest such deposit in the state. Total production from the Hillsboro District amounts to over $14.5 million dollars, including copper, silver, gold, lead, zinc, and manganese.

Farther south, the Lake Valley Mining District was discovered in 1878 by George W. Lufkin and Chris Watson, who were cowboys and part-time prospectors. Initial samples showed remarkably high silver contents. Over the next 75 years, the Lake Valley mines produced nearly 6 million ounces of silver and over 20,000 tons of manganese from carbonate-hosted silver-manganese replacement deposits in the Mississippian Lake Valley Limestone. Silver assays as high as 20,000 ounces per ton were common, and high-grade silver pockets were found throughout the district. The famous Bridal Chamber Mine contained a large concentration of nearly pure cerargyrite (silver chloride), 200 feet long by 25 feet thick. Most of the silver mines were played out by the late 1880s, although some sporadic mining continued into the 20th century. Manganese was still exploited there as a strategic material during World War II. The New Mexico Abandoned Mine Lands Program subsequently closed most of the old workings.

—Virginia T. McLemore, David J. McCraw, and Robert Eveleth

Loadout facility in Lake Valley where high-grade silver deposits were recovered from Mississippian Lake Valley Formation rocks.

If You Plan to Visit

The Lake Valley village and mine site, a ghost town, is now managed by the Bureau of Land Management and is open all year. Hillsboro is sparsely populated but does have museums that provide further information on the history of this area. A self-guided walking tour of the townsite is available, beginning at the schoolhouse museum. The tiny village of Nutt has the only other facility along the route, the Corner Bar (formerly the "Middle of Nowhere Bar").

To get to the north end of Lake Valley Back Country Byway, use the Hillsboro exit (Exit 63) off I–25 and then NM 152 west towards Hillsboro. To access the south end of the byway, take NM 26 southwest from Hatch or northeast from Deming to the town of Nutt and then turn west on NM 27. For more information:

Bureau of Land Management
Las Cruces District Office
1800 Marquess Street
Las Cruces, NM 88005-3370
(575) 525-4300

At right, the former Lake Valley schoolhouse is now the Visitors center. In the background, Monument Peak, also known as Lizard Mountain (elevation 5,682 feet), is composed of Rubio Peak Andesite.

Fort Selden State Monument and Leasburg Dam State Park

Fort Selden State Monument and Leasburg Dam State Park are both located near the Rio Grande, west and northwest of the town of Radium Springs, about 14 miles north of Las Cruces. Fort Selden was constructed in 1865 on the Camino Real de Tierra Adentro (Royal Road of the Interior Lands) that connected Mexico City to northern New Mexico. Soldiers stationed at the fort protected travelers from Apache raiders and bandits. Fort Selden was the last dependable place that settlers and traders could find forage and water for their livestock before heading north along the Jornada del Muerto. During the time of military occupation between 1865 and 1891, the Rio Grande was very close to the fort. The river has since shifted to the west.

Leasburg Dam was built in 1909 to divert water for agricultural needs along this section of the Rio Grande. Camping, hiking, wading in the river, and bird watching are common activities at the state park. Several small warm springs related to geothermal activity in this area are found on both sides of the river below the dam.

OPPOSITE: Steaming warm springs emerge in the river bed below Leasburg Dam. The springs are most visible during the winter when the river level is low and the air temperature is below freezing.

Remains of the foundation of the jail at Fort Selden, built of limestone and rhyolite from the nearby Robledo Mountains. The parade ground is behind the jail.

Geography and Regional Geology

Both facilities offer good views of the northern end of the Robledo Mountains, an uplifted block within the Rio Grande rift that rises 2,000 feet above the valley floor. Paleozoic and Eocene sedimentary rocks, cut by an Eocene rhyolite sill, are exposed in this mountain range. Sedimentary rocks in the Robledo Mountains generally tilt toward the south; thus, the oldest rocks (Ordovician) are found at the north end. The northern tip of the range, visible from Fort Selden, is composed of Permian sedimentary rocks that are down-faulted against Ordovician sedimentary rocks along an east-trending fault. The Prehistoric Trackways National Monument and the Robledo Mountains Wilderness Study Area are located in the southern part of the range.

Visible to the east of I–25 are the Doña Ana Mountains, which preserve a deeply eroded Eocene caldera, an explosive supervolcano about 35 million years old. Pennsylvanian through Eocene sedimentary rocks in this mountain range were disrupted by caldera collapse and associated igneous intrusions.

The Rock Record

Shallow- to marginal-marine sedimentary rocks preserved in the northern Robledo Mountains include Ordovician to Pennsylvanian beds dominated by carbonates (dolomite and limestone). The carbonates are overlain by red Permian river deposits that interfinger with marine limestones. Limestones are intruded by rhyolite sills that, in places, are up to 1,000 feet thick. Lookout Mountain in the northern Robledo Mountains served as an observation post for Fort Selden and is made up of a thick rhyolite sill that was emplaced about 35 million years ago.

Geologic History

The oldest rocks in Radium Springs area are the Eocene Palm Park Formation and a rhyolite sill found near Leasburg dam. These rocks formed during an episode of regional volcanism that also produced the volcanic rocks in the Organ Mountains to the east and the Mogollon–Datil volcanic field to the northwest. Crustal stretching associated with the development of the Rio Grande rift followed the volcanism, ultimately creating the mountains and valleys of today.

Geologic Features

The only bedrock exposed in this area is in Leasburg Dam State Park. The west side of the dam abuts a rhyolite sill intruded into the Eocene Palm Park Formation. Examples of a well-developed chilled margin of the intrusion are visible along the trail.

Photograph of fractured rhyolite which is the geothermal reservoir at Radium Springs.

The intrusive contact between the rhyolite and the Eocene Palm Park Formation is exposed along a trail at Leasburg State Park. The arrow highlights the fine-grained, chilled margin of the intrusion. A GPS unit is shown for scale.

The Palm Park Formation is primarily a sedimentary unit of andesitic debris shed from stratovolcanoes about 38 to 36 million years ago. An outcrop of Palm Park Formation can be seen in the bed of the Rio Grande about 0.5 miles downstream of the dam.

The state park and fort are mostly built on a late Pleistocene river terrace known as the Leasburg terrace, which formed about 15,000 years ago. The terrace contains silt and clay used for the adobe bricks in Fort Selden. The adobe bricks were not of high quality, so most of the fort walls are badly weathered. Limestone and rhyolite blocks from the Robledo Mountains were used in the construction of the jail. Prisoners scratched markings into the limestone blocks, but the harder rhyolite remains unblemished.

Natural hot springs seeping from the fractured rhyolite intrusion north of Leasburg Dam were used for bathing and medicinal purposes for many years. The "Radium Springs" do not actually contain radium. The name was used to tempt visitors to experience the benefits of

North-south oriented geologic cross section through the north end of the Robledo Mountains to the west of Fort Selden State Park and Leasburg Dam State Park showing the sill that forms Lookout Peak.

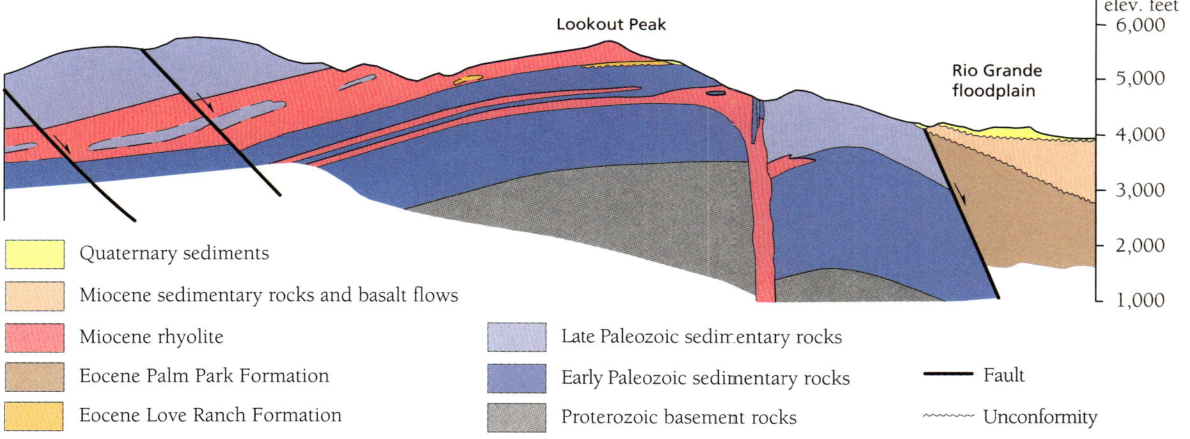

drinking and bathing in the water, as radium was then considered to be healthful. Within Leasburg Dam State Park, several minor hot springs are present. On very cold days, steam emanates from them.

In 1987, Alexander Masson built a large greenhouse complex just north of Leasburg Dam that is heated by geothermal water. The more than 150 people now employed there grow flowering plants. Originally, the greenhouse was heated by two shallow wells less than 350 feet deep, which provided 165°F water from fractures in the rhyolite. Later, a deeper well (800 feet deep) was drilled through the rhyolite intrusion and the Palm Park Formation into the under-lying fractured Hueco Limestone aquifer. Water from that depth has a temperature of 199°F.

—*Shari A. Kelley*

If You Plan to Visit

To reach Fort Selden State Monument, take the Radium Springs exit on I–25 (Exit 19; about 16 miles north of Las Cruces), and turn west/southwest onto Fort Selden Road, and drive for about 1.5 miles. The Monument and its Visitor Center parking lot are on the right-hand (north) side of the road.

Fort Selden State Monument
1280 Fort Selden Road
Radium Springs, NM 88054
(575) 526-8911
www.nmhistoricsites.org/fort-selden

To reach Leasburg Dam State Park take the Radium Springs exit on I–25 (Exit 19; about 16 miles north of Las Cruces), and turn west/southwest onto Fort Selden Road, and drive for about 0.8 miles. Turn right on Leasburg Park Road to get to the park's Visitor Center and campgrounds. Alternatively, you can continue on Fort Selden Road, drive past Fort Selden State Monument to just before the junction with NM 185. There, take Leasburg Dam Road/Doña Ana Road north to the park's lower picnic facilities, river, and dam overlook.

Leasburg Dam State Park
12712 State Park Road
Radium Springs, NM 88054
(575) 524-4068
www.emnrd.state.nm.us/spd/leasburgdamstatepark.html

Mesilla Valley Bosque State Park
NEW MEXICO STATE PARKS

Mesilla Valley Bosque State Park is located on the southwestern outskirts of Las Cruces, just west of the town of Mesilla along the west bank of the Rio Grande. It is a good place to visualize the dynamic nature of the river during the last 5 million years. A 2.3-mile-long hiking trail winds through the bosque and along ditch banks, offering good views of the nearby Rio Grande and the Organ Mountains in the distance. Birds and wildlife are readily seen along the trail. Educational displays in the Visitor Center describe the formation of the Rio Grande rift and the paleontology of the Permian redbeds at the Prehistoric Trackways National Monument. Another display shows how the Rio Grande main channel shifted westward between 1850 and 1885.

The Visitor's Center entrance and patio offer an inviting space for picnics and weddings.

The Rock Record and Geologic History

The only "outcrops" are in arroyos on the west side of the park. Tan sands with thin, flat bedding and interspersed thin gravel lenses belonging to the Santa Fe Group form a low bench. The textures and clasts in the gravels suggest that this unit was deposited by a river. The surface developed on top of this fluvial deposit formed during a time when the drainage level in this area was about 20 feet higher than it is today.

Geologists have been examining the ancestral Rio Grande deposits in the Las Cruces area since the 1950s. The groundbreaking work that came out of these early studies has been applied to investigations of other ancient arid river systems in rifts around the globe. Scientists have mapped the distribution of the deposits and have dated fossils, volcanic ash, and magnetization preserved in these rocks to piece together an amazing story (see the introduction to this part of the book).

The Rio Grande first arrived in southern New Mexico less than 5 million years ago after the river filled a playa lake system in the southern Albuquerque Basin just north of Socorro, causing the river to spill southward. The few remnants of 5 to 3.6 million-year-old ancestral Rio Grande deposits that are preserved in southern New Mexico suggest that this early river closely followed the modern course between Socorro and Las Cruces. Between Las Cruces and El Paso, the river emptied into a series of lakes and playas known as Lake

Cabeza de Vaca that formed in the southern Mesilla basin, the Hueco basin, and basins in west Texas and northern Mexico.

As time passed, the river valley widened considerably until river sediments were deposited across a 25-mile-wide swath at the latitude of Las Cruces. The river continued to empty into Lake Cabeza de Vaca at this time. By about 2.6 million years ago, the river had filled the broad fault-bounded basin in the Las Cruces region and spilled east through Fillmore Gap (between the Organ and the Franklin mountain ranges) into the southern Tularosa and the Hueco basins. Ancestral Rio Grande deposits containing a 2 million-year-old volcanic ash bed overlie Lake Cabeza de Vaca sediments in the southeastern Hueco Basin, indicating that the lake had filled with sediments by about 2.25 million years ago. The river flowed into west Texas toward Big Bend after 2.25 million years, and finally flowed into the Gulf of Mexico sometime between 1.5 and 0.8 million years.

Broad, regional-scale sediment deposition by the ancestral Rio Grande ended at about 800,000 years ago. The elevation of the highest sediment aggradation is marked by the La Mesa surface, which is about 550 feet above the modern floodplain. The long interval of filling of the basins of the Rio Grande rift was replaced by episodic downcutting that created the relatively narrow river valley that we see today. Downcutting was in progress by 740,000 years ago, as recorded by deposition of Bishop volcanic ash within the valley. A basalt flow erupted from the Santo Tomas volcano about 6 miles south of the park and flowed off of the La Mesa surface onto an inset surface 115 feet above the modern floodplain.

At least five episodes of downcutting and partial backfilling are preserved in the Las Cruces area. These backfilling episodes produce terraces, although only one terrace surface is visible in the park. Downcutting occurred during times of Pleistocene glaciation when climates were less arid, large amounts of water flowed in the river, and sediment loads were low. During the warmer interglacial times, river discharge was low, and sediment loads were high, causing backfilling and terrace formation.

—Shari A. Kelley

Exposures of ancestral Rio Grande deposits on the west side of the park.

If You Plan to Visit

From Las Cruces, take the East University Avenue exit (Exit 1) from I–25 and head west on that road, which becomes West University Avenue; turn right on Avenida de Mesilla; in about 0.5 miles turn left on Calle del Norte and drive for about 2 miles; then turn left at South Fairacres Road, which leads directly into the park. For more information:

Mesilla Valley Bosque State Park
5000 Calle del Norte
Mesilla, NM 88046
(575) 523-4398
www.emnrd.state.nm.us/spd/mesillavalleystatepark.html

Organ Mountains–Desert Peaks National Monument

Aguirre Springs and Dripping Springs Recreation Areas

Organ Mountains–Desert Peaks National Monument is a large, diverse and recently established park. It has four separate geographic areas: the Organ Mountains, the Doña Ana Mountains, the Potrillo Mountains, and the "Desert Peaks." This chapter deals only with the Organ Mountains. The geology of the Doña Mountains area is briefly discussed in the chapter on Fort Selden–Leasburg Dam. The Potrillo Mountains include the remote Potrillo volcanic field and are covered in the Kilbourne Hole chapter. The "Desert Peaks" includes the Robledo Mountains and Sierra de las Uvas. The Robledo region is described in the chapters titled "Robledo Mountains" and "Prehistoric Trackways" (a separate, but contiguous national monument).

Rising nearly a mile above Las Cruces is the jagged crest of the Organ Needles, the backbone of the Organ Mountains. Few skylines in southern New Mexico are as inspirational to artists, climbers, and outdoor enthusiasts. Although views of the Organ Mountains reflecting the glow of the setting sun can induce a sense of tranquility, the scene here 36 million years ago was far from peaceful. Instead, three caldera-forming super-eruptions rocked the region, marking the beginning of a volcanic episode that would forever change the face of southwestern New Mexico. Following these explosive eruptions, the Organ caldera was ripped apart by large-scale faulting, providing geologists with a unique opportunity to study both the erupted volcanic rocks and the crystallized magma chambers beneath them.

An early name for the Organ Mountains was La Sierra de Soledad or "the mountain of solitude." Their dramatic peaks and spires, reminiscent of a pipe organ, were an essential landmark along El Camino Real between the 15th and 18th centuries. Sporadic mining of precious metals and minerals occurred in the Organ Mountains during the 18th and 19th centuries. In the late 19th century, Dripping Springs Resort

OPPOSITE: The name "Organ Mountains" is thought to originate because of the similarity of the many spires and needles, such as these, to the pipes of an organ.

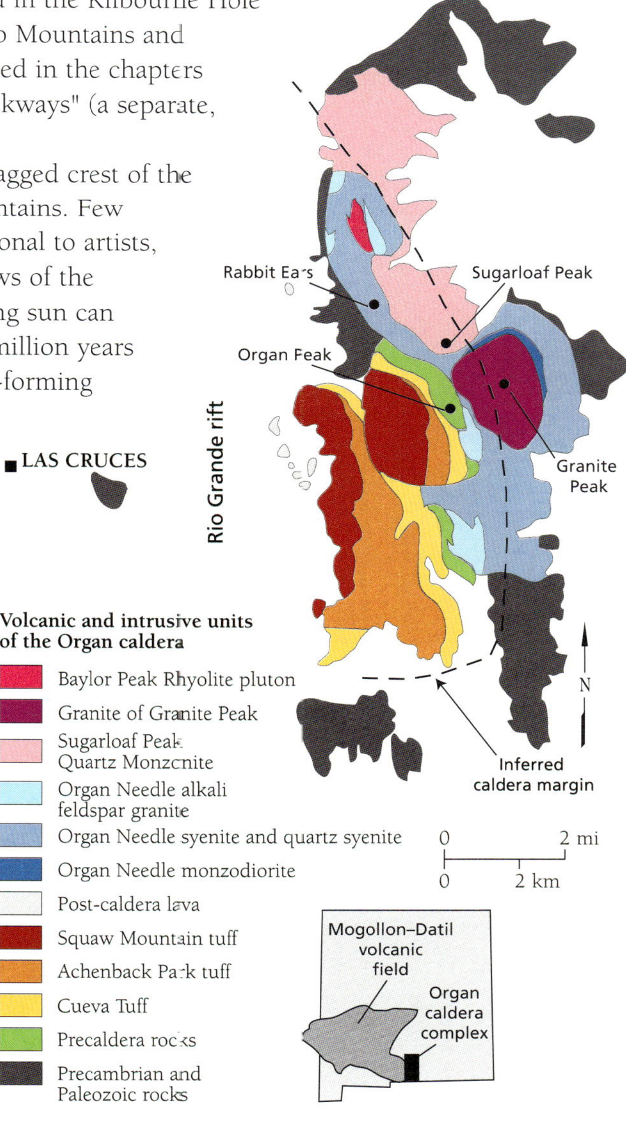

was established as a reclusive mountain getaway. It was later converted to a sanatorium. The Organ Mountains were designated as part of the Organ Mountains–Desert Peaks National Monument in 2014.

Regional Setting

The Organ Mountains are located in a complex zone of overlapping geologic regions. The Organ caldera marks the southeastern margin of the Mogollon–Datil volcanic field. Superimposed on the eastern margin of this field is the Rio Grande rift, the north-south trending series of down-dropped basins related to east-west stretching of the crust. The Organ Mountains are part of a 150-mile-long block of the Earth's crust that was uplifted during rifting. The San Andres Mountains to the north and Franklin Mountains to the south are parts of this same west-tilted fault block. The Organ Mountains uplift is sandwiched by two basins, the Tularosa Basin to the east and Jornada del Muerto to the west.

Northern Organ Mountains skyline as seen from Leasburg Dam State Park.

The Rock Record

The majority of the rocks exposed in the Organ Mountains are intrusive or volcanic in origin, the latter including both ash-flow tuffs and lavas. The Organ caldera erupted three major ash-flow tuffs about 36 million years ago. Tuffs are composed mostly of volcanic ash (tiny particles of volcanic glass) deposited from an incandescent cloud of gas and ash as it flowed rapidly across the land surface. Lens-shaped features in the tuffs are flattened pumice, indicating that the ash was hot enough to fuse together after deposition and form a welded tuff. The three tuffs are exposed along the western flank of the mountain and dramatically thicken westward from less than one thousand feet in the east to nearly 10,000 feet in the west. The wedge-shape geometry of the tuffs resulted from deposition within a trapdoor-style caldera, formed when the western part of the circular caldera subsided more deeply along faults than did the eastern side. Similar to most calderas, lavas erupted prior to and after the caldera formed. The best lava

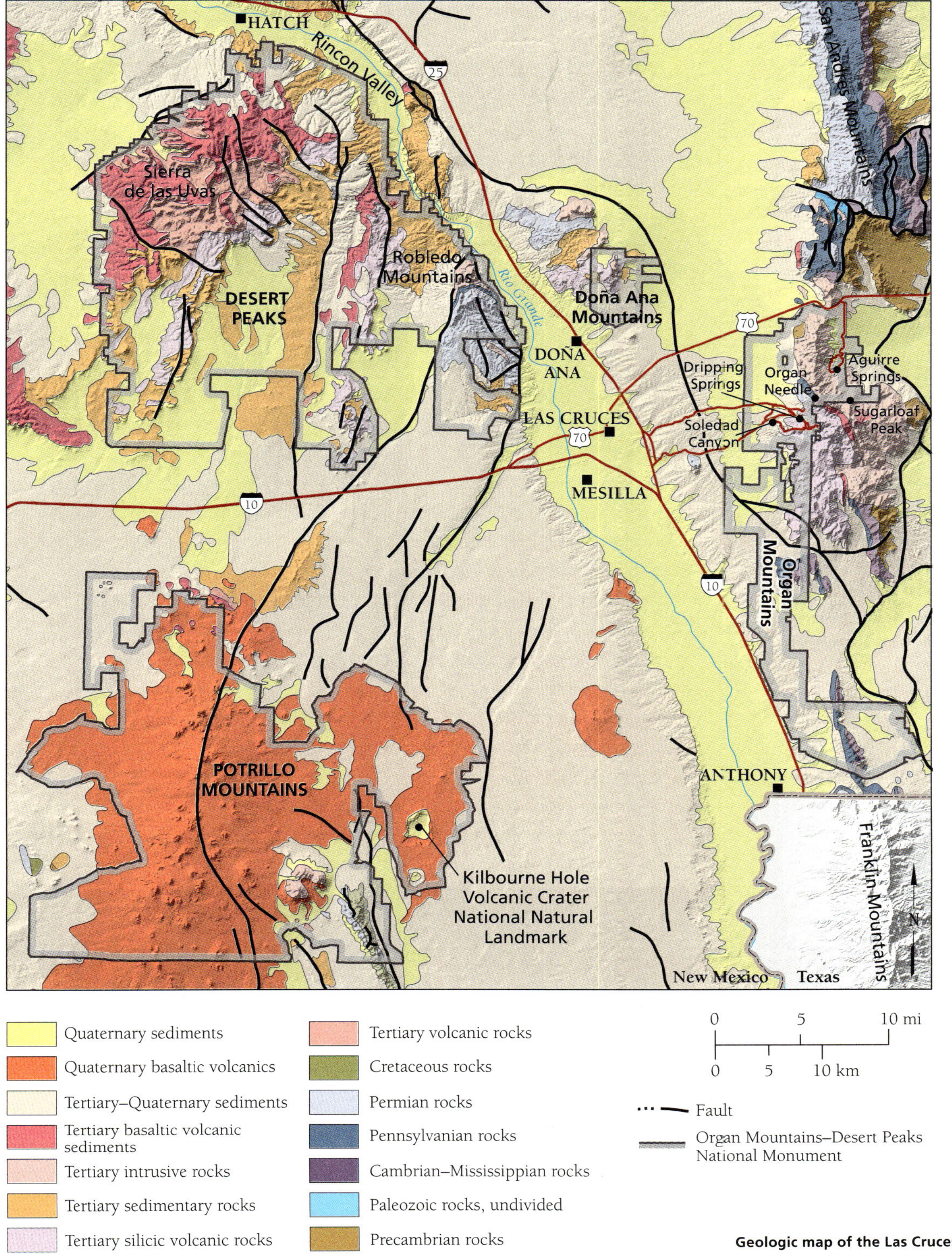

Geologic map of the Las Cruces area.

exposures are near the head of Fillmore Canyon, close to the crest of the mountain.

World-class examples of granitic intrusions are found in the Organ Mountains. Granite forms when silica-rich magma cools and crystallizes completely within the crust. Crystals in granite range in size from a fraction of an inch to several inches. Typically, smaller crystals form when magma cools quickly. Granite contains mostly quartz and feldspar, with lesser amounts of mica and hornblende. Some granites of the Organ Mountains were emplaced as magma chambers that were themselves the sources of the ash-flow tuffs. Other granites are younger, recording the slow cessation of magmatism. Together, this group of intrusions is called the Organ batholith. The Organ batholith is similar to the tip of an iceberg in that most of it is hidden in the subsurface. Distinguishing the intrusive rocks from volcanic rocks here is easy; the intrusions typically weather into serrated spires that dominate the northern part of the range, whereas the volcanic rocks usually weather into rounded massifs that dominate the southern range.

Rocks not related to the Organ caldera are also exposed in the Organ Mountains. The oldest rocks in this area are 1.4 billion-year-old granites. These are texturally similar to the younger intrusive rocks but are typically darker in color and more eroded. Exposed along the western flank of the Needles are variably altered Paleozoic sedimentary rocks, chiefly limestones, sandstones, and siltstones. The rift-related Pliocene–Pleistocene Camp Rice Formation is the youngest sedimentary unit in the area. Thin layers of Pleistocene volcanic ash in the upper part of the Camp Rice Formation include those from the

Examples of volcanic and intrusive rocks of the Organ caldera. The volcanic rock (left) contains few crystals and is mostly made of ash and pumice. The intrusive granite (right) is completely crystalline.

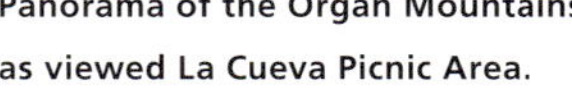

Panorama of the Organ Mountains as viewed La Cueva Picnic Area.

Yellowstone caldera in Wyoming and the Long Valley caldera in California. These far-travelled ashes are a testament to the explosive power of these modern calderas, which were much like the long-extinct one exposed in the Organ Mountains.

Geologic History

The earliest geologic history of the Organ Mountains region is characterized by multiple mountain-building events and deposition of enormous volumes of sediments during Precambrian time. Between 1.7 and 1.1 billion years ago, volcanic islands and continents collided with the fledgling North American continent to build what is now much of the Southwest. Silicic magmas, such as those preserved as Precambrian granites in the Organ Mountains and in the Sandia Mountains, were intruded into the crust and crystallized during this time, contributing to the regional metamorphism of the earlier volcanic and sedimentary deposits.

Following a long episode of regional uplift and erosion, vast Paleozoic seas and rivers deposited many thousands of feet of sediment between 500 and 250 million years ago. The Laramide Orogeny thrust both the Precambrian and Paleozoic rocks upward, deforming the once horizontal sedimentary rocks into a series of folds and fault-bounded blocks during the Late Cretaceous and Early Cenozoic (about 75 to 45 million years ago).

This was followed by three caldera-forming eruptions in the Organ Mountains about 36 million years ago. Geologic mapping suggests that each successive caldera was superimposed on the previous one, thereby obscuring the structure of all but the youngest caldera. Following the caldera formation, lava flows erupted for at least 1 million years. As the

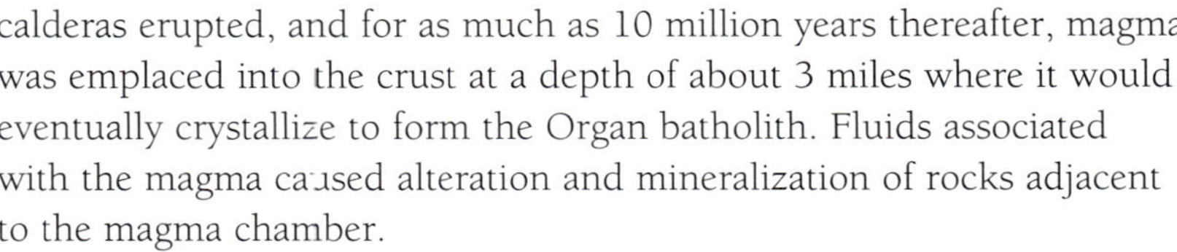

calderas erupted, and for as much as 10 million years thereafter, magma was emplaced into the crust at a depth of about 3 miles where it would eventually crystallize to form the Organ batholith. Fluids associated with the magma caused alteration and mineralization of rocks adjacent to the magma chamber.

Rift-related faulting, basin subsidence, and sedimentation are the youngest events in the Organ Mountains area. Faulting and basin development likely began during the waning stages of magmatism, as some of the intrusions appear to have been emplaced into fault zones. The combination of faulting, westward tilting of the fault block, and subsequent erosion has exposed the intrusive rocks beneath the caldera.

Geologic Features

DRIPPING SPRINGS AND SOLEDAD CANYON—Intra-caldera tuffs (tuff that accumulated within the caldera) are well exposed at both Dripping Springs and in Soledad Canyon. The towering 2,000-foot walls that surround the ruins of Dripping Springs resort are composed almost entirely of intra-caldera Squaw Mountain tuff, deposited during a single eruption about 36 million years ago. The terrain north of Soledad Canyon is also composed of Squaw Mountain tuff, but the hills to the south are mostly intra-caldera Achenback Park tuff. Thin sediments between the tuffs were deposited between eruptions.

LA CUEVA—The intracaldera Cueva Tuff is exposed at the La Cueva shelter. The lower part visible here, in contrast to the younger tuffs, contains an abundance of rock fragments that are mostly volcanic in origin. The rock fragments were likely incorporated into the tuff when an extremely violent explosion initiated the caldera-forming eruption. Amazing exposures of various parts of the Cueva Tuff are also found near Peña Blanca, south of La Cueva.

Waterfalls in the Dripping Springs Recreational Area. The rocks exposed here are almost entirely intra-caldera Squaw Mountain tuff.

AGUIRRE SPRINGS—Intrusive rocks related to the Organ caldera are exposed along the east slopes and crest of the Organ Mountains. The highest peaks are a part of the Organ batholith and are composed of a rock called the Organ Needle Quartz Monzonite, which is similar to granite. The quartz monzonite also forms the iconic Rabbit Ears pinnacles. The age and chemistry of the quartz monzonite are similar to the Squaw Mountain tuff, suggesting a genetic relationship.

The 4-mile-long Pine Creek Loop Trail provides easy access to the Sugarloaf Peak Quartz Monzonite, which was emplaced 1 to 2 million years after the caldera eruptions. The crystals in this intrusion are very similar in size to one another, although some zones have larger crystals set in matrix of smaller crystals, a texture called porphyritic. Also seen in this intrusion are baseball- to beach-ball-sized masses of darker rock. These features record the mixing of hot (2,100°F) basaltic magma into the quartz monzonite magma. This mixing added new heat to the quartz monzonite magma, slowing its cooling and possibly triggering smaller eruptions.

Several debris-flow deposits are exposed near Aguirre Springs. In August 1991, heavy monsoon rains triggered a debris flow that roared down Anvil Creek and crossed the Loop Road just east of the Pine Creek Trailhead. Most of the debris-flow material has been cleared away, but remnants can be seen from the road. Parallel levees composed of large boulders delineate the margins of the flow.

EXFOLIATION DOMES—Sugarloaf Peak, south of Aguirre Springs, and San Agustin Peak, just north of San Agustin Pass on NM–70, display rounded outcrops that result from a type of physical weathering common in granitic rocks called exfoliation. Granites are emplaced deep within the crust and are under pressure from the surrounding rocks.

Sugarloaf Peak viewed from Aguirre Springs campground. The rounded nature of this peak is because of the weathering process called exfoliation.

The jagged peaks of the northern Organ Mountains are composed of intrusive rocks or fossil magma chambers. The two prominent peaks near the center are called "The Rabbit Ears".

When eventually exposed at the surface, decreased confining pressure causes the granite to expand, creating a series of concentric fractures parallel to the surface termed exfoliation joints. Exfoliation joints are subsequently enlarged by a variety of processes, including freezing/thawing of water and thermal stresses caused by daily or seasonal temperature cycles. These processes incrementally enlarge the joints and pry the intervening curved slabs of rock apart, eventually producing rounded knobs like those of Sugarloaf Peak and San Agustin Peak.

VIEWSCAPE FROM CREST OF THE NEEDLES—Superb views of south-central New Mexico, Texas, and northern Mexico are a worthwhile reward for those willing to make the rugged hike and climb to the Organ Needle summit. The endless array of peaks and valleys are characteristic of the southern Rio Grande rift. On a clear day, the vista stretches for at least 100 miles and includes many locations discussed in this book, such as Sierra Blanca and White Sands National Park to the northeast, Kilbourne Hole to the southwest, the Florida Mountains, Rockhound State Park, Cookes Peak to the west, and the Robledo Mountains and Mogollon–Datil volcanic field to the northwest.

—*Matthew J. Zimmerer*

If You Plan to Visit

Aguirre Springs Campground and Trails

This site lies on the east side of the Organ Mountains. From Las Cruces, take the US 70 exit off I-25 (Exit 6) and go 14 miles east on US 70. Then turn south on the Aguirre Springs Road (marked by a small sign 1.1 miles east of San Agustin Pass). It is about 6 miles to the campground and at mile 4 the road becomes a winding one-way loop with steep grades.

Dripping Springs Recreation Area and A. B. Cox Visitor Center

(including La Cueva Trail, Filmore Trail, and trails to Cox Ranch and Van Patten Mountain Ranch)

The area is located approximately 10 miles east of Las Cruces, on the west side of the Organ Mountains. From Exit 1 on I–25, make a left onto University Avenue/Dripping Springs Road and follow the road east to the end at the Visitor Center. Alternatively, this area can be accessed from the north and east from US 70 by taking the NASA Road/ Baylor Canyon Road Exit off that highway (about 11 miles northeast of Las Cruces and just west of the village of Organ). Go south on Baylor Canyon Road until its T-junction with Dripping Springs Road; make a left and then follow that road east to the Visitor Center. For more information:

Bureau of Land Management
Las Cruces District Office
1800 Marquess Street
Las Cruces, NM 88005-3370
(575) 525-4300

Dripping Springs Visitor Center
(575) 522-1219
www.blm.gov/nm/st/en/prog/blm_special_areas/national_monuments/
omdp_national_monument.html

Prehistoric Trackways National Monument
BUREAU OF LAND MANAGEMENT

The Prehistoric Trackways National Monument covers over 5,200 acres in the rugged and barren southern Robledo Mountains, just northwest of Las Cruces. One of New Mexico's newest national monuments, it was created by an act of Congress in 2009. There, layers of red sandstones and siltstones were deposited along an early Permian shoreline about 290 million years ago, and these layers preserve the world's best fossil record of early Permian footprints and other trace fossils.

Geography and Geologic Province

The Prehistoric Trackways National Monument is located in the southern Robledo Mountains, a complex, south-dipping fault block of Paleozoic sedimentary rocks. The youngest Paleozoic rocks are exposed in the area of the monument. To the east is the Mesilla Basin of the Rio Grande rift, a structural depression filled with Miocene–Pleistocene sedimentary rocks that were deposited by the ancestral Rio Grande and its tributaries. The basin to the west of the Robledo Mountains has a nearly flat, undissected floor (For more details, see the chapter about the Robledo Mountains).

Rock Record

Paleozoic beds exposed in the Prehistoric Trackways National Monument belong to the early Permian Hueco Group, a succession of mostly marine limestones and shales about 1,700 feet thick. Most Hueco strata are beds of gray limestones and tan or gray layers of shale that commonly contain limestone nodules. The Hueco Group contains myriad fossils, including the shells of marine organisms and the petrified trunks of trees that drifted offshore.

The Prehistoric Trackways National Monument is located in the southern part of the Robledo Mountains, just northwest of Las Cruces.

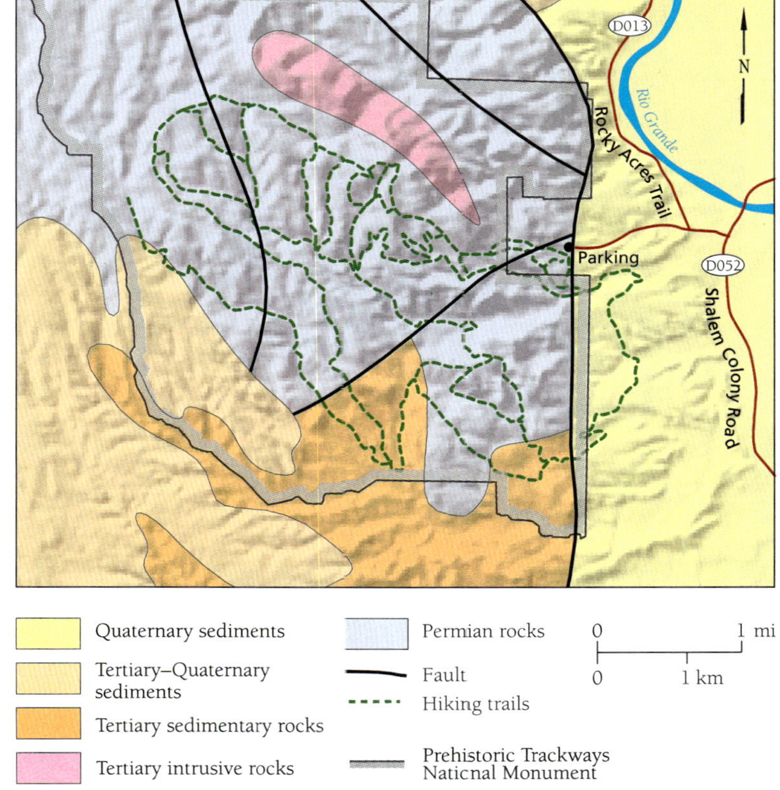

In the Prehistoric Trackways National Monument, fossil logs found in marine strata of the Hueco Group represent trees that died and floated offshore and were buried by sediment on the early Permian sea floor. In the distance, the red rocks that contain trackways can be seen.

The upper part of the Hueco Group in the Robledo Mountains is an interval about 380 feet thick with layers of red sandstones and siltstones interbedded with the gray limestones and shales. These strata, termed the Robledo Mountains Formation, are the bedrock of much of the Prehistoric Trackways National Monument. The early Permian layers preserve fossil tracks and trails of the many land animals that lived near the Hueco seacoast. They also contain fossil impressions of the foliage of land plants. These fossils are the only direct evidence of the organisms that lived along the tropical shoreline of southern New Mexico, 290 million years ago.

In the southernmost part of the Prehistoric Trackways National Monument, some Tertiary-age rocks are exposed. They mostly belong to the Eocene Palm Park Formation and consist of gray, green, purple, and maroon mudstones, sandstones, and conglomerates, as well as some debris-flow breccias. These rocks consist of sediments that were eroded from the southeastern margin of the Mogollon–Datil field, a major volcanic field active about 40 to 25 million years ago in southwestern New Mexico and adjacent parts of Arizona. Other Tertiary rocks in the Prehistoric Trackways National Monument have small outcrop areas and mostly are younger conglomerates, rhyolites, and basalts.

Bedrock in the Prehistoric Trackways National Monument is locally obscured by younger sediments, mostly alluvium associated with the Rio Grande and its tributaries that was deposited during the last few million years.

Geologic History

During the Carboniferous, about 330–320 million years ago, the continents came together to form the single supercontinent called Pangea, which remained amalgamated until the Late Triassic. Collision of the continents that created Pangea produced a series of geologic uplifts across much of the southern and western United States that geologists refer to as the Ancestral Rocky Mountains.

During the early Permian, in the last stage of development of the Ancestral Rocky Mountains, New Mexico was located near the equator along part of the western shoreline of Pangea. A tropical sea, the Hueco seaway, covered southern New Mexico. Its approximately east-west-trending shoreline advanced and retreated from near Las Cruces to as far north as Derry (south of Truth or Consequences), driven in part by sea-level cycles caused by waxing and waning of glaciers in the southern hemisphere.

The shallow, sunlit seafloor of the Hueco seaway was populated by brachiopods, bryozoans, and crinoids as well as some bivalves (clams) and gastropods (snails). Locally, algae grew in thickets that trapped carbonate mud and formed mounds on the seafloor. Nautiloid and ammonoid cephalopods, as well as sharks, swam the Hueco seaway and were top predators. When the Hueco seaway lapped northward across

Impressions of the foliage of walchian conifers are the most common fossil plants found in the Prehistoric Trackways National Monument.

the area of the Prehistoric Trackways National Monument, carbonate mud and clay accumulated to form limestones and shales. When the sea retreated to the south, red sand and silt were deposited on extensive river floodplains and shoreline tidal flats.

North of the Robledo Mountains, the Abo Formation is a thick (up to 1,200 feet) succession of red sandstones and mudstones. It is the landward equivalent of the marine Hueco Group strata. Thus, while the Hueco seaway covered much of southern New Mexico, the Abo Formation was being deposited by rivers that drained the Ancestral Rocky Mountains in northern and central New Mexico. These rivers built

Some characteristic trace fossils from the Prehistoric Trackways National Monument and reconstructions of the trace makers. From left to right: footprints of large, sail-backed pelycosaurian reptile (*Dimetropus*); tracks of a horseshoe crab (*Kouphichnium*); and the unusual corkscrew-shaped burrow called *Augerinoichnus*.

a vast floodplain across much of central New Mexico as they flowed southward to the Hueco seashore near the present-day latitude of Las Cruces. This floodplain was a hot and humid region with a monsoonal climate, where months of rain alternated with months of dry weather.

The seasonally arid Abo floodplain was shrouded in a dense vegetative cover of conifers and seed ferns. Fossilized impressions of the fronds of the coniferous tree, *Walchia*, are especially commonly found in this unit. Scorpions, beetles, and other arthropods lived in the underbrush, hunted by small amphibians and reptiles. Larger reptiles, notably the sail-backed *Dimetrodon*, were top predators. On the tidal flats, various arthropods, including primitive insects (such as beetles and silverfish), spiders, and scorpions searched for food. Small, salamander-like amphibians and lizard-like reptiles hunted the arthropods. Even larger predators, especially the 3- to 6-feet-long pelycosaurs, fed on the amphibians and the smaller reptiles.

These animals left footprint impressions in the sediment and other trails that paleontologists refer to as trace fossils, evidence of ancient animal behavior. At many places in the Prehistoric Trackways National Monument, red layers of sandstones and siltstones contain thousands of such trace fossils. Probably, seasonal flooding and sediment deposition from the rivers onto the tidal flats contributed to the preservation of so many trace fossils. Thus, a footprint left in soft mud when the climate was wet hardened under dry conditions and then was rapidly buried during the next flood.

Trace fossils from the Prehistoric Trackways National Monument document life on the ancient tidal flats, from jumping insects to rapidly running primitive reptiles. In terms of abundance and diversity, the monument contains one of the most extensive trace fossil records known anywhere from any geologic time period, and it certainly is the most extensive of all Permian trace fossil localities.

Geologic Features

The eastern edge of the Prehistoric Trackways National Monument is at an abandoned flagstone quarry known as the Community Pit. The Community Pit exposes fresh outcrops of the Hueco Group that includes numerous trace fossils in red sandstone layers and an abundance of brachiopod fossils in gray shale layers.

The east-west-trending arroyo just north of the Community Pit is the trail to the main Permian tracksite in the Prehistoric Trackways National Monument. A hike of about 1 mile up this arroyo (to the west) takes you to the site, where the Bureau of Land Management installed interpretive signage, and many footprints and other trace fossils can be seen in place. Just north of the trail, and just west of the Community Pit, an impressive, 7 million-year-old basaltic igneous plug is visible.

Complex faulting and igneous intrusions have fractured the Hueco Group strata into many broken blocks of rock that form the rugged canyons and steep ridges that characterize the landscape of the Prehistoric Trackways National Monument. These features can be examined by walking into any canyon in the Monument. Some areas can also be reached via off-highway vehicles and mountain bike trails.

—Spencer G. Lucas

Additional reading

Traces of a Permian seacoast: Prehistoric Trackways National Monument, Spencer G. Lucas, New Mexico Museum of Natural History and Science, Albuquerque, 2011.

If You Plan to Visit

From the north: Turn off I–25 at the Doña Ana Exit 9 onto NM 320 (Thorpe Road) and go about 2 miles west to US 85. Turn right (north) for about 0.5 mile. Turn left (west) on Shalem Colony Trail. Go about 1.5 miles until you cross over the Rio Grande and take a right onto Rocky Acres Trail (D013). Go approximately 0.25 mile, and then turn left onto the only unpaved road that goes west, Permian Tracks Road. Cross over a cattle guard and continue west to the BLM parking area.

From the south (Las Cruces): Travel west on Picacho Avenue (US 70). Turn right (north) onto Shalem Colony Trail at the stop light. Go about 5.5 miles. Prior to reaching the Rio Grande, you will come to a county road, Rocky Acres Trail. Turn left (west) here. Go approximately 0.25 mile and then turn left onto the only dirt road that goes west, Permian Tracks Road. Cross over a cattle guard and continue west to the BLM parking area. For more information:

Bureau of Land Management
Las Cruces District Office
1800 Marquess Street
Las Cruces, NM 88005
(575) 525-4300
www.blm.gov/visit/ptnm

Robledo Mountains and Broad Canyon Trail System

This outcrop of the Hueco Group strata in one of the canyons in the Prehistoric Trackways National Monument shows the layers of red sandstones and siltstones that contain the world's most significant record of early Permian trackways.

The Robledo Mountains Off-Highway Vehicle (OHV) Trail System is a network of trails in the southern Robledo Mountains northwest of Las Cruces, which includes both OHV and mountain bike trails. Trails within the system lie within both the Prehistoric Trackways National Monument and the Organ Mountains–Desert Peaks National Monument.

In the Robledo Mountains, the trails are dominated by enormous rocks, making the terrain exceptionally challenging for riders. The area also includes the "SST" (Shane's Super Trail) mountain bike trail, which is only open to non-motorized uses. North of the Robledo Mountains is the less rugged OHV Trail System in Broad Canyon.

Geography and Geologic Province

A complex fault block approximately 10 miles long (north to south) by 4 miles wide, the Robledo Mountains lie along the western side of the Rio Grande Valley, just northwest of Las Cruces. The mountain range is within the Rio Grande rift and the adjacent Basin and Range physiographic province, which includes much of southwestern New Mexico. The Robledo Mountains primarily expose Paleozoic sedimentary rocks that are extensively fractured but generally tilt to the south. In the southern part of the range, volcaniclastic beds of the Eocene Palm Park Formation are exposed.

The basin east of the Robledo Mountains is the Mesilla Basin of the Rio Grande rift, a late Cenozoic structural feature filled with Miocene–Pleistocene sedimentary rocks, most of which were deposited by the ancestral Rio Grande and its tributaries. The modern Rio Grande and its tributaries have extensively dissected this basin during the past 800,000 years. The basin to the west of the Robledo Mountains has a nearly flat, undissected floor.

The Broad Canyon area to the north of the Robledo Mountains is also part of the Rio Grande rift and the Basin and Range province. There, Broad Canyon cuts through Tertiary volcanic rocks exposed in two mountain ranges, the Cedar Hills and the Sierra de las Uvas.

Rock Record

Most of the Paleozoic strata in the Robledo Mountains, and all that are exposed in the Robledo Mountains OHV Area, belong to the early Permian Hueco Group, a succession of mostly marine limestones and shales about 1,700 feet thick. The Hueco Group strata contain a diverse assemblage of fossils. Limestones and shales contain shells of marine animals and the petrified trunks of trees that drifted offshore. Red siltstones and sandstones abound in tracks and trails of ancient arthropods, crustaceans, amphibians, and reptiles. They also contain impressions of plant foliage from the forested lands that grew just north of the Hueco seaway.

In the Broad Canyon area, most of the exposed bedrock consists of Oligocene volcanic rocks of the Bell Top Formation (about 35 million years old) overlain by the Uvas Basaltic Andesite (about 27 million years old). These tuffs and lavas are overlain by conglomerate beds of the Miocene Rincon Valley Formation.

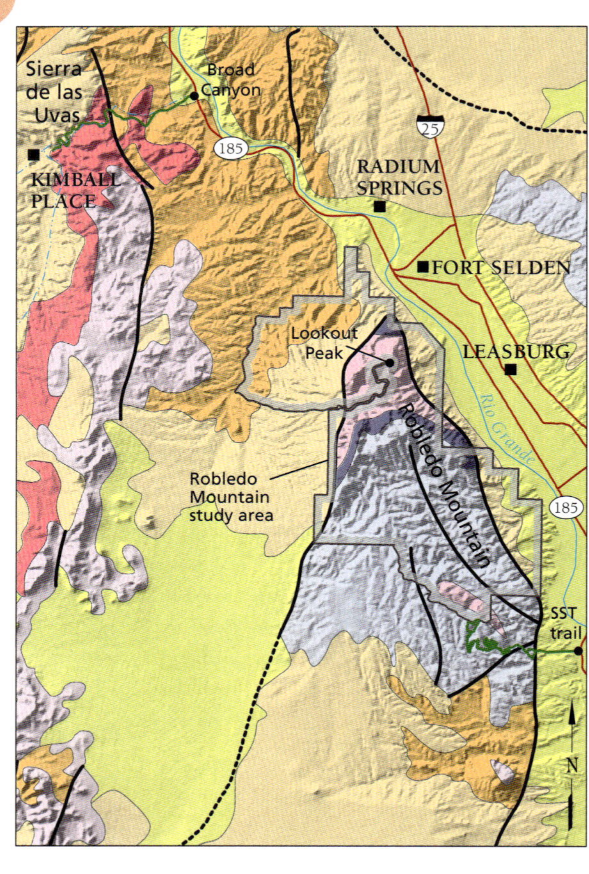

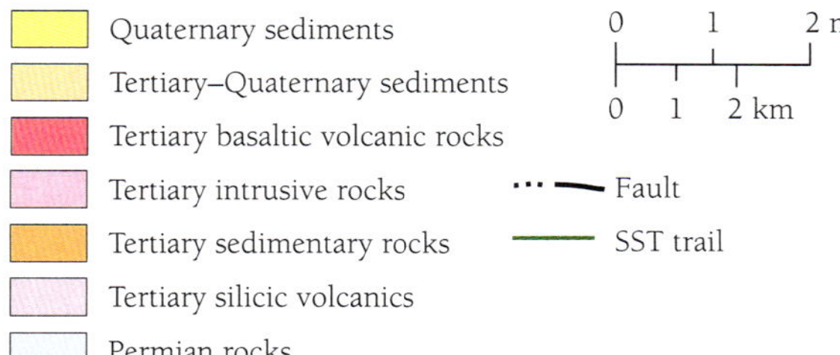

Geologic History

During the early Permian, all of the world's continents were united in a single supercontinent called Pangea. What is now New Mexico was along the western shoreline of Pangea nearly at the equator. A warm shallow sea covered southern New Mexico, with a shoreline that fluctuated from near Las Cruces northward for about 35 miles. North of that shoreline, rivers flowed south from the Ancestral Rocky Mountains toward the seaway.

Many plants and animals lived along the seacoast and on inland river floodplains to the north. Conifer forests covered much of the landscape. Various arthropods, including primitive insects (such as beetles and silverfish), spiders, and scorpions searched the sandflats for food. Small, salamander-like amphibians and lizard-like araeoscelid reptiles hunted the arthropods. Even larger predators, especially the pelycosaurs, fed on the amphibians and the smaller reptiles.

The northwestern face of the Robledo Mountains exposes hundreds of feet of limestone-dominated Hueco Group strata that were deposited by the tropical sea that covered much of southern New Mexico during the early Permian.

After the early Permian, no rock record was preserved in the Robledo Mountains until the middle Eocene—a gap (unconformity) of about 230 million years. Beginning about 40 million years ago, volcanism dominated the landscape of southwestern New Mexico. The eruptions of large volcanoes, and the rivers that flowed from them, deposited the Palm Park Formation.

Subsequent explosive eruptions during the Oligocene produced extensive ash-flow tuffs, such as those of the Bell Top Formation, which are well exposed in the Broad Canyon OHV. Lavas and other tuffs of the Uvas Basaltic Andesite were erupted during the Oligocene. By Miocene time, huge alluvial fans filled ancient valleys eroded into the volcanic rocks within a subsiding rift basin in the Broad Canyon area, producing the conglomerates of the Rincon Valley Formation.

Geologic Features

Complex faulting and some igneous intrusions have fractured the Hueco Group strata into many broken blocks of rock that form the rugged canyons and steep ridges of the Robledo Mountains landscape traversed by the OHV and mountain bike trails.

The Broad Canyon Trail is about 9 miles long and takes you through Broad Canyon, which contains boulders of Oligocene igneous rocks up to 3 feet in diameter. The canyon walls feature Oligocene volcanic rocks and, locally, Miocene conglomerates.

—Spencer G. Lucas

If You Plan to Visit

The Robledo Mountains OHV area and SST Trail are in the Robledo Mountains about 4 miles northwest of Las Cruces. From Las Cruces, take North Valley Drive (NM 185) north to Shalem Colony Trail, then go west on Shalem Colony Trail across the Rio Grande, turn right on Rocky Acres Trail (D013). Go approximately 0.25 mile, and then turn left onto the only unpaved road that goes west, Permian Tracks Road. Cross over a cattle guard and continue west to the BLM parking area. Signs for extreme four-wheel drive trails are south of the main road, which goes west toward an abandoned flagstone quarry (the Community Pit). Take the left fork in the arroyo bottom, climb the hill, and go up the canyon to the mountain bike trail and additional extreme four-wheel drive routes.

The Broad Canyon Trail is about 20 miles northwest of Las Cruces. To reach the trailhead, drive north on I–25 from Las Cruces and take Exit 19. Turn left (west) onto NM 157 (Fort Selden Road), then right (northwest) onto NM 185. Travel approximately 6.8 miles and turn left on the dirt road into Broad Canyon.

Bureau of Land Management
Las Cruces District Office
1800 Marquess Street
Las Cruces, NM 88005
(575) 525-4300
www.blm.gov/visit/robledo-mountains-ohv-trail-system

Kilbourne Hole Volcanic Crater National Natural Landmark

BUREAU OF LAND MANAGEMENT

Kilbourne Hole, located in a remote part of southernmost New Mexico, is a stunning example of a maar, a type of volcanic crater formed when hot, basaltic magma interacts explosively with near-surface groundwater during eruption. This volcanic crater is part of the Potrillo volcanic field, on the western margin of the Rio Grande rift, and is unusual in containing material derived, from the upper mantle. Brilliant green peridot (olivine) crystals and other minerals from Kilbourne Hole are of great interest to collectors and geologists alike—the former for jewelry and the latter for a better understanding of the Earth's mantle. Kilbourne Hole is a wonderful place to visit a young, well-exposed maar volcano.

Geography and Geologic Province

Kilbourne Hole is located in the Mesilla Basin, one of a string of broad, sediment-filled rift basins between central Colorado and northern Mexico. Erupted through the basin deposits are the basaltic lavas of the Potrillo volcanic field, which hosted countless small eruptions between

Panoramic aerial view of Kilbourne Hole, a maar volcano crater.

about 1.2 million and 20,000 years ago. The Potrillo volcanic field owes its origin to crustal thinning caused by continental extension in the Rio Grande rift.

The Rock Record

The oldest rock exposed at Kilbourne Hole is the Pliocene–Pleistocene Camp Rice Formation, a sedimentary unit deposited within the Mesilla Basin by local streams and the ancestral Rio Grande. The Camp Rice Formation is exposed as pale pinkish-orange sediments on the lower walls of the crater. Overlying these sediments is a flat-lying, dense, basaltic lava flow, called the Afton Basalt, which was erupted from a small vent northeast of Kilbourne Hole. The Afton Basalt is 90,000 years old. Overlying the Afton Basalt or, in some places, directly above the Camp Rice Formation is a pale-orange chaotic deposit containing large blocks of Afton Basalt. This deposit is an explosion breccia related to the earliest stages of eruption at Kilbourne Hole. The breccia contains mantle xenoliths (literal translation "strange rocks") for which Kilbourne Hole is famous. Overlying the explosion breccia is a sequence of complexly cross-bedded, gray pyroclastic surge deposits (that is, deposits produced by a fluidized mix of gas and rock fragments). These rocks form an outward-sloping apron around the crater. The Kilbourne Hole eruption occurred around 47,000 years ago and probably lasted between 1 and 10 years.

Geological History

Kilbourne Hole is a classic example of a maar crater, formed during a explosive eruption caused by hot basaltic magma interacting with

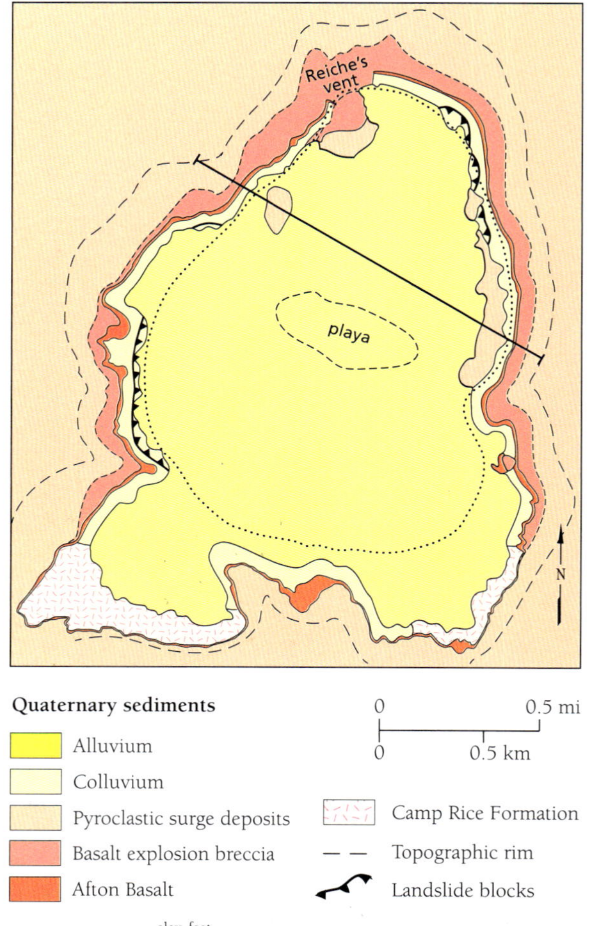

Simplified geologic map and cross section of Kilbourne Hole and surrounding area.

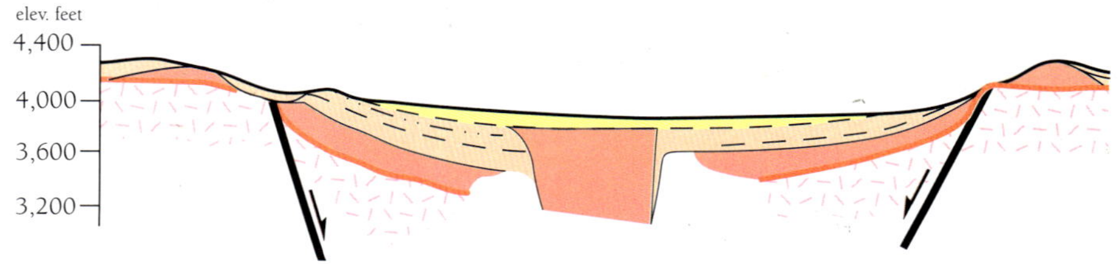

groundwater at depth. The interaction of hot magma with groundwater created a steam explosion that excavated a deep, cone-shaped crater. This "excavation" stage of the eruption resulted in the emplacement of the chaotic explosion breccia described above. The explosion breccia contains a wide range of xenoliths (rock fragments not part of the original magma but rather torn from surrounding rocks and carried from various depths within the Earth's crust and mantle). Most notable of these are the xenoliths known as peridotites, which contain green olivine, black pyroxene, and apple-green chrome diopside. These peridotites were carried upward by magma from more than 30 miles depth in the Earth's mantle. The geochemistry of these xenoliths has been extensively studied to learn more about this inaccessible part of the Earth (the deepest hole ever drilled, for comparison, only reached a depth of 7.5 miles). The green olivine is valued by collectors as the semi-precious gemstone peridot.

Overlying the explosion breccia is a beautiful sequence of complexly bedded gray surge deposits. The surge deposits, which are about 100 feet thick in places, formed from a series of small explosive eruptions, mostly triggered by hot magma interacting with water in the previously excavated vent. Low-density, fast-moving turbulent clouds of debris, composed of mixed fragments of fresh magma, Afton Basalt, and Camp Rice sediments, deposited the cross-bedded surge beds. Some of these beds were deposited from a wet eruption cloud, as shown by the presence of accretionary lapilli, which are pea-sized balls of wet ash with a concentric growth structure similar to hailstones. Also visible are bedding sags caused by the impact of large blocks of Afton Basalt. These blocks were blown high into the sky by the eruption and then fell into the wet surge deposits.

Geologic Features

A short walk from the southern parking lot will take you to the crater rim, where the entire maar crater is visible. Deposits of pale pink-orange Camp Rice Formation low on the crater wall are capped, in places, by the dense Afton Basalt lava flow. Overlying the basalt, explosion breccias are locally present, overlain by a thick blanket of gray surge deposits. To examine the lava, explosion breccias, and surge deposits, hike to the east-northeast around the inside of the crater wall.

Peridotite xenolith erupted from the Kilbourne Hole maar volcano. The main mineral in this xenolith is olivine, which is bright green. The black mineral is pyroxene, and the rare, apple-green phase is chromium diopside.

The large block is a fragment of older Afton Basalt that became caught up in the explosive eruption cloud that formed the Kilbourne Hole surge deposits. These blocks are called "ballistics", because after being thrown upward by the explosive eruption, they would arc back down towards the surface, and would be moving rapidly when they hit the ground. The strong deformation of the surge beds around the ballistic block provides evidence that the volcanic deposit was wet, because dry surge beds typically do not deform as dramatically when impacted.

Crater wall of the Kilbourne Hole maar crater, showing the Camp Rice formation, overlain by Afton basalt, capped by gray pyroclastic surge deposits related to the maar crater eruption.

Accretionary lapilli and bedding sags are present in surge beds. As you continue to the northeast, the explosion breccia becomes thicker, but is difficult to observe because it does not form cliffs but rather is exposed on a gentle slope. After traveling about 1 mile around the crater, you will see green to black peridotite xenoliths, along with xenoliths of other rock types, lying on the crater slopes. Because Kilbourne Hole is a Bureau of Land Management site, you may collect xenoliths or other rocks.

—*Nelia W. Dunbar*

If You Plan to Visit

The rim of Kilbourne Hole can be accessed via northern and southern routes—both require driving on rough, unpaved roads. Stay on roads because of soft sand in some areas. The floor of Kilbourne Hole is private property, but visitors may drive around the rim and hike on the rim and inner slopes. No facilities are available. Rattlesnakes are found in the area, so take care when hiking.

Driving directions for southern rim access: Take I–10 south from Las Cruces or north from El Paso to Exit 155 (Vado Drive). Take NM 227 west 3 miles to NM 28. Take NM 28 south (left) 2 miles to Doña Ana County Road (CR) B008. Take CR B008 west (right) for 11 miles to CR B004 and turn south (left). Drive 6 miles to railroad tracks. Turn left and cross the railroad. Turn west (right) on CR A017 and drive for 7 miles to CR A011. Turn west (right) and proceed 8 miles to Kilbourne Hole. Kilbourne Hole is on the right, past the big tan dirt bank.

Directions to the north rim and Hunt's Hole (a similar crater) are available at the website address given below. For more information:

Bureau of Land Management
Las Cruces District Office
1800 Marquess Street
Las Cruces, NM 88005-3370
(575) 525-4300
www.blm.gov/visit/kilbourne-hole-volcanic-crater

THE PERMIAN BASIN

This region, which includes the Guadalupe Mountains and adjacent areas to the north and east, has spectacular rock outcrops and caves that appeal to a wide range of visitors. The Permian Basin has world-class energy and mineral resources that lie beneath roughly 75,000 square miles of southeastern New Mexico and west Texas These features are the reason we made this region a separate section of the book, despite the fact that it is actually part of the much larger Great Plains province.

The Permian Basin is a huge, complex sedimentary basin that formed about 320–250 million years ago. Its Pennsylvanian and Permian strata now lie mostly beneath the surface. An exception is the Guadalupe Mountains, which expose the northwestern edge of the Delaware Basin, the westernmost part of the greater Permian Basin. The Guadalupe Mountains offer some of the finest exposures in the world of an ancient reef, and other rocks associated with it. As a consequence, this area has been officially adopted as the international standard for rocks of Middle Permian age (termed the Guadalupian Epoch) and has become a pilgrimage site for geologists from around the world.

OPPOSITE: **Pumpjacks and a storage tank in a Permian Basin oil field.**

Landsat satellite image of the Guadalupe Mountains and surrounding areas. The bright red line marks exposures of the Capitan reef at the edge of the Northwest Shelf. Green spots at lower left are irrigated crop circles.

Regional Setting

The Guadalupe Mountains of west Texas and New Mexico represent a transition zone between the block-faulted terrain of the Basin and Range province (or the eastern flank of the Rio Grande rift) to the west and the stable Great Plains to the east and north. As geologic transitions go, it is remarkably sharp. The steep, mile-high western escarpment of the Guadalupe Mountains was formed by numerous Late Cenozoic normal faults that uplifted the range and tilted its strata gently to the east while down-dropping the Salt Flat graben to the west. In contrast, the topography of the eastern Guadalupe Mountains, resulted primarily from erosion that exhumed the seaward margin of the Permian Reef.

Traveling south on US 62/180 from Whites City down to El Capitan, the distinctive peak that marks the southern tip of the Guadalupe Mountains,

El Capitan, the distinctive peak at the southern tip of the Guadalupe Mountains. The rounded peak to the right is Guadalupe Peak, the highest point in Texas (8,751 feet). Sandstones of the Brushy Canyon and Cherry Canyon formations, seen in the foreground, were deposited in the deep Delaware basin. The lighter colored rocks of the main peaks are dolomitic limestones of the Capitan Formation along the basin margin.

you are essentially passing through a 250 million-year-old landscape. The road traverses and steps down through a series of progressively older seafloor deposits of the Delaware Basin. The massive upper parts of the cliffs to the west mark the Capitan reef, and the slopes below are those down which reef debris tumbled and slid into the basin. The flat-lying rocks at the very top of the plateau, and in the canyons that dissect the mountain front, expose the backreef deposits of the Northwest Shelf (a shelf is part of the edge of a continent submerged by relatively shallow ocean water). The whole spectrum is remarkably unaltered from Permian time—as though someone recently drained the water away leaving a stranded ancient landscape behind. However, it was the relatively recent erosion and dissolution of highly soluble, post-reef evaporite beds that created this illusion.

Although the parks in the Guadalupe Mountains proper (Carlsbad Caverns National Park, Guadalupe Mountains Scenic Byway, and Sitting Bull Falls Recreation Area) provide spectacular outcrops, other localities in and near the Pecos Valley, especially Bottomless Lakes State Park, provide additional excellent exposures of Permian rocks in the region. As seen on the geologic map, there are few outcrops of Permian rock in the area east of the incised Pecos River valley because they are covered with younger Mesozoic and Cenozoic deposits of the High Plains. These relatively thin fluvial, alluvial, and eolian deposits, described in more detail in Part 6 of this book, conceal vast underlying Permian strata. These buried units are among the thickest accumulations of Permian sedimentary deposits in the world (locally up to 17,000 feet thick).

Geologic History

The region has a sedimentary rock record, documented by drill holes, that extends back to early Paleozoic time. Thick Ordovician to Early Pennsylvanian limestones, dolomites, sandstones, and shales reflect primarily marine deposition in a stable continental margin setting. Subsequent geologic features of the Permian Basin, now mainly concealed under High Plains deposits, were formed during the Ancestral Rocky Mountains deformation, a mountain-building and basin-forming event related to the collision of South and North America. This slow collision and mountain building created part of the supercontinent of Pangea, roughly 320 to 270 million years ago.

Most of the Paleozoic strata exposed in south-eastern New Mexico are of Middle and Late Permian age, about 270–250 million years old. The Middle Permian was a time of extensive carbonate reef and bank growth. Late Permian time saw the precipitation of thick evaporites (mainly gypsum and halite) that filled and capped remaining basins. Those deposits mark the end of the long history of Paleozoic marine sedimentation in this area and

Reconstruction showing the collision zone between North America and South America in Late Pennsylvanian time (ca. 300 million years ago). The collision zone is marked by a dashed line. The location of New Mexico is shown. Major uplifts of the Ancestral Rocky Mountains lie just to the north.

ERA	PERIOD	EPOCH	NORTHWEST SHELF			DELAWARE BASIN	
PALEOZOIC	Late	Lopingian	Rustler Formation				
			Salado Formation				
						Castile Formation	
	Middle Permian	Guadalupian	Artesia Group	Tansill Formation	Capitan Reef*	Bell Canyon Formation	Delaware Mountain Group
				Yates Formation			
				Seven Rivers Formation		Cherry Canyon Formation	
				Queen Formation			
				Grayburg Formation			
			San Andres Formation			Brushy Canyon Formation	
	Early	Cisuralian	Yeso	Victorio Peak		Cutoff Formation / Bone Spring Formation	
			Abo Formation			Hueco Group	
	Older Paleozoic strata						

Stratigraphic column for the Guadalupe Mountains area. The white areas on the diagram represent missing section due to non-deposition or erosion.

*lower portion of reef section includes Goat Seep Reef.

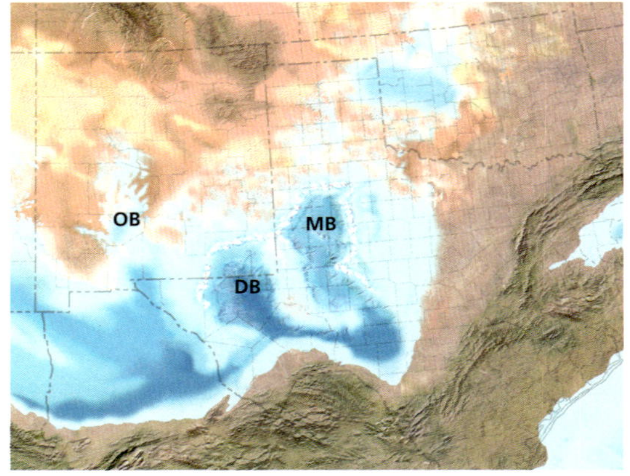

288

also reflect the progressively increasing aridity that marked Permian time in this area.

Thin Mesozoic and Cenozoic deposits subsequently blanketed the area during a long period of tectonic stability and slow, mostly non-marine sedimentation. Tectonic stability ended in middle Cenozoic time (about 25 million years ago), when the region began to experience periods of uplift, erosion, and extensional deformation in its western parts. Faulting uplifted the Guadalupe Mountains and led to development of spectacular caves, including Carlsbad Cavern.

Reconstruction of the southwestern United States about 278 million years ago. The Midland Basin (MB), Delaware Basin (DB) and Orogrande Basin (OB) were the main depositional centers at that time.

A comparison of the Permian Basin region in early Middle Permian (top) and late Middle Permian time (bottom) showing the filling of the Midland Basin between those two intervals. The Delaware Basin remained but was filled during the Late Permian. Hachures mark areas of present-day exposures of the Permian reef and associated rocks.

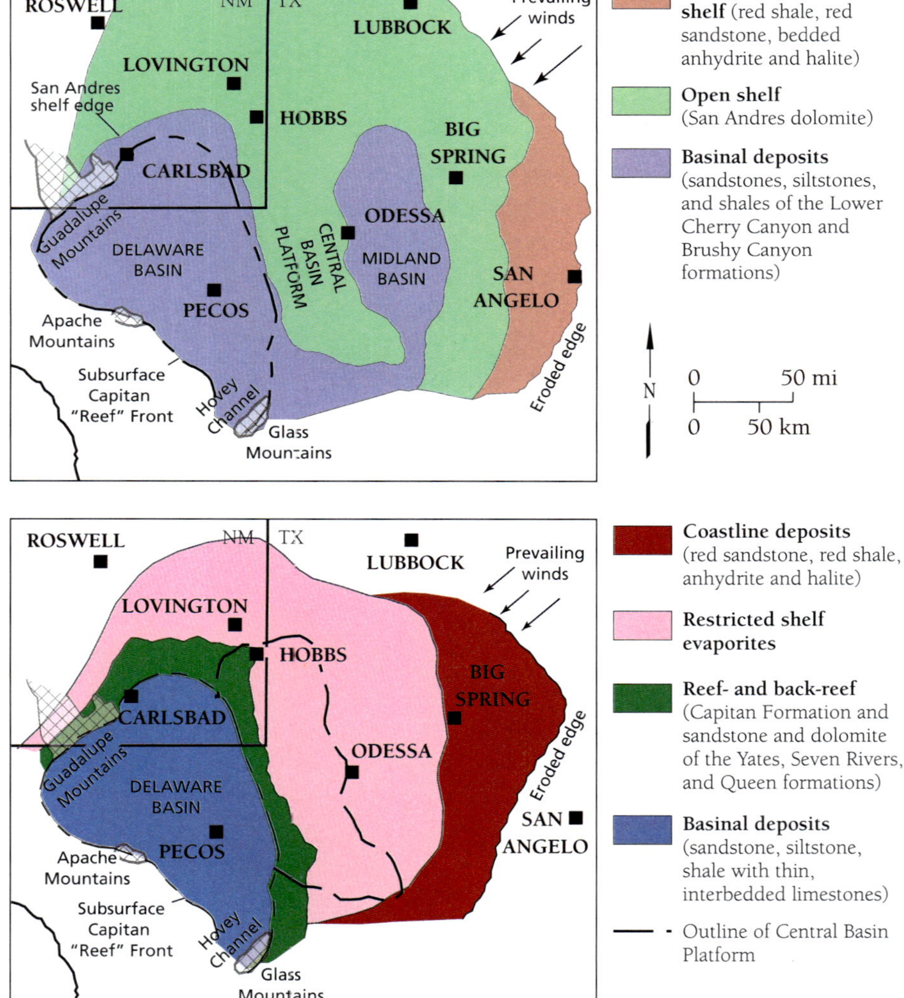

The Rock Record

The Middle Permian rocks of the Guadalupe Mountains and adjacent areas were deposited during the last stage of Ancestral Rockies deposition. But what a spectacular ending it was! Two large and deep basins existed in the region about 270 million years ago, at the start of the Middle Permian—the Midland Basin of west Texas and the larger Delaware Basin of west Texas and New Mexico. By the end of Middle Permian time (about 260 million years ago), only the Delaware Basin remained active. By the close of Permian sedimentation (about 250 million years ago), both basins were filled to the brim.

Most of the strata exposed in the Guadalupe Mountains were deposited landward of the Capitan Reef on a continental shelf so broad that at times it extended all the way to northern New Mexico and was repeatedly flooded by marine waters during Middle Permian time. Those intervals of flooding, however, were interspersed with repeated episodes of sea-level fall and shelf exposure. The sea-level variations

Idealized cross section of the Guadalupe Mountains and adjacent Delaware Basin. Cisuralian (Lower Permian) rocks are exposed mainly on the west side of the mountains. More widespread Guadalupian (Middle Permian) units include predominantly basinal sandstones and siltstones (Delaware Mountain Group); shelf dolomites, evaporites and redbeds (San Andres Formation and Artesia Group); and reef, forereef-slope and basinal limestones and dolomites (Capitan Reef). The Lopingian units (Castille and Salado formations) are Upper Permian evaporites that filled and overtopped the basin.

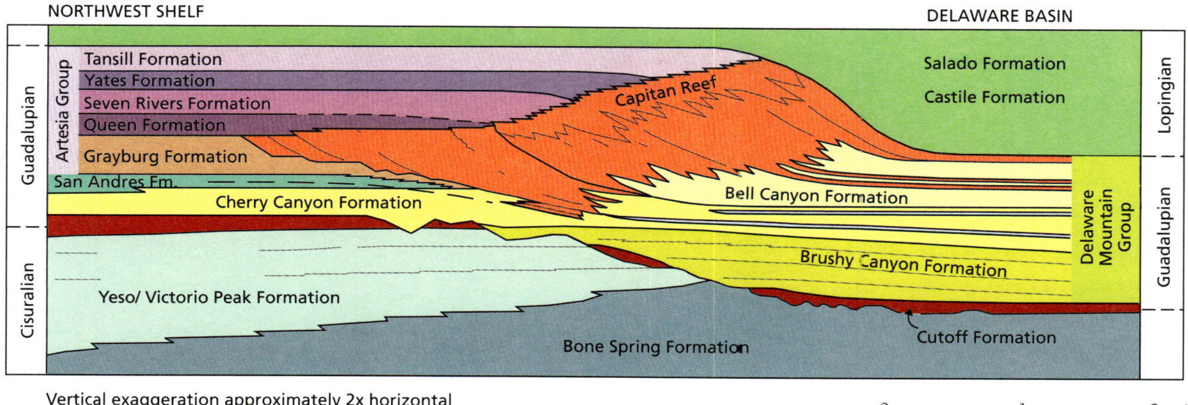

Vertical exaggeration approximately 2x horizontal

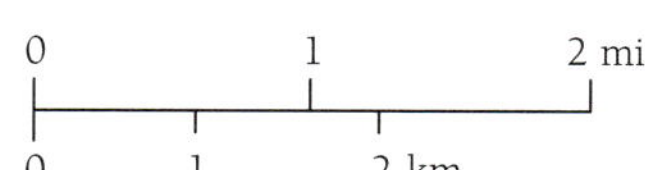

were driven by waxing and waning glacial ice on the southern part of Pangea, during the last part of one of the great "icehouse" intervals in geologic history (Late Mississippian to Middle Permian time).

During warm interglacial periods, when sea levels were high, the Northwest Shelf hosted islands and lagoons, as well as reefs and backreef shoals at the shelf margin. Shelf deposits formed during these high-stand intervals consist largely of limestone and dolomite (seen at almost all the parks covered in this part of the book), with additional gypsum and anhydrite evaporite beds that formed in hypersaline lagoons and in coastal areas (best seen at Bottomless Lakes State Park). During oft-repeated times of low sea level (glacial intervals), the shelf areas were drained of marine waters. During those times, dunes and

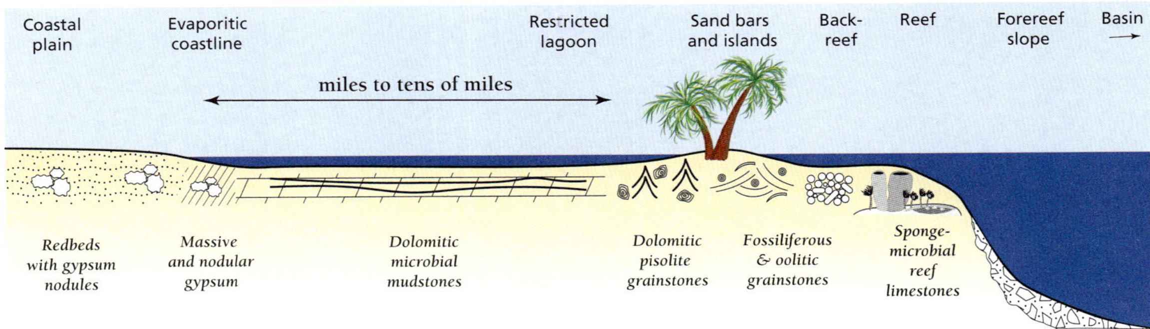

290

Idealized view of environments of highstand sedimentation on the Northwest Shelf and adjacent Delaware Basin. These patterns persisted through much of the Middle Permian, although the width of this spectrum varied from just a few miles to several tens of miles at various times.

sheets of wind-blown sand extended all the way to the shelf edge, and some of the sand and silt was blown and/or washed into the basin.

The Delaware and Midland basins, which typically had water depths of 1,000 to 2,000 feet, accumulated varied materials washed or blown off the shelves. These consisted of quartz-feldspar sands during sea-level lowstands and reworked carbonate sediments during highstands. A major change in that style of deposition occurred at the end of Middle Permian time as the climate grew warmer and more arid. As the North American part of Pangea drifted northward during Early and Middle Permian time, glaciation waned globally and the rising Appalachian and Ouachita mountains formed rain shadows to the east. In addition, ocean water inflow to the Delaware Basin may have become more restricted over time. As a result, the youngest deposits in the basin consist of more than 3,000 feet of Upper Permian evaporites (mostly gypsum and halite) that were associated with an abrupt end to the growth of the Capitan Reef.

Geologic Features

This area has an astounding variety of geologic features and rock types, and most fit into a well-defined pattern that existed through most of Middle Permian time. The main "factory" for sediment formation was at or near the shelf margins where warm waters, strong wave action, and the availability of nutrients allowed growth of high-productivity reefs and banks. The biota in Permian reefs was quite different from modern ones, in part because scleractinian corals (the main framework of modern reefs) did not yet exist. Instead, calcareous sponges, bryozoans, green and red algae, crinoids, mollusks,

The interlayers of organic carbon-rich calcium carbonate (dark) and gypsum (light) in the Upper Permian basin-filling Castile Formation. Each pair of layers is interpreted to be an annual deposit. Finger tip at lower left corner provides scale.

brachiopods, and microbes predominated in the Permian reefs. These organisms were cemented together to form huge volumes of limestone. Some of that limestone was retained in the reef environment, but most of it tumbled, slid or was otherwise transported downslope, contributing to a gradual filling of the basins and expansion of the shelves. During Capitan Reef deposition, the position of the shelf edge shifted laterally between 2 to 10 miles basinward, depending on location. The basinward migration of the shelf edge can be seen especially well in Walnut and Slaughter canyons in Carlsbad Caverns National Park and in McKittrick Canyon in Guadalupe Mountains National Park.

During times of high sea level, the Northwest Shelf hosted a spectrum of depositional environments landward of the Capitan Reef. This spectrum, extending from reef to mainland shoreline, included

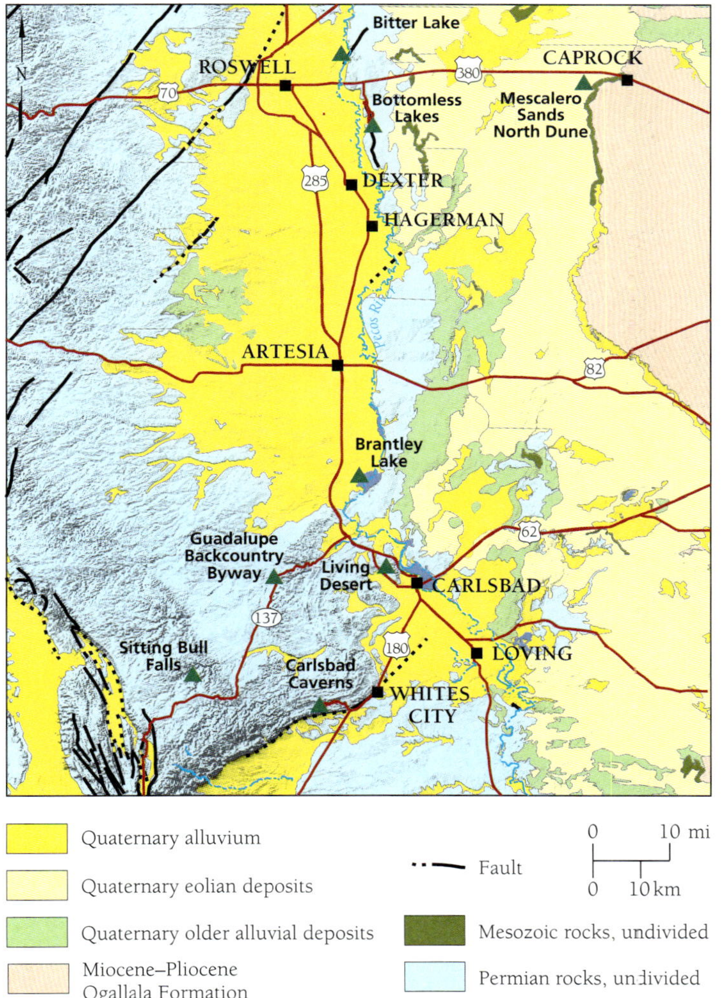

Generalized geologic map of the Guadalupe Mountains and southern Great Plains area of southeastern New Mexico. Parklands discussed in this book are marked with green triangles.

submerged banks and barrier islands, muddy carbonate lagoons, coastal areas with saline lakes (salinas), tidal flats (sabkhas), and coastal plain redbeds. Middle Permian back-reef deposits (the back reef is the part closest to shore) are well exposed at several localities covered in this book, including Walnut Canyon in Carlsbad Caverns National Park, Bottomless Lakes State Park, and the Guadalupe Back Country Byway.

The thick, Late Permian evaporite deposits of the Castile and Salado formations that formed the late-stage filling of the Delaware Basin are, in themselves, unusual and interesting geologic features. The exposed Castile deposits, in particular, show more than 175,000 thin, paired layers of calcium carbonate (calcite) and calcium sulfate (gypsum/anhydrite) that have been interpreted as annual deposits (varves) reflecting seasonal variations in water salinity. If those cycles are indeed annual, then the roughly 1,500 feet of Castile evaporite formed in about 175,000 years, very rapid deposition by geological standards. The overlying Salado Formation reflects even more arid conditions and consists of rapidly deposited cyclic deposits of halite (salt) and various potash minerals.

Because evaporite minerals are highly soluble in water, they are preserved in surficial deposits only in arid areas such as southeastern New Mexico. Although the varved Castile beds are not exposed in any of the public lands covered in this book, excellent roadside exposures are found between New Mexico mile markers 2 and 6 on US 62/180 (south of Whites City and just north of the Texas border).

Drilling rig used for oil and gas exploration in the Delaware Basin east of Carlsbad.

Geologic Resources of the Permian Basin—Oil and Gas

A total of 6.2 billion barrels of oil has been produced in New Mexico since the first oil discovery in 1924 and approximately 95 percent of this oil came from the Permian Basin. Annual New Mexico oil production peaked at 129 million barrels in 1969 and then declined to 75 million barrels by 1980. From 1980 until 2008, the decline became more gradual as a combination of new discoveries, new technologies, and increased use of enhanced recovery techniques brought additional oil into production.

By 2008, oil production in New Mexico began to increase. This increase, slow at first but then reaching gains of about 20 percent per year between 2012 and 2015, was due to the employment of horizontal

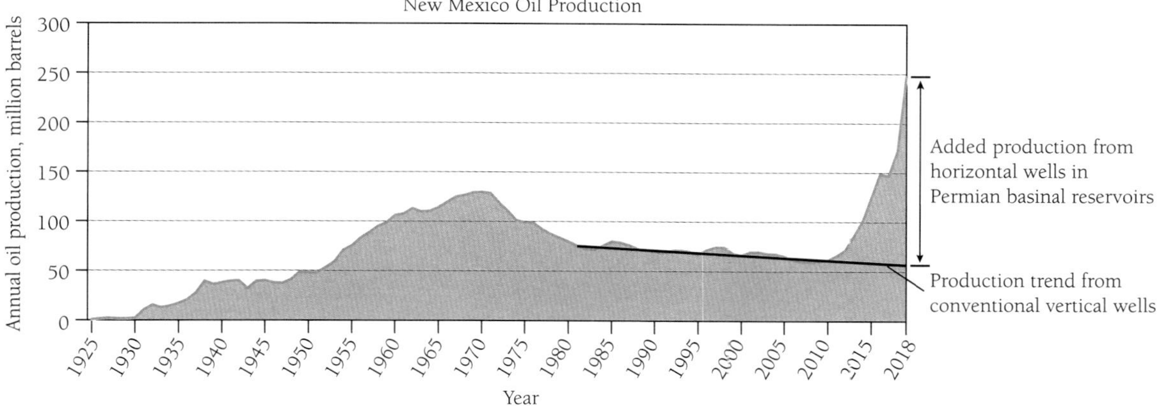

drilling. A "horizontal well" is drilled vertically downward and then is curved into the target zone until it is horizontal. The well is then drilled laterally within the oil-bearing layer for distances up to two miles. In most cases, the reservoir rock is then hydraulically fractured to increase the flow rate of oil into the well. The increased flow allows for production of oil from low permeability (or "tight") shale reservoirs that would otherwise not have been commercially viable.

The resultant dramatic increase in New Mexico oil production reached 146.8 million barrels in 2015, with Permian-age reservoirs in southeastern New Mexico contributing 124.5 million of that total. In addition, a remarkable 479 million barrels of oil were produced in 2015 from the adjacent Texas part of the Permian Basin.

The increased production of New Mexico's Permian Basin oil in the past decade came mostly from basinal siltstone, shale, and limestone reservoirs, typically at depths of 8,000 to 12,000 feet, rather than from traditional shelf deposits. For comparison, in 2000, 78 percent of Permian Basin production from Permian age reservoirs came from shelf and shelf-margin strata with 22 percent from basinal units; by 2015 those numbers had reversed, with only 30 percent of production from shelf settings and 70 percent from basinal ones.

Although oil production dominates, the New Mexico part of the Permian Basin also produced 558 billion cubic feet of natural gas during 2015, 36 percent of all natural gas produced in New Mexico. Historically, 60 percent of natural gas production in the New Mexico part of the Permian Basin was obtained from Permian reservoirs, and this percentage has increased in recent years as co-produced gas from Permian basinal oil reservoirs has increased. Older and more deeply buried units continue to produce gas, the deepest coming from Silurian carbonates at 17,750 foot depth. The oldest gas-producing reservoirs are in Ordovician strata of the Ellenburger Formation.

Overall, despite the recent huge increases in oil production in other areas, the Permian Basin of west Texas and southeastern New Mexico remains the single most productive oil province in the United States. It is currently responsible for 20 percent of total U.S. oil production. Recent estimates of undiscovered oil and gas in basinal shales indicate that remaining reserves in this basin are enormous—among the largest in the world.

Geologic Resources of the Permian Basin—Potash

Upper Permian potash deposits in the Salado Formation constitute another major resource in the New Mexico portion of the Permian Basin. Potash is the common industrial term for soluble minerals in which potassium is combined with other elements. Potassium is the third most widely used fertilizer nutrient after nitrogen and phosphorus, and agricultural use accounts for more than 90 percent of total potash consumption. Potash has been produced commercially in southeastern New Mexico since 1931, primarily through the mining of two minerals—sylvite (potassium chloride) and langbeinite (a potassium-magnesium sulfate). The Carlsbad district is the largest potash producing area in the U.S., and New Mexico is the top-ranked state in national potash production with sales of more than 200 million dollars in 2018.

Intrepid Mining LLC and Mosaic Co. operate mines in the Carlsbad district with production from depths of 800–1,500 feet. The estimated potash reserves in the district are more than 550 million short tons. Potash is also being produced by solution mining at the HB Solar Solution mine. There, water is injected into the old underground workings (formerly the Eddy Mine) and potash-rich solutions are extracted. The HB Solar Solution Mine has 5 million short tons of proven and probable reserves and an estimated mine life of 28 years.

Upper Permian salt deposits (halite, or sodium chloride) of the Delaware Basin also host the nation's only deep geologic repository for nuclear waste at the Waste Isolation Pilot Plant (WIPP) site.

A "continuous miner" at the Intrepid underground potash mine. The broken ore drops into the hopper below and is then pulled onto a conveyor belt. The reddish potash ore in the walls and ceiling show tool marks left by the cutter.

The underground facility stores transuranic waste from the nation's nuclear defense program. These thick and stable salt deposits are well suited for disposal of such waste because the pressure of more than 2,000 feet of overlying rock causes the salt to flow plastically, thereby sealing the waste from surface contact. Operated by the U.S. Department of Energy, the facility is not open to the general public, but an exhibit at their Carlsbad office building (located at 4021 National Parks Highway) features a brief documentary on the project and numerous displays.

—Peter A. Scholle, Ron Broadhead, and Virginia McLemore

Additional Reading

Classic Upper Paleozoic reefs and bioherms of west Texas and New Mexico: Field guide to the Guadalupe and Sacramento Mountains of west Texas and New Mexico, P.A. Scholle, R.H. Goldstein, and D.S. Ulmer-Scholle, Open File Report 504, New Mexico Bureau of Geology and Mineral Resources, 2007.

New Mexico potash—past, present, and future, J.M. Barker and I. Gundiler. New Mexico Earth Matters, New Mexico Bureau of Geology and Mineral Resources, 2008.

Carlsbad Caverns National Park

NATIONAL PARK SERVICE

Carlsbad Caverns National Park (CCNP) is located in southeastern New Mexico about 20 miles south of Carlsbad. It was initially designated a national monument in 1923, was elevated to a national park in 1930, and was recognized by United Nations Educational, Scientific and Cultural Organization as a World Heritage Site in 1995. Its 46,766 acres include 120 known caves, the two largest of which are Carlsbad Cavern and Lechuguilla Cave, with total passage lengths of 32 miles and 143 miles, respectively, which places them among the world's longest. Lechuguilla (open only to experienced researchers) is the second deepest limestone cave in the United States at 1,604 feet. More importantly than cave size is the great variety and beauty of the formations and the complexity of the processes that formed them.

In conjunction with the nearby Guadalupe Mountains National Park in west Texas, CCNP also preserves access to some of the world's finest exposures of Middle Permian rocks. The Capitan reef, its landward equivalents on the Northwest Shelf, and its deeper water equivalents in the Delaware Basin (described in the introduction to this part of the book) can all be seen in their original context despite more than 250 million years of subsequent burial and erosion. It is impossible to describe all the features of this area in the space available, so we have focused on four especially significant areas: Walnut Canyon, which hosts the entry road to Carlsbad Cavern and exposes rocks formed in reef, near back-reef, and shelf environments; Carlsbad Cavern, which is the most visited attraction in the park; Slaughter Canyon, which shows a spectacular cross section of rocks from deep forereef to shallow shelf environments; and Rattlesnake Springs, located in a basinal setting with a lush flora and diverse fauna supported by groundwater sourced from the Guadalupe Mountains.

It is also worth inquiring at the Visitor Center for availability of other cave tours or hikes. Such tours, offered to small groups and led by Park Service guides, are sometimes available for otherwise off-limits parts of Carlsbad Cavern, as well as Spider Cave and Slaughter Canyon Cave.

OPPOSITE: **Stalagmites (which grow upward) and stalactites (which hang down like icicles) in Carlsbad Cavern.**

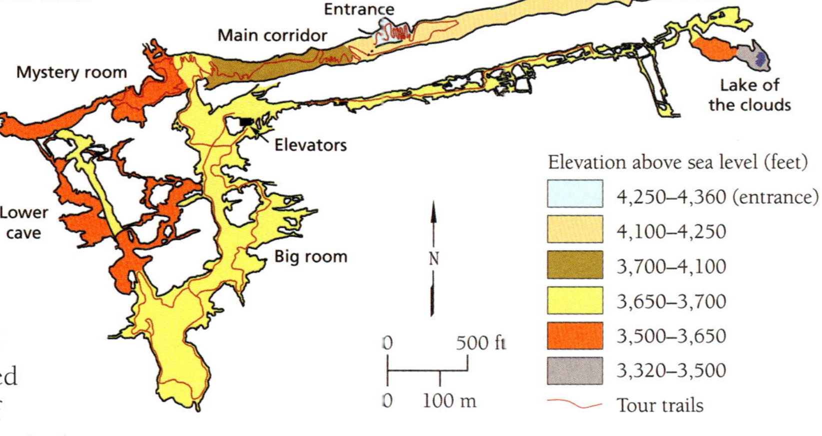

Modified map of Carlsbad Cavern showing the elevation. Note the rectilinear pattern of cave rooms and passages that reflect the influence of regional fracture patterns on water flow and limestone dissolution.

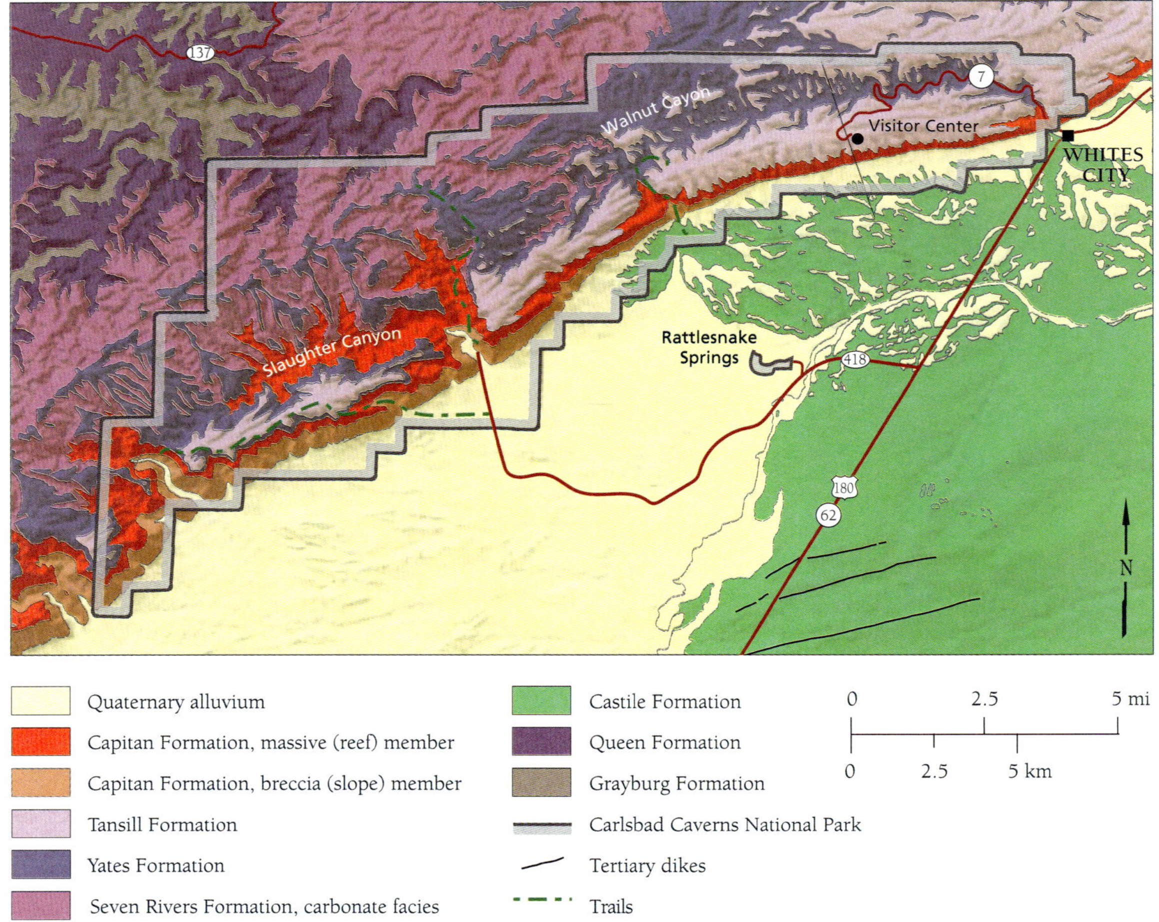

Geologic map of Carlsbad Caverns National Park and surrounding areas.

The Rock Record

All the exposed strata in the park, other than Quaternary alluvium, are of Middle Permian age and were formed in platform-interior to platform-margin settings associated with the Northwest Shelf and its transition to the Delaware Basin. The Capitan Formation (or Capitan Limestone) is the formal name applied to the reef (massive limestones and dolomites) and forereef formed on the ocean-ward side of the reef deposits (steeply dipping brecciated and bedded limestones and dolomites). The Capitan reef and slope units, shown on the accompanying geologic map and cross section, form both the eastern escarpment of the Guadalupe Mountains and the eastern edge of the main part of this park. The forereef rocks are interlayered downslope (to the southeast) with basinal rocks of the Bell Canyon Formation. Basinal rocks are visible only in a few places in CCNP (primarily in Slaughter and Rattlesnake canyons) but are better exposed in and near Guadalupe Mountains National Park to the south.

The strata that lie immediately to the northwest of the Capitan reef (in particular, the Tansill, Yates, and Seven Rivers formations) all represent back-reef and shelf units deposited in a variety of environments including banks, shoals, islands, lagoons, and coastal plains (see diagram in the prior overview chapter). Those environmental patterns persisted during dozens of sea-level highstands when the shelf was submerged under shallow marine waters. The highstand rocks are well bedded and include carbonate (mainly dolomite), evaporite (mainly gypsum) and, far to the west, coastal redbeds (evaporitic siltstones). They alternate with thinner, but very widespread, yellow-brown siltstones and sandstones that reflect interspersed lowstand periods when wind-blown silt and sand dunes spread across the then-dry shelf.

Cenozoic uplift and gentle eastward tilting of the Guadalupe Mountains, coupled with subsequent erosion, has resulted in a situation in which progressively older rocks are exposed toward the west across the shelf (see cross section). The Tansill Formation, the youngest unit in the Artesia Group, is found only on the easternmost part of the platform (including at the top of the section in Walnut and Slaughter Canyons and at Carlsbad Cavern). The Yates Formation is found throughout much of the platform area in the park, and Seven Rivers strata are widespread in the western parts of CCNP. Still farther west, mainly outside CCNP but well exposed along the Guadalupe Back Country Byway and along the road into Sitting Bull Falls, one finds the lower parts of the Artesia Group—the Queen and Grayburg formations.

Elongate, chambered, calcareous sponges in the Capitan reef. The dark material surrounding them consists of microbial crusts and marine cement. The coin shown for scale is about an inch in diameter.

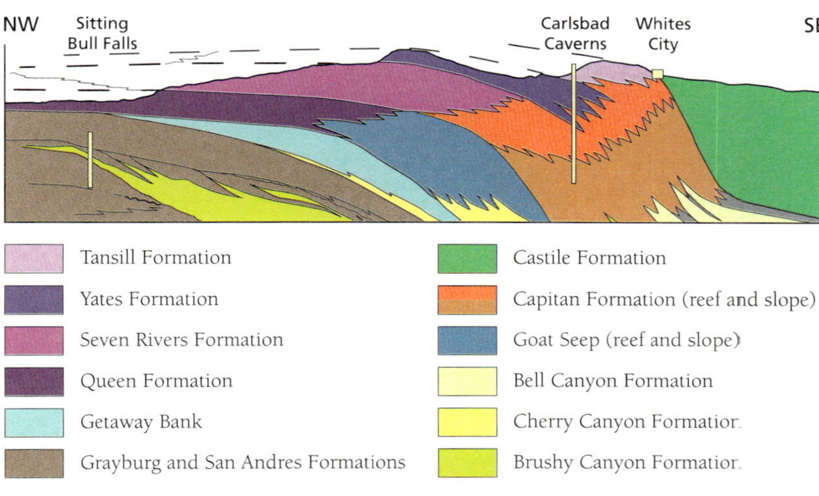

Tansill Formation

Yates Formation

Seven Rivers Formation

Queen Formation

Getaway Bank

Grayburg and San Andres Formations

Castile Formation

Capitan Formation (reef and slope)

Goat Seep (reef and slope)

Bell Canyon Formation

Cherry Canyon Formation

Brushy Canyon Formation

Generalized geologic cross section of Permian-aged strata based on well logs, drawn just north of Carlsbad Caverns National Park (Dark Canyon Arroyo). The vertical range is about 2,200 feet and the horizontal distance is approximately 20 miles. Yellow bars represent the exposed rock record.

Walnut Canyon (main park entrance road)

The area near the entrance gate to the park is the best place to see the Capitan reef and the transition from reef to immediate back-reef settings. Unfortunately, the finest exposures of the reef lie just outside the park, on private lands. You should get permission for access (inquire in Whites City) and some fees may be charged. Please remember that collecting of fossils or rock samples is not permitted in the National Park or on these private lands.

The main area of reef exposure is on a large spur of rock south of the park entrance road, a few hundred feet west of Whites City, between Walnut Canyon and the next small valley to the south (Bat Cave Draw). Fore-reef slope deposits are not visible here as they are covered by Upper Permian Salado and Castile basin-filling evaporites. Farther south along the mountain front, these evaporites have been dissolved and the fore-reef deposits are more prominent.

The structures of the reef here are obscured by weathering. The key to recognizing the reef is to find areas where the rocks have been acid-etched by geologists. Along the eastern side of the Walnut Canyon spur you can see a variety of platy and finger-like sponges, phylloid (leaf-like) algae, clusters of a bright-white, extinct organism of uncertain affinities (*Shamovella/Tubiphytes*), a few solitary horn corals, and abundant microbial crusts (*Archaeolithoporella*). These organisms were the main reef builders of that time.

Within the reef framework one can see cavities filled with carbonate mud (generally light gray) and abundant marine cement (dark brown to nearly black). This cement is calcium carbonate (mainly aragonite, now calcite) that precipitated in cavities and on the seafloor simultaneously with reef growth. In places, more than 50–70 percent of the rock consists of such cement, and its formation during reef growth helped to strengthen and harden the reef sediment into rock at the seafloor. Because cementation happened primarily in shallow waters, deeper sediment remained soft. This allowed settling and compaction that led to

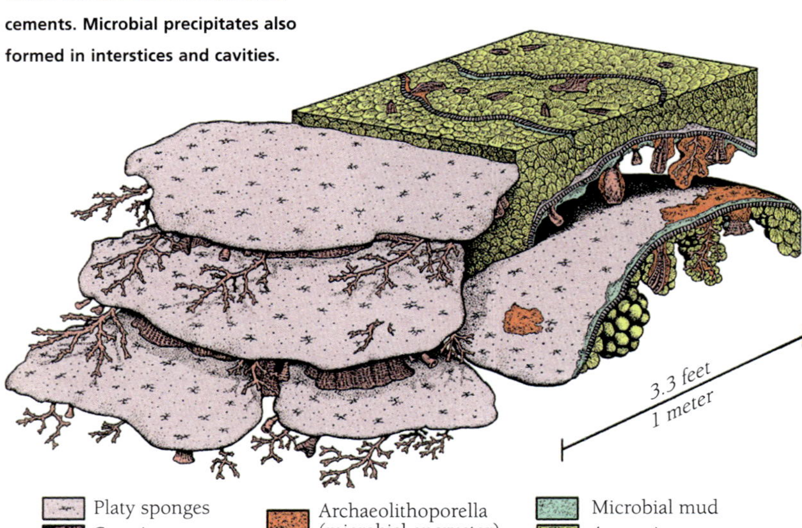

A generalized representation of organisms within the Capitan reef showing large, platy sponges forming the main structural framework. Smaller cryptic (cavity-dwelling) organisms colonized protected areas within the framework and marine cements. Microbial precipitates also formed in interstices and cavities.

early seafloor fracturing of the reef. This locality is an excellent place to see such linear fractures, some tens of feet long, which generally trend parallel to the reef front. Those open fractures too were filled with marine cements in most places.

It is worthwhile to walk southward across the nose of the spur, and then westward a few hundred feet up Bat Cave Draw to an area of large, smooth, steep rock exposures. The slope leading to the crest of the spur has wonderful exposures rich in sponge remains, including some large platy forms up to six feet across. With care, one can see cavities that existed below these sponges and a rich fauna of organisms, including other sponges, that lived in those sheltered sites (called "cryptic fauna"). The largest of these cavities are several feet in diameter and are filled with multiple generations of marine cement, with radiating clusters ("sprays") of long, thin crystals.

Returning to Walnut Canyon, one can walk up-canyon for a few hundred yards, now within CCNP. The gradual change in the rocks from massive to bedded reflects the transition from Capitan reef to grainy Tansill near-back-reef deposits. The back-reef strata are composed of small round grains of green algae, gastropods (snails), bivalves (clams), fusulinid foraminifera (protozoans) and other organisms along with coated grains (ooids and pisoids). The higher cliffs on the north side of the road expose mainly the Tansill Formation. Three thin sand-stones at the cliff base mark the contact with the underlying Yates Formation. Those sandstones are sometimes covered by stream sediment near the mouth of the canyon but are always visible in roadside outcrops as one drives up-canyon to the Caverns Visitor Center.

On the twisting seven-mile drive, the road turns and parallels the reef front, so one stays in rocks that were formed in roughly the same depositional environments. These envi-ronments include near-back-reef subtidal areas and islands/shoals. Excellent examples of shallow-water deposits are visible at many of the turnouts along the road, and some of the best examples of pisoids are found near the Hairpin Turn about 4.8 miles up the road from the park entrance. Pisoids are grains ranging in size from BBs to golf balls with concentric coatings of precipitated calcium carbonate. These beds are clearly related to shoals and islands that separated the biologically diverse reef and near-back-reef areas to the east from the biologically barren, hypersaline lagoons farther west on the shelf. Despite understanding that these grains formed in a shelf-crest setting, geologists still disagree about the exact origin of the pisoids. Some think that they formed as spring deposits or caliche crusts (lithified soil) on

A large former cavity (about 3 feet across) in the Capitan reef, filled with several generations of early marine cement (initial dark cement followed by a more coarsely crystalline later stage).

An example of a pisolite deposit near the uppermost part of the Yates Formation. Pisoids are defined as coated grains larger than 0.08 inches (2 mm) in diameter. In this example, the originally aragonitic grains and cements have been replaced by dolomite. Here, the largest grain is approximately the size of a U.S. nickel.

exposed barrier islands; others see these grains as current-formed or microbially-generated nodules in shallow-water settings. Pisoid-bearing rocks, and their associated "tepee structures" (layers buckled up by growth of marine cements), can also be seen in the upper parking lot at the Cavern Visitor Center and at the entrance amphitheater of the Cavern. Additional outcrops of mid-shelf Tansill and Yates rocks can be seen along the Walnut Canyon Desert Drive that loops from near the Visitor Center to the Hairpin Turn on the park entry road.

On a clear day, the Visitor Center area offers a superb view to the southwest along the Capitan reef front all the way to El Capitan. The flat country to the southeast is underlain by exhumed deposits of the deep Delaware Basin. The elevation difference between the Cavern parking lot and the valley floor to the southeast approximates the Middle Permian water depth in the Delaware Basin.

Steep descent through the natural entrance passage of Carlsbad Cavern.

Carlsbad Cavern

With their great expanse of limestone, the Guadalupe Mountains seem ideal for cave formation. But this process requires acidic water that is able to dissolve carbonate rock, and in this dry climate the water seeping into the ground has little capacity for dissolution. The cave's natural entrance is located in an odd position—at the top of a ridge. The shallow gully leading into the entrance seems inadequate to deliver the huge amount of water needed to form the cave. The answer to this puzzle becomes clear during the cave tour.

As one enters the cavern through the Tansill Formation, the trail descends from the entrance and leads to a vast corridor that stretches out of sight in both directions. The aromatic Bat Cave branches left, but the tour turns right, into the Main Corridor. Its floor gradually drops until the ceiling looms more than 200 feet above. The lower walls are composed of the massive Capitan Formation with layered shelf rocks (Yates Formation) visible in the upper walls. This passage opens into an astonishing gulf. The winding trail descends gently, but the original explorers, more than a century ago, had to drop straight down the steep slope by hanging on to fencing wire.

There is no evidence that descending water might have formed the cave—no channels carved by flowing water; no rounded boulders carried by streams. Could the water have risen from below and exited to the surface through springs above the current natural entrance? If so, what was its source? More about this later.

The trail passes around a giant iceberg-shaped rock, then climbs
into the Lunch Room, which offers food, rest rooms, and elevators for
eventually returning to the surface. Beyond stretches the Big Room,
America's largest cave room and the second largest in the northern
hemisphere. The trail rambles around its perimeter for about a mile.
The floor is fairly flat, with only a few local high points. This chamber
was formed mainly in the reef and fore-reef rock of the Capitan
Formation. At the western end of the Big Room is a 90-foot overlook
into the Lower Cave. Beyond, the arched ceiling rises to nearly 300 feet
high. At the southern end of the Big Room is "Bottomless Pit" (the
bottom is 130 feet below). This is one of several pits that descend far
below the main levels, but they simply terminate in fissures too narrow
to explore.

SPELEOTHEMS—A major attraction of the cave is its variety of spectacular
speleothems ("cave formations"). Most were formed after the cave
ceased enlarging. They include stalactites that hang from the ceiling
and stalagmites—both formed by dripping water that traveled from
the surface through pores and fissures in the rock. Near the surface,
groundwater acquires carbon dioxide gas from the soil and forms weak

The narrow northern end of the
Big Room. Note the arched cross
section (approximately 50 feet high),
relatively flat floor, and speleothems.

carbonic acid, which allows it to dissolve limestone. But when groundwater enters the cave, where the carbon dioxide gas concentration is lower, much of the gas escapes, causing limestone to be deposited as calcite speleothems. Rims of calcite form around the shores of pools. In drier areas, where there is seepage through pores, speleothems are deposited by a film of evaporating water. These have no gravitational orientation, but instead form knobs, erratic fingers, or needle-like crystals of aragonite. Microbial activity can also influence speleothem growth. In places, they form web-shaped or irregular finger-like deposits of calcite (biothems).

CLUES TO CAVE ORIGIN—The smooth walls and arched ceilings give the impression that acid-rich water once filled the entire cave, dissolving all surfaces simultaneously. Appearances can be deceptive! On the floor of the Big Room are large blocks of gypsum, clearly visible just beyond the overlook into Lower Cave as white rocks with vertical holes dissolved by dripping water. The gypsum blocks contain ghost-like shapes that mimic the textures in the limestone, including the outlines of fossils. This gypsum has replaced some of the limestone cave walls and thus formed at the same time and rate that the limestone was dissolving. This is the work of sulfuric acid, which forms in water when hydrogen sulfide (the "rotten egg" gas, H_2S) reacts with oxygen.

A rift in Left Hand Tunnel, which originally delivered water and hydrogen sulfide to the main cave level (depth approximately 50 feet).

Most hydrogen sulfide forms deep beneath the surface where gypsum reacts with organic compounds such as petroleum. Both are abundant in this region. This gas is extremely soluble and is carried in solution by flowing groundwater. Where hydrogen sulfide encounters oxygen, the two gases react to form sulfuric acid, which easily dissolves limestone and forms caves.

At each phase of cave development, inflowing groundwater apparently covered large parts of the cave floor at or near the water table. This was the ideal way to introduce hydrogen sulfide into the system. But oxygen could enter only from the overlying land surface. Mixing of these two gases was needed. Apparently a small amount of hydrogen sulfide escaped into the cave air but was quickly absorbed by moisture films and droplets on the cave walls and ceilings. These offered a tremendous surface area for the absorption of both gases and the dissolution of limestone (or, in places, its replacement by gypsum). The reaction to form sulfuric acid is aided by microorganisms that derive energy from it.

As it dissolved the limestone, the acid was replenished by incoming hydrogen sulfide and oxygen. The reaction with limestone was vigorous enough to produce a crust of gypsum on the dissolving surface. The porous gypsum was able to transmit acid toward the walls and the dissolved components back out. The gypsum crusts expanded and thus were poorly bonded to the walls. These crusts commonly peeled off and fell into acidic waters at the floor level where they were recrystallized into thick deposits of gypsum. Much additional gypsum was dissolved and carried out of the cave by water flow. These processes have ceased in the cave, but dripwater continues to carry away some of the gypsum in solution. However, the process can be viewed, with caution, in still-active hydrogen sulfide-rich caves elsewhere in the world.

Although sulfuric acid is no longer active in the cave, corrosion continues by condensation of water from air currents. Moist air rises from pools, and water condenses on cooler surfaces higher in the cave. Droplets and thin films of water contain no dissolved material and thus readily dissolve limestone. Dome-shaped ceiling pockets, corroded walls, and asymmetrically dissolved speleothems indicate patterns of air movement. Where dry air moves through the cave it causes evaporation, and clusters of crystals form on the bedrock or speleothem surfaces.

Sulfuric acid also altered certain minerals in the cave to new minerals. An example is alunite (an aluminium potassium sulfate). Limestone neutralizes acid rapidly, so alunite cannot form in typical cave water. But in air, sulfuric acid droplets can condense on clay and alter it to alunite, as observed in active hydrogen sulfide caves. Carlsbad must have been actively enlarging when its alunite formed, but it could not have been water-filled. This point is important in the following section.

HOW OLD IS THE CAVE?—Tiny amounts of radioactive material in speleothems can be analyzed to obtain dates with great accuracy. Although some small precursor cavities are lined with large calcite crystals having radiometric dates up to 90 million years, typical speleothems in Carlsbad are a maximum of about 800,000 years old. Speleothems can form at any time after the cave itself, so their ages do not tell us

Lake of the Clouds, at the eastern end of Left Hand Tunnel, is a former source of rising hydrogen sulfide at the lowest point in Carlsbad Cavern. Today it has no accessible floor opening.

305

Spongework ("Boneyard") near the Lunch Room was not formed by groundwater. Instead, such features are enlarged mainly in air by thin films and droplets of water condensed from the atmosphere. This water absorbs hydrogen sulfide and forms sulfuric acid, which etches the walls in bizarre patterns.

the cave age. But the alunite described above formed during the main sulfuric acid development of the cave. Radiometric analysis of potassium in the alunite has been used to determine the true age of the cave during the final phases of dissolution. This technique was first developed in the 1990s in Guadalupe caves. These ages (with 2–3 percent uncertainty) show a systematic decrease from the highest Guadalupe caves to the lowest—from about 12 million years (late Miocene) to 4 million years ago (early Pliocene). Carlsbad's three main levels have dates of approximately 6, 5, and 4 million years. They also reveal the evolution of the Guadalupe Mountains and the record of past uplift, erosion, and hydrology.

Development of the Carlsbad Cavern apparently took place in the following stages:

1) About 6 million years ago, deep sulfide-rich groundwater rose to the water table at the level of the Main Corridor (approximately 4,100–4,200 feet above present sea level). Escaping hydrogen sulfide gas mixed with atmospheric oxygen to initiate dissolution of the passage. Groundwater flowed eastward to discharge at a spring at the edge of the Guadalupe escarpment. Most cave enlargement took place above the water table. The entrance passage probably began to form as an outlet for hydrogen sulfide. This passage is now quite large because it acted as a conduit for rising hydrogen sulfide throughout all later stages of cavern development. As the erosional base-level and the water table dropped as a result of regional uplift, passages began to develop at lower levels.

2) The Big Room and Left Hand Tunnel next began to form at approximately 3,650 feet (modern elevation) about 5 million years ago. This level was apparently stable for a long period. Water containing hydrogen sulfide entered via Bottomless Pit and fissures in Left Hand Tunnel. As described above, much enlargement was caused by sulfuric acid introduced to the cave atmosphere during these inputs. The spring outlet remained in same area as before, but at a lower elevation.

3) The Lower Cave began to form about 4 million years ago as the water table continued to fall. Hydrogen sulfide-bearing groundwater entered from several pits at approximately 3,350 feet, and also at Lake of the Clouds at 3,300 feet. Lake of the Clouds now appears to be perched 200 feet above the modern water table. The present water table in Lechuguilla Cave, 3.5 miles west of the Carlsbad entrance, is at 3,150 feet.

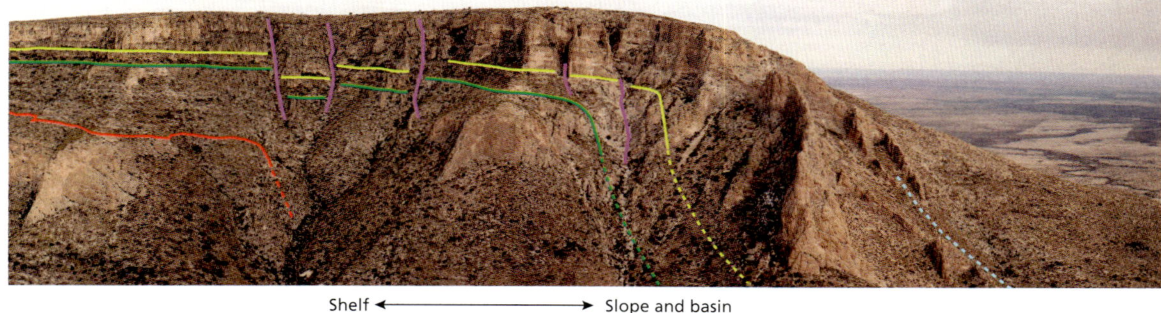

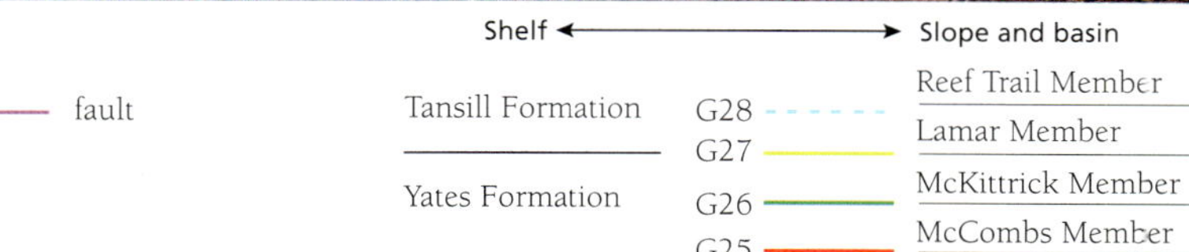

Slaughter Canyon

Slaughter Canyon provides the most southerly access to the Capitan reef in CCNP. Along with McKittrick Canyon in Guadalupe Mountains National Park, it has one of the world's most spectacular cross sections through an ancient reef and its associated landward and seaward beds. A trail up to Slaughter Canyon Cave gives the best view of the reef (reservations and fees are required for the cave, but not for the trail). The trail extends several miles up-canyon, showing further reef and back-reef exposures. The Slaughter Canyon cave tour starts with a 600-foot climb but provides the reward of exploring a cave with only rustic trails. Exotic cave formations, relics of early mining of bat guano, and a sense of history are all part of the adventure.

The photograph above shows the view from near Slaughter Canyon Cave. The massive rocks in the canyon wall are mainly cemented reef limestone that formed at the transition from the flat-lying shelf beds (left) to the fore-reef slope (right). Colored lines on the panorama mark approximate time lines and show that the reef and associated deposits built upward and outward (basinward) through time.

Trails to Slaughter Canyon Cave and others heading up the canyon floor give views of the fore-reef rubble that lay seaward of the Capitan reef. Blocks, some up to house-size, are composed of reef limestone that slid or rolled down-slope during growth of the reef. Longer-distance transport involved submarine debris flows and turbidity currents. During sea-level lowstands, when the shelf was dry and the reef had probably ceased to grow, sand dunes and intermittent streams transported quartz-feldspar sands across the shelf and downslope into the basin. Those basinal sandstones now host significant hydrocarbon deposits in the subsurface of the Delaware Basin.

A panoramic image showing the north wall of Slaughter Canyon. The red, green, and yellow lines show the position and shape of the Capitan reef and basin margin through time (where dashed, the lines denote approximate boundaries). A consistent pattern of deposition is shown throughout the 3–4 million-year period represented here. Flat-lying shelf rocks grade laterally into massive Capitan reef limestones and then into steeply-bedded fore-reef rubble beds that thin and flatten eastward into the Delaware Basin. Slope and basin units shown in the legend are named members of the Bell Canyon Formation. G25 to G28 represent four of the many time slices through the reef.

The longer North Slaughter Canyon trail follows the canyon floor and ultimately provides access to shelf rocks after passing through fore-reef and reefal deposits.

Rattlesnake Springs

Rattlesnake Springs, an isolated unit of CCNP, is located in the valley of the upper Black River, a tributary to the Pecos River. The water supplied by the spring supported homesteading and farming that began in the 1880s. The site, acquired by the Park Service in 1934, remains an important source of water to nearby ranches. Water is pumped from the springs to the top of the Guadalupe Escarpment to supply the Carlsbad Cavern Visitor Center. The springs also help sustain many animals, including frequently sighted wild turkeys. The area is well known as one of the richest bird habitats in southeastern New Mexico, with over 350 documented species.

The water from Rattlesnake Springs is of good quality, with a discharge rate of about 1,900 gallons per minute. The water flows from Quaternary conglomerate (exposed in bluffs along the Black River at Washington Ranch, about one mile northeast of the springs). The Quaternary conglomerate in this portion of the Black River valley is composed of limestone pebbles and cobbles derived from the Capitan reef and near-backreef units of the Guadalupe Mountains a few miles north and west of Rattlesnake Springs. The conglomerate is well-cemented by calcium carbonate.

The conglomerate behaves as an aquifer in the subsurface. Its recharge takes place in the upper reaches of the alluvial fan at the mouth of Slaughter Canyon. The conglomerate is hydraulically isolated from the underlying Capitan reef aquifer, in which water levels are more than 300 feet lower. Groundwater flows from the mouth of Slaughter Canyon through solution conduits in the conglomerate, with discharge occurring at Rattlesnake Springs and as baseflow directly into the Black River.

A visit to Rattlesnake Springs and Slaughter Canyon also provides an opportunity to view rocks of the Delaware Basin by traveling south on US 62/80. On the New Mexico stretch of the road, one can see several

Diversion channels at Rattlesnake Springs with adjacent cottonwoods and marshland habitat.

outcrops of some of the final-stage basin-filling evaporites—the laminated gypsum beds of the Upper Permian Castile Formation. Further south, in Texas, one can see progressively older (Middle Permian) rocks representing both highstand (limestone) and lowstand (sandstone and siltstone) basinal deposition.

—Peter A. Scholle, Arthur N. Palmer, and Lewis Land

Additional Reading

Geology of Carlsbad Cavern and other caves in the Guadalupe Mountains, New Mexico and Texas, C.A. Hill, Bulletin 117, New Mexico Bureau of Geology and Mineral Resources, 1987.

Wild turkeys a-courtin' along the road near Rattlesnake Springs.

If You Plan to Visit

Carlsbad Caverns National Park's only entrance road is NM 7. Turn north from US 62/180 at Whites City, New Mexico, which is 20 miles southwest of Carlsbad, New Mexico and 145 miles northeast of El Paso, Texas. It is 7 miles from the park entry gate at Whites City to the Visitor Center and cavern entrance. To access Rattlesnake Springs or Slaughter Canyon take US 62/180 south from Whites City for 5.5 miles. Then turn right (west) on CR 418 and follow the signs to Rattlesnake Springs (total distance about 2.5 miles) or continue on CR 418 following signs to Slaughter Canyon (roughly an additional 8 miles).

It is worth inquiring at the Visitor Center for availability of other cave tours or hikes. Such tours, offered to small groups and led by Park Service guides, are sometimes available for otherwise off-limits parts of Carlsbad Cavern, as well as Spider Cave and Slaughter Canyon Cave. For more information:

Carlsbad Caverns National Park
3225 National Parks Highway
Carlsbad, NM 88220
(575) 785-2232
www.nps.gov/cave

This recreation area in the Lincoln National Forest has spectacular cliffs and features a year-round spring-fed stream, waterfalls (the largest of which has a 150-foot drop), and many refreshing clear-water wading pools. There are picnic areas and several hiking trails, including T–68 and T–68A that lead to the spring source of the waterfalls. The Apache name for the area was "*gostahanagunti*," which means hidden gulch, and the presence of flakes of worked chert, grinding holes (bedrock mortars), and nearby pictographs testify to long-standing Native American use of this wildlife-rich oasis in the desert.

Regional Setting

Sitting Bull Falls is located in the Guadalupe Mountains just to the west of a major fold (the Huapache monocline) in the regionally nearly flat-lying Permian shelf to shelf-margin strata. The falls are about 12 miles northwest of the younger, late Middle Permian shelf margin reef exposed in the Guadalupe Mountains (see Carlsbad Caverns National Park chapter). The change in location of the shelf-edge

Sitting Bull Falls Recreation Area with parking lot and picnic shelters at lower left. Dark rocks in right-center foreground are tufa deposits produced by carbonate-rich water flowing over waterfalls that shifted position through time. Higher cliffs in background consist of bedded San Andres and Grayburg carbonate rocks in the upper third of the cliffs, above yellow-brown Cherry Canyon sandstones in the lower slopes. Dashed line shows shelf edge "rollover" bedding (flat on the shelf at right and getting progressively steeper on the slope).

between Sitting Bull Falls and Carlsbad Caverns reflects basinward expansion of the shelf over roughly 10 million years.

Basin and Range faulting during the Late Cenozoic uplifted and tilted the Guadalupe Mountains eastward. Uplift allowed erosion and cave formation to sculpt this narrow canyon, which was then partially refilled by tufa (porous travertine) precipitated from spring waters.

The Rock Record

Access via the northern route from Carlsbad is mainly through shelf strata of the Artesia Group, primarily the Seven Rivers Formation (see Guadalupe Back Country Byway chapter). The broad, largely featureless area one crosses midway through this trip is the Seven Rivers Embayment, where dissolution of lagoonal and shoreline evaporites of the Seven Rivers Formation has occurred. The last, very curvaceous stretch of access road on CR 409 takes you past the sheer cliffs of Last Chance and Sitting Bull canyons, which are cut into San Andres, Grayburg, and Cherry Canyon formations.

In addition to Permian rocks, Sitting Bull Falls also has interesting recent deposits. The waterfalls are located in a narrow canyon fed by the upstream escape of groundwater from caves and springs in thick Permian limestones and dolomites. The emerging water has precipitated enormous volumes of calcium carbonate tufa, a process that continues to this day.

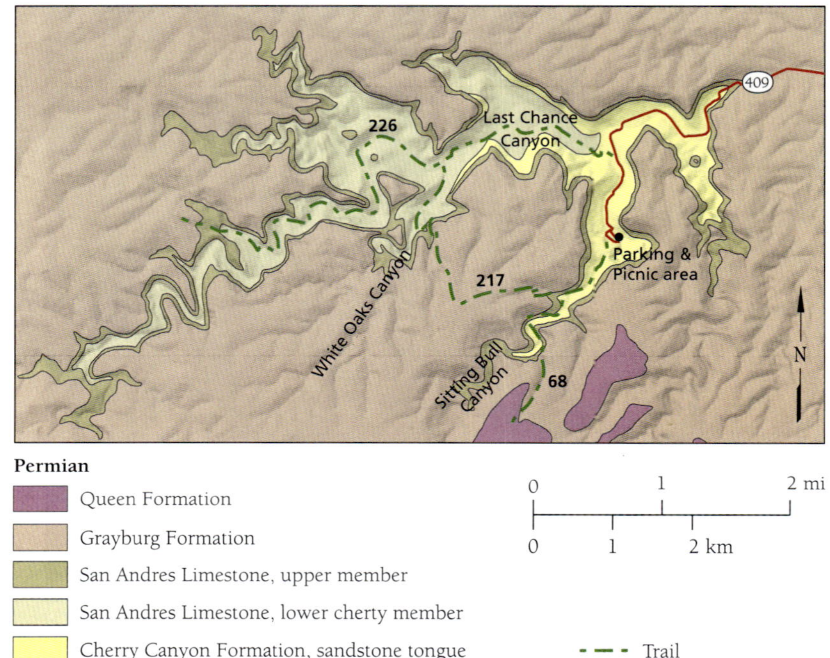

Geologic map of the Sitting Bull Falls Recreational area and Last Chance Canyon.

Geologic History

In Middle Permian time, this area was part of a broad shelf and basin complex that extended across much of southeastern New Mexico. Highlands created by the Ancestral Rockies deformation during Pennsylvanian to Early Permian time were deeply eroded by Middle Permian time, yet they were still shedding substantial amounts of sand and silt. The thick, yellowish-brown sands of the Cherry Canyon Formation exposed in the lower half of the canyon walls record this sediment influx. The steep bedding reflects the transport of those sands downslope from the Northwest shelf toward the Delaware Basin to the east. The overlying limestones of the Upper San Andres and Grayburg formations record times of higher sea level when sands were trapped along distant shorelines to the west and north, and carbonate-producing organisms flourished on the shelves. The irregular thickness of these limestone and dolomite beds may be due to small patch reefs and/or tidal channels on the shelf margin. A walk up Last Chance Canyon allows one to see the toe of the Cherry Canyon slope as well as slightly older carbonate rocks deposited by debris flows. These outcrops are near the road, just downhill from the Sitting Bull Falls area, or via trail T–226 farther up the canyon.

This area was eventually overlain by hundreds of feet of additional Permian shelf strata and thin Triassic and Cretaceous deposits. About 60 million years ago, a large fold (the Huapache monocline) formed just west of Sitting Bull Falls during Laramide crustal shortening. Beginning in the Miocene, Permian rocks were again exposed by Late Cenozoic uplift and erosion. During that stage, canyons were cut, ground-water flow produced caves, and spring-related tufa deposits began to form.

A now abandoned waterfall area showing the sheeted structure of the tufa deposits. As tufa builds up in any given area, the waterfall shifts to lower-lying sites thereby distributing the tufa over large areas and producing broad, rocky terraces.

Geologic Features

Perhaps the most impressive feature at Sitting Bull Falls is the Quaternary tufa. The entire hillside immediately behind the parking lot, shelters and picnic sites is composed of modern tufa that precipitated on the long-lived waterfalls in this area. Water that emerges from upstream caves and springs is saturated with calcium carbonate. As the water evaporates, warms, and degasses during flow, calcite (calcium carbonate) precipitates to form drapery-like sheets of rock. Plants and algae also become encrusted with calcite, and when the organic matter decomposes, it leaves behind the hollow molds that are visible in the tufa. You can still see these processes at work in the modern falls where it is possible to walk up behind some of the tufa sheets. A hike to the top of the falls and then upstream to the cave source of the waters allows observation of stream tufa formation.

—*Peter A. Scholle*

Sitting Bull Falls waterfall and splash pool. Note the numerous shield-like tufa projections and attached plants over which the waters flow and precipitate calcium carbonate.

If You Plan to Visit

Take US 285 north from Carlsbad for about 12 miles and turn west onto NM 137. Continue on NM 137 for about 20 miles until you reach CR 409. Turn right (west) on CR 409 and continue to the Sitting Bull Falls Recreation Site at the end of the road. This route follows the Guadalupe Back Country Byway (see next chapter in this book).

Alternatively, the park can be accessed by going south about 9 miles from Carlsbad on US 62/180 and turning west on the Dark Canyon road (CR 408). Follow CR 408 (not always well maintained due to flash floods and not well sign-posted) for about 24 miles to its T-junction with NM 137; turn left (south) and go 3 miles on that road and then right (west) on CR 409 to its termination at Sitting Bull Falls.

On both routes, drivers should watch out for cattle (open range) and high-speed traffic associated with oil-field and quarry operations. For more information:

Lincoln National Forest
Guadalupe District
5203 Buena Vista Drive
Carlsbad, NM 88220
(575) 885-4181
www.fs.usda.gov/recarea/lincoln/
recarea/?recid=34238

Guadalupe Back Country Byway
BUREAU OF LAND MANAGEMENT

The Guadalupe Back Country Byway begins at the intersection of US 285 and NM 137 (Queen Highway) about 12 miles north of Carlsbad. The Queen Highway is one of the earliest roads built by the Civilian Conservation Corp (CCC) in New Mexico, connecting US 285 with the Queen, El Paso Gap, and Dog Canyon areas of the western Guadalupe Mountains. The 27-mile paved road passes from subdued desert topography with mesquite, creosote bush, and cholla cactus, past dry arroyos and steep rocky cliffs covered in prickly pear, ultimately transitioning to the less vegetated gypsum hills and bluffs of the Seven Rivers Embayment. Driving along the byway, visitors may observe pronghorn and mule deer, as well as a variety of birds, including hawks, quail, roadrunners, and turkey vultures.

The Bureau of Land Management describes the Back Country Byway as a "working landscape," consistent with its mandate for multiple use of the public lands it oversees. Gravel quarries, oil and gas operations, and livestock grazing occur along the Guadalupe Back Country Byway. Watch for cattle, as much of the area is open range.

The discussion of geologic features references mileage from the northeastern (Carlsbad) end of the byway. Visitors may wish to zero their trip odometers when turning off of US 285 and onto NM 137.

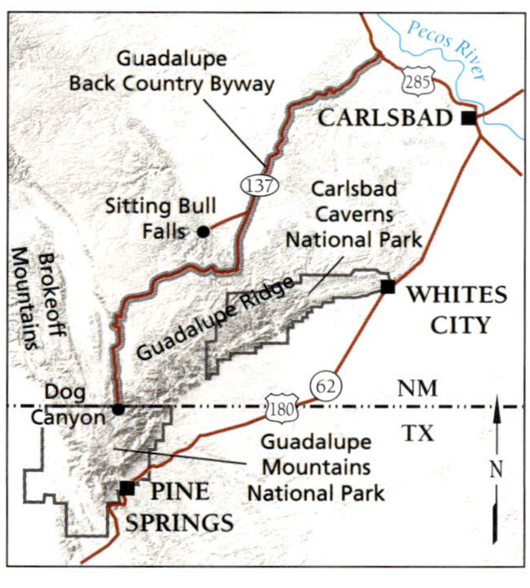

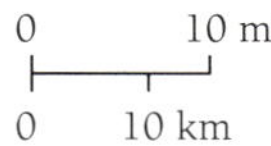

Regional Setting

The Guadalupe Back Country Byway begins in the lower Pecos Valley of southeastern New Mexico, a few miles west of Brantley Lake and ends in the high plateau of the Guadalupe Mountains. The geologic setting of the Byway is on the Permian Northwest Shelf, an Ancestral Rockies platform that lay adjacent to the deeper Delaware Basin to the east (discussed in more detail in the Introduction to Part 5). A substantial section of sedimentary rocks extends across the Northwest Shelf for several hundred miles into northern New Mexico, thinning from south to north.

The Rock Record

With the exception of some Quaternary terrace gravels and tufa deposits, essentially all the rocks exposed along (or visible from) the Guadalupe Back Country Byway are from the Middle Permian Seven

Rivers, Queen, and Grayburg formations (about 270–260 million years old). Over the full course of the Byway, the Seven Rivers Formation, the main exposed unit, changes from cliff-forming shelf dolomites to interbedded gypsum and reddish-brown mudstones reflecting more restricted shelf to shoreline settings. Older Early to Middle Permian rocks can be accessed from the Byway by taking a side-trip to the Sitting Bull Falls Recreation Area.

Geologic History

During Middle Permian time, the Delaware Basin was a semi-enclosed body of water with restricted circulation, similar to the modern-day Black Sea. The basin was rimmed by the Capitan Reef. Remnants of the reef are still visible as a prominent limestone escarpment along the southeast flank of the Guadalupe Mountains, south of Carlsbad. Behind the reef, fine-grained sediments were deposited in the shallow waters of a broad lagoon that extended for hundreds of miles across the Northwest Shelf. In the near backreef section of the lagoon, those sediments consisted of lime mud and sand composed of calcium carbonate. Farther north and west from the reef, gypsum, silt, and mud were deposited in an arid coastal setting characterized by tidal flats and ephemeral streams. The carbonate sediments of the near backreef now make up the dolomite beds of the Artesia Group exposed in the cliffs and canyons of the Guadalupe Mountains. Evaporitic sediments of the far backreef hardened to form interbedded gypsum and reddish-brown mudstone.

Roadcut with "crinkle beds" of probable microbial origin in Seven Rivers dolomite.

Geologic Features

The first few miles of the Guadalupe Back Country Byway crosses a low-relief landscape largely covered by Quaternary river terrace gravels. At mile 4.5, the route begins to follow Rocky Arroyo, and the first Permian bedrock outcrops, dolomites of the Seven Rivers Formation, are exposed in roadcuts partially covered with ocotillo shrubs and prickly pear cactus. At mile 4.9, wavy bedding, or "crinkle beds" can be observed in the roadcuts a few feet above the road. These features represent carbonate structures built up by alternating laminations of carbonate sediment and microbial mats in a shallow lagoonal setting.

Tufa deposits are also extensively exposed along this portion of Rocky Arroyo. The tufa outcrops are mainly Pleistocene spring deposits formed during wet interglacial climate intervals. The tufa has a brown, spongy appearance in the outcrops, strikingly different from the much older, white to tan beds of dolomite. The spongy texture is caused by plant-fragment casts preserved in the tufa.

At mile 5.8, Seven Rivers dolomites are exposed in a steep cliff to the left above a ranch house. A short distance farther, just before crossing Rocky Arroyo, caves can be observed that have formed within massive tufa deposits on the right shoulder of the road.

At mile 7.3, the route begins a traverse through the transition from near-backreef dolomites to far-backreef gypsum and red mudstones within the Seven Rivers Formation. This lithologic transition occurs over the course of just one mile as the route continues west. A short distance beyond these roadcuts, the route enters the Seven Rivers Embayment. The gypsum and mudstone in this area are more soluble, softer, and more easily eroded than the cliff-forming dolomites, resulting in more subdued topography. In places one can see rubbly (brecciated) dolomites that formed where underlying gypsum beds dissolved and the overlying dolomite collapsed into the solution caves.

At mile 10.3, the route passes a distinctive peak known as the Teepee to the right. This conical hill is made up of mudstones and gypsum and is capped by more erosion-resistant dolomites of the Seven Rivers Formation. The Indian Basin field, with pumpjacks and storage tanks visible from the road, has produced over 100 million barrels of oil and 3-trillion-cubic feet of gas from Pennsylvanian sandstones and dolomites since its discovery in 1963.

The byway continues for another 10 miles past cone-shaped hills and bluffs formed by erosion of gypsum and mudstone within the Seven Rivers Embayment. At mile 20, the route intersects Dark Canyon Road (NM 408), an alternative return route to US 62/180 and

The Teepee with exposures of Seven Rivers mudstones, evaporites, and dolomites.

Carlsbad. Although not part of this scenic byway, the Dark Canyon route has excellent exposures of gypsum, siltstone, and carbonate rocks of the Seven Rivers and Yates formations. In addition, one of the northernmost exposures of the Capitan reef is visible in low outcrops at the mouth the canyon.

The Back Country Byway continues southwest on NM 137 to the Sitting Bull Falls turnoff at mile 22.6. At this point, visitors may choose to continue south on NM 137 toward the small community of Queen in the high Guadalupe Mountains. This route crosses the Huapache monocline (a large fold), leaves the Seven Rivers Embayment, and passes through the older dolomites of the Grayburg and Queen formations.

Alternatively, one can turn right onto NM 409 toward Sitting Bull Falls. At mile 26.5, the route descends into Last Chance Canyon past dolomite cliffs of the Queen and Grayburg formations. The Guadalupe Back Country Byway ends at mile 26.9, at the boundary of Lincoln National Forest. Visitors are encouraged to continue another 3 miles to the Sitting Bull Falls day-use area, where the Grayburg, San Andres, and Cherry Canyon formations are exposed in the tufa-draped cliffs of Last Chance Canyon. Sitting Bull Falls is described in more detail in a separate chapter of this book.

—*Lewis A. Land*

If You Plan to Visit

The entrance to the Guadalupe Back Country Byway is located about 12 miles north of Carlsbad and 23 miles south of Artesia, New Mexico, at the intersection of US 285 and NM 137. Turn west onto NM 137 to begin your trip, which ends 27 miles to the southwest at the boundary of the Lincoln National Forest-Guadalupe Ranger District. The southern end of the byway can be accessed with some difficulty from the west and northwest via unpaved and relatively poorly maintained dirt roads not shown on map in this chapter. For more information:

Bureau of Land Management
Carlsbad Field Office
620 East Greene St.
Carlsbad, NM 88220
(575) 234-5972
**www.blm.gov/nm/st/en/prog/recreation/carlsbad/Guadalupe_Backcountry_
Byway.html**

Living Desert Zoo and Gardens State Park

Living Desert Zoo and Gardens State Park is located in the Ocotillo Hills, on the northern outskirts of Carlsbad, New Mexico. The zoo has exhibits of more than 40 species of animals and hundreds of plants native to the Chihuahuan Desert. The ridges and canyons around the park expose Middle Permian rocks characteristic of the northern Guadalupe Mountains. There is no camping in the park, but picnic areas and hiking trails are available.

The zoo is structured around a self-guided, 0.9-mile trail that introduces visitors to a variety of habitats of the Chihuahuan Desert, including sand hills, gypsum hills, desert uplands, and arroyos. The tour also includes a walk-through aviary and a prairie dog village. Most of the animals in the zoo were injured or abandoned, and are unsuitable for return to the wild.

Regional Setting

The Park is located on the Northwest Shelf of the Delaware Basin, an ancient, large, sediment-filled depression of the Earth's crust in southeastern New Mexico and west Texas, discussed in more detail in the introduction to this part of the book.

Prairie dogs (genus *Cynomys*) are herbivorous burrowing rodents native to the grasslands of North America. Despite the name, they are not actually canines.

The Mexican gray wolf (*Canis lupus baileyi*) is the rarest subspecies of gray wolf in North America. In 2017, there were only 114 wolves remaining in the wild.

Contact between the upper Yates sandstone (buff) and overlying Tansill dolomite (gray).

The Rock Record

Sedimentary rocks exposed in the vicinity of the Living Desert Zoo and Gardens State Park consist of the Yates and Tansill formations of the upper Artesia Group. The Tansill Formation is predominantly dolomite. Dolomite also makes up most of the underlying Yates Formation, although it also contains sandstone beds, especially at the top and base of the formation.

Geologic History

During Middle Permian time about 270–260 million years ago, the Delaware Basin was rimmed by the Capitan reef, the fossil remnants of which are exposed as a limestone escarpment along the southeast flank of the Guadalupe Mountains. Behind the reef, fine-grained lime mud and sand were deposited in a broad, shallow lagoon, in the near backreef. The park is located in such a near-backreef setting, in what was an intertidal to shallow subtidal, restricted marine environment. That environment is distinguished from the more arid coastal setting of the far backreef, represented by beds of gypsum and mudstone, such as those exposed in the cliffs of the McMillan Escarpment, north of Carlsbad, and at Bottomless Lakes State Park.

During periods of high sea level, sediments deposited in the near backreef were predominantly shallow-marine lime muds and carbonate sands. These sediments eventually became lithified and were replaced by dolomite. During episodes of low sea level, the lagoon and shelf areas were exposed and wind-blown quartz sands were deposited on the shelf and in the Delaware Basin. Those sands now make up the sandstone beds of the Yates Formation, as well as other sandstones within the Artesia Group.

Geologic Features

Living Desert Zoo and Gardens State Park is located on the Tracy Dome, known to local residents as C-Hill. In 1926, the Ohio Oil Co. drilled a 5,805-foot-deep dry hole, the Tracy #1, on this structure.

In outcrops along the access road from US 285, Tansill dolomite beds are inclined approximately 30 degree northeastward, toward the Pecos River. Approximately two-thirds of the way up the hill, the contact between the Tansill Formation and the underlying upper Yates sandstone is exposed in a roadcut on the right.

Small and poorly accessible outcrops of Tansill dolomite can be observed in various areas within the park, particularly within the hooved-animal enclosures. The walls of the canyons bordering the zoo provide far better exposures of the Yates and Tansill formations, their bedding sharply defined by vegetation. The Ocotillo Nature Trail (1.8 mile roundtrip) descends past outcrops of Yates and Tansill dolomites that are deeply sculpted by small-scale weathering features known as rillenkarren (small-scale solution rills). Plant life is typical of the northern Chihuahuan Desert. An indicator species for the Chihuahuan Desert, ocotillos are abundant and display lush green leaves and bright red flowers in the spring and during the summer monsoon. The nature trail ends at the parking lot behind New Mexico State University–Carlsbad.

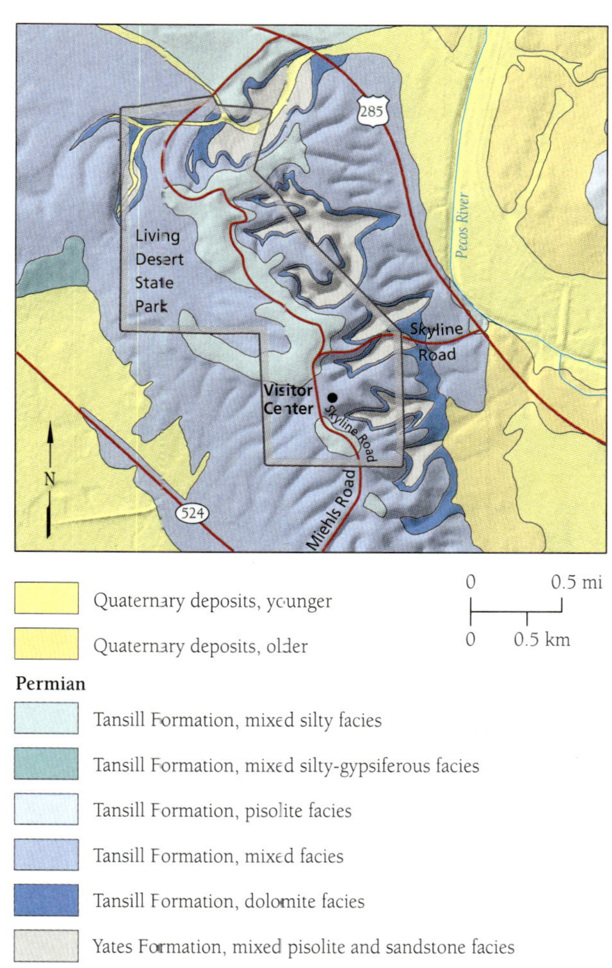

Geologic map of Living Desert Zoo and Gardens State Park.

—*Lewis A. Land*

If You Plan to Visit

The Park is located on the north side of Carlsbad, New Mexico. From US 285/West Pierce Street, turn west onto Skyline Road. Follow Skyline Road to the crest of C-Hill and bear left (south). The Park is located on the east side of Skyline Road. For more information:

Living Desert Zoo and Gardens State Park
1504 Miehls Drive North
Carlsbad, NM 88220
(575) 887-5516
www.emnrd.state.nm.us/SPD/livingdesertstatepark.html

Ocotillo flowering in the spring at Living Desert Zoo and Gardens State Park.

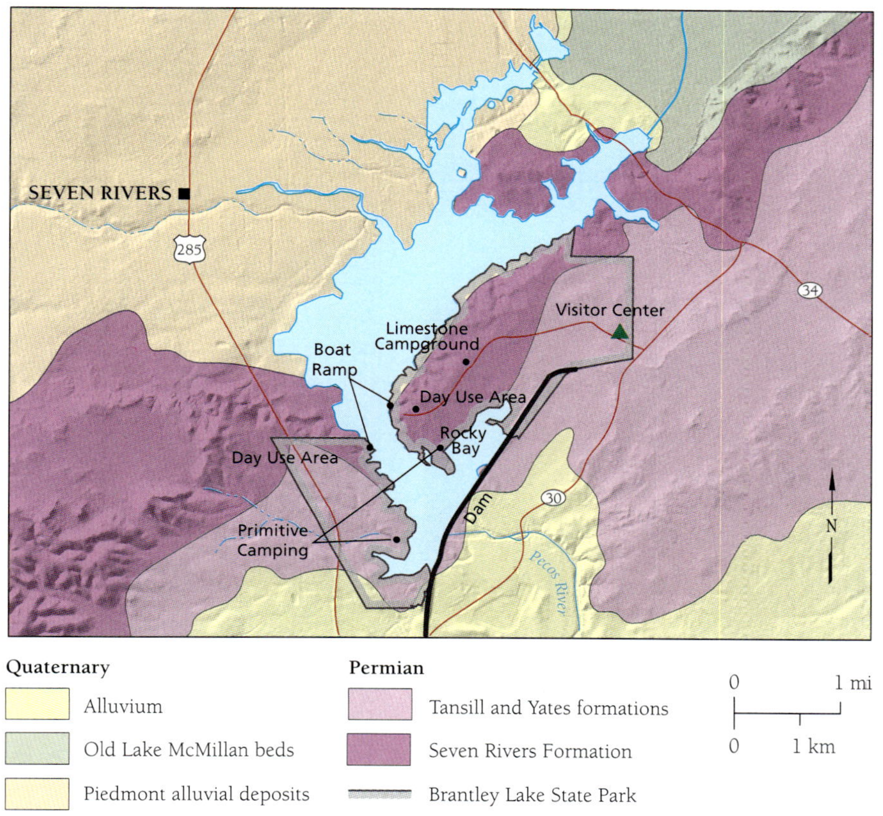

Geologic map of
Brantley Lake State Park
and surrounding areas.

Quaternary

Alluvium

Old Lake McMillan beds

Piedmont alluvial deposits

Permian

Tansill and Yates formations

Seven Rivers Formation

Brantley Lake State Park

0   1 mi

0   1 km

Brantley Lake State Park is located about 12 miles north of the city of Carlsbad. The lake itself is an artificial impoundment on the Pecos River created by the construction of Brantley Dam. It is the southernmost lake in the state and is stocked with a variety of game fish including bass, walleye, catfish, bluegill, and crappie. The park offers developed campsites, boat ramps, a 2.6-mile hiking trail, and primitive camping along both shores of the lake. The principal functions of the lake are flood control, water storage for the Carlsbad Irrigation District, and recreation.

Regional Setting

The park is situated in the lower Pecos Valley of southeastern New Mexico, on the Northwest Shelf of the Permian Delaware Basin, a large sedimentary basin in southeastern New Mexico and west Texas (see discussion in the introduction to this part of the book). A thin section of sedimentary rock stretched northward from the reef-rimmed margin of the basin, extending across the Northwest Shelf for hundreds of miles.

The Rock Record

Bedrock exposed at Brantley Lake State Park and beneath the lake consists mostly of dolomite of the Seven Rivers Formation, in the middle of the Artesia Group (see stratigraphic diagram in the introduction to this part of the book). There are limited exposures of gypsum of the overlying Yates Formation along the east side of the park near the Visitor Center. Much of the northwest shore of the lake consists of Quaternary alluvial deposits.

Geologic History

During Middle Permian time, the Delaware Basin was a semi-enclosed body of water with restricted circulation. The basin was rimmed by the Capitan reef. Behind the reef, fine-grained sediments were deposited in the shallow waters of a broad lagoon that extended across the Northwest Shelf. In the near-backreef waters of the lagoon, these sediments consisted of mud and sand composed of calcium carbonate. Farther north and west, gypsum, silt, and mud were deposited in restricted lagoonal to arid coastal settings characterized by ephemeral streams and evaporitic tidal flats. Those restricted-shelf rocks are exposed along the McMillan Escarpment a short distance northeast of the park.

An outcrop of Seven Rivers Formation along the McMillan Escarpment.

Engineering Impacts of Geologic Setting

Long before the completion of Brantley Dam in 1987, Lake McMillan was created in 1894 with a dam at the base of the McMillan Escarpment, just upstream from what is now Brantley Lake. Lake McMillan was constructed on evaporitic rocks of the Seven Rivers Formation, at a time when geologic assessments of dam sites were rarely conducted.

When Lake McMillan was completed, it was reported to be "the largest artificial reservoir in America and… one of the greatest in the world." However, because of the soluble nature of the underlying gypsum bedrock, Lake McMillan almost immediately began experiencing leakage problems through sinkholes and solution channels that formed in the lakebed. In the early twentieth century, a dike was constructed along the southeast shore of the lake, near the base of the McMillan Escarpment, in an attempt to isolate the worst areas of sinkhole formation. Those efforts had only limited success.

Deposition of suspended sediment steadily decreased the storage capacity of Lake McMillan, until by the early 1940s less than half of its original capacity remained. Dredging the lake was considered, but that plan was discarded when it was realized that removal of accumulated sediment would increase leakage through sinkholes in the underlying gypsum bedrock. Fortunately, the rate of storage-capacity loss decreased by the mid-1940s because stands of invasive salt cedar became established in bottom lands upstream from Lake McMillan, trapping much of the suspended sediment on the river flood plain above the lake.

Plans for construction of a new dam began as early as 1905. The U.S. Geological Survey recommended that the new reservoir be located on the less soluble dolomite of the Seven Rivers Formation. Because these beds contain less evaporite, they are more resistant to dissolution and formation of sinkholes. The U.S. Bureau of Reclamation began

View of Brantley Dam from Brantley Lake showing central concrete spillway section and earth-fill wings to the east and west.

Large dolomite clast from the Seven Rivers Formation displaying rillenkarren formed by modern dissolution.

construction of Brantley Dam in 1984, and in 1991 McMillan Dam was intentionally breached, allowing Lake McMillan to drain into Brantley Lake.

Geologic Features

The principal feature of Brantley Lake State Park is the lake itself. The Limestone Campground on the eastern shore of the lake is located on outcrops of dolomite of the Seven Rivers Formation, which also is exposed at the primitive camping area on the western shore of the lake. In many areas, the dolomite exposures display striking small-scale weathering features, including rillenkarren, or small dissolution channels that form on carbonate rock by flowing water. The hiking trail from the campground to the Visitor Center crosses a poorly-exposed formation contact between Seven Rivers dolomite and gypsum of the Yates Formation. The more subdued topography along the trail closer to the Visitor Center is the result of the soluble nature of the gypsum bedrock, in contrast to more resistant dolomite at the Limestone Campground.

The Seven Rivers Formation carbonate-to-evaporite transition is exposed at the old McMillan Dam site and along the McMillan Escarpment northeast of the park. Visitors can walk onto the breached dam and observe interbedded dolomite and gypsum in cliffs along the Kaiser Channel of the Pecos River below the dam. The undulating character of the outcrop is result of dissolution of gypsum in the shallow subsurface and sagging of overlying strata. This location also provides a view of the dry lakebed, the McMillan Escarpment along the eastern shore of the lake, and the man-made dike at the base of the escarpment.

—*Lewis A. Land*

If You Plan to Visit

Brantley Lake State Park is located about 12 miles north of Carlsbad, New Mexico. From US 285, turn east onto CR 30/Capitan Reef Road, cross the Pecos River below Brantley Dam, then turn west (left) onto Brantley Lake road to the Visitor Center.

To visit the old dam site, turn left when exiting Brantley Lake State Park onto CR 30/Capitan Reef Road, and drive 1.7 miles to the T-junction with CR 34. Turn left onto CR 34/Lake Road and continue for one mile to the end of pavement. For more information:

Brantley Lake State Park
33 East Brantley Lake Rd.
Carlsbad, NM 88221
(575) 457-2384
www.emnrd.state.nm.us/SPD/brantleylakestatepark.html

Folds in the Seven Rivers Formation north of Brantley Lake State Park.

Bottomless Lakes State Park is located about 15 miles southeast of Roswell on the Seven Rivers Escarpment, a topographic feature that marks the steep eastern margin of the Pecos River Valley. Established in 1933, Bottomless Lakes was the first state park in New Mexico. The principal attractions are eight sinkhole lakes formed in gypsum and mudstone exposed along the Seven Rivers Escarpment. These features are some of the largest of the many sinkholes and other karst features in the lower Pecos Valley, including those found at Bitter Lake National Wildlife Refuge northeast of Roswell. The term "karst" refers to landforms that result from dissolution of soluble rock such as limestone or gypsum by circulating groundwater. Sinkholes such as the ones found at this park are regarded by many geoscientists as the diagnostic landform of a karst landscape.

The sinkholes (or *cenotes*) of the lower Pecos Valley are unusual in that they occur in a semiarid setting, where mean annual evaporation exceeds precipitation by a factor of seven or more. The Pecos Valley sinkholes are also distinctive because they occur in an area of groundwater discharge, whereas most sinkholes form in groundwater recharge areas. The sinkholes at Bottomless Lakes are fed by submerged springs discharging from an underlying artesian aquifer, and they illustrate the fundamental role that karst processes played in shaping the morphology of the region.

Regional Setting

Bottomless Lakes State Park is located on the Northwest Shelf of the Delaware Basin, described in more detail in the introduction to this part of the book. The shelf extends from the reef-rimmed northern edge of the Delaware Basin for hundreds of miles to the north and northwest.

The Rock Record

Bedrock exposed in the cliffs of the Seven Rivers Escarpment at this park consists of gypsum and reddish-brown mudstone of the Middle Permian Seven Rivers Formation (part of the Artesia Group). In some areas in and adjacent to the park, the gypsum bedrock is overlain by a thin layer of Quaternary sand and gravel. These young sediments were deposited by the ancestral Pecos River over the past several hundred thousand years as the river meandered across its floodplain.

OPPOSITE: Wavy-bedded dolomites and redbeds in the Seven Rivers Formation resulting from dissolution of interbedded evaporites.

Satellite view of Bottomless Lakes State Park area. Lea Lake is the dark feature near the bottom of the image. The chain of smaller sinkhole lakes is visible in the upper center of the view.

Geologic History

During Middle Permian time, fine-grained sediments were deposited in the shallow waters of a broad lagoon that extended for over 100 miles across the Northwest Shelf. Sediments deposited in a lagoon that formed behind the Capitan reef were originally mud and sand composed of calcium carbonate. These deposits later were replaced by dolomite (calcium magnesium carbonate), forming the beds of the Artesia Group in the Guadalupe Mountains. Farther to the north, the environment of sediment formation passed transitionally from the relatively unrestricted waters of a shallow lagoon to an arid coastal setting characterized by ephemeral streams and tidal flats. The sediments that were deposited likewise changed from carbonates in the open lagoon to reddish-brown mud and gypsum in the more evaporitic environments of the interior Northwest Shelf, where the Bottomless Lakes sinkholes are found.

Probably during late Miocene time, an ancestral lower Pecos Valley began to form by sinkhole creation and solution-subsidence (karst processes) in the carbonate and evaporite rocks of the Artesia Group. At this time, the upper Pecos River was a separate drainage system flowing southeast through the Portales Valley. As the lower Pecos River extended farther to the north by headward erosion, it eventually captured the upper Pecos River near Fort Sumner during the Pleistocene, at which time the modern drainage system was established.

The Pecos River originally passed to the west of Roswell, but the channel migrated eastward due to continued uplift of the Sacramento Mountains, and tilting of the land surface to the east. As the river migrated eastward, it eroded and undercut strata of the Artesia Group, a process that continues to this day, giving the Roswell-to-Carlsbad reach of the Pecos Valley a distinctly asymmetric character. Much of the eastern margin of the valley is defined by steep gypsum bluffs of the Seven Rivers Formation, in contrast to the gently-sloping west side, which is cut on the resistant limestone of the San Andres Formation.

Hydrology

Bottomless Lakes State Park is located at the downstream end of a regional aquifer system in the Roswell Artesian Basin. The artesian

Geologic map of Bottomless Lakes State Park and surrounding areas.

Quaternary

- Alluvium
- Piedmont alluvial deposits

Permian

- Artesia Group
- Bottomless Lakes State Park
- ▲ Campground

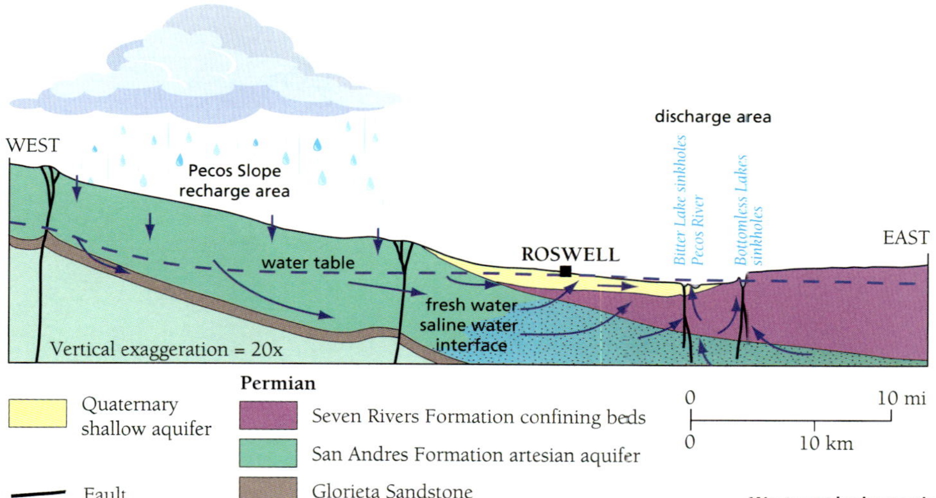

West-east hydrostratigraphic section of Roswell Artesian Basin, showing groundwater recharge and discharge areas, regional groundwater flow paths, and the freshwater-saltwater transition zone.

aquifer is hosted by the San Andres Formation, which is exposed on the Pecos Slope west of Roswell. Recharge to the aquifer occurs in the Sacramento Mountains and on the Pecos Slope, where water carried eastward by ephemeral streams infiltrates into the subsurface. The groundwater then flows through the tilted rocks and passes beneath gypsum beds of the Seven Rivers Formation a few miles west of Roswell. The Seven Rivers Formation acts as the upper confining unit for the aquifer, causing it to become pressurized, thus contributing to artesian flow from springs and irrigation wells in the basin. All of the sinkholes in the park are fed by springs discharging from the artesian aquifer. Discharge has caused subsurface dissolution of gypsum within the Seven Rivers Formation, localized subsidence, and upward propagation of collapse chimneys, which ultimately formed the sinkholes.

In the early 20th century, several of the lakes overflowed into wetlands east of the Pecos River, but the progressive decline in hydraulic pressure in the artesian aquifer due to irrigation-water withdrawals caused lake levels to fall, so that now only Lea Lake overflows. A comparison of water levels in the artesian aquifer with discharge from Lea Lake demonstrates the relationship of water levels in the Bottomless Lakes sinkholes to groundwater circulation within the Roswell Artesian Basin. Water levels are lowest in the summer, when the aquifer is heavily pumped for irrigation, and rise significantly in winter months when irrigation decreases. Discharge from Lea Lake closely mimics the seasonal water-level cycle in the artesian aquifer.

Water in all the lakes is brackish, with high sulfate concentrations, mainly due to prior passage of the water through subsurface gypsum

View south to Mirror Lake, showing local bedding inclination in wall of sinkhole.

beds of the Seven Rivers Formation. Lea Lake, the largest and deepest of the sinkhole lakes, has a salinity of about 9,500 parts per million. Lazy Lagoon has the highest salinity in the park, around 38,000 parts per million, greater than the salinity of seawater.

Geologic Features

The most distinctive features at Bottomless Lakes State Park are the eight gypsum sinkhole lakes that formed within and at the base of the Seven Rivers Escarpment. The sinkholes occur along the east side of the Pecos River floodplain, adjacent to an abandoned former channel of the Pecos River, which now flows about 2 miles west of the park. The gentle eastward regional dip of the Artesia Group rocks is locally reversed along the escarpment, where strata in the Seven Rivers Formation incline steeply and abruptly to the west by as much as 40 degrees. This change results from subsurface dissolution of gypsum in the vicinity of the sinkholes and consequent slumping of overlying beds.

The sinkholes in the park are best observed along the lower road on the west side of the park. Entering the park from the north, take the right fork and begin a descent into the Pecos Valley. The first of the lakes encountered from this direction is Lazy Lagoon, which is actually three separate sinkholes that occupy an abandoned channel of the Pecos River. Most of the lake is very shallow, and it dries up during the summer irrigation season, with the exception of the three deep sinks. A turnoff to the right provides access to a short trail with a scenic overlook of Lazy Lagoon. This overlook also provides an excellent view to the north of large mudstone-gypsum blocks that have slumped off the escarpment above the road, an indication of mass-wasting processes that continue to erode the steep eastern margin of the Pecos Valley.

Continuing south toward the Visitor Center, several dry sinkholes can be observed to the left, high on the escarpment. The next sinkhole lake encountered is Cottonwood Lake, about 30 feet deep and located next to the Visitor Center parking lot. The Visitor Center includes several exhibits that explain the origin of the sinkholes at the park and provide three-dimensional imagery of the lakes. A trail leads from the

Lea Lake, the largest lake in this park, with outcrops of Seven Rivers Formation along the Seven Rivers Escarpment at left.

Visitor Center parking lot south along the base of the escarpment to Mirror Lake, which is actually two sinkholes that have merged. The remaining sinkholes, from north to south, are Inkwell Lake, Figure 8 Lake, Pasture Lake, Lost Lake, and Lea Lake, the largest and deepest of the lakes in the park, with a maximum water depth of about 80 feet. A ninth sinkhole, Dimmit Lake, is located on private land south of the park boundaries.

Lea Lake is the only lake in the park where swimming and non-motorized boating is permitted. Lea Lake is also the only sinkhole that continuously overflows, with discharge emerging from several spring outlets at the base of the escarpment along the eastern margin of the lake. This robust outflow contributes to the clarity of the water in Lea Lake, making it a popular destination for scuba divers.

Although the Bottomless Lakes sinkholes probably formed during the Pleistocene, catastrophic solution collapse processes are still active along this portion of the lower Pecos Valley. The remains of a rockslide that occurred in 1975 can still be seen on the eastern margin of Lea Lake across from the beach and picnic area. Large open fissures and slump blocks are present along the Seven Rivers Escarpment, indicative of ongoing mass-wasting processes. These can be observed from the scenic overlook above Lea Lake.

— Lewis A. Land

If You Plan to Visit

Bottomless Lakes State Park is located about 15 miles east of Roswell, New Mexico. Drive east from Roswell on US 380/East 2nd St., cross the Pecos River and ascend the Seven Rivers Escarpment, then turn south on CR 409/Bottomless Lakes Road. For more information:

Bottomless Lakes State Park
545 A Bottomless Lakes Rd.
Roswell, NM 88201
(575) 624-6058
www.emnrd.state.nm.us/SPD/bottomlesslakesstatepark.html

Bitter Lake National Wildlife Refuge
U.S. FISH & WILDLIFE SERVICE

Morning sunrise over an impoundment flooded for waterfowl in the
Pecos floodplain.

The Bitter Lake National Wildlife Refuge covers about 38 square
miles in central Chaves County, east and northeast of Roswell. Situated
on the Pecos River where the Chihuahuan Desert meets the Great
Plains, the refuge supports an amazing diversity of flora and fauna,
including numerous migratory waterfowl in the winter months.
Established in 1937, the refuge includes the new Joseph R. Skeen
Visitor Center, which sits on the Orchard Park terrace overlooking
floodplain wetlands and the redbed escarpment on the eastern skyline.

The Joseph R. Skeen Visitor Center on a terrace overlooking the Pecos River floodplain.

Beginning at the Visitor Center are a bike trail, a short walking trail, and an 8-mile auto-tour loop. The loop traverses a Pecos Valley terrace overlooking the floodplain for the first 4 miles. The remainder explores floodplain environments, from an abandoned Pecos River meander oxbow lake to the low-lying wetlands. The more adventuresome visitors can explore the refuge's north tract, which includes the Salt Creek Wilderness. Here, one can hike to the mouth of Salt Creek at the Pecos River and continue onto some of the refuge's many sinkholes, including the scenic Ink Pot.

Regional Geologic Setting

Bitter Lake lies at the northern end of the Roswell Artesian Basin in the lower Pecos Valley. Although the Pecos floodplain makes up the majority of the refuge, the western margin is composed of terraces

Bitter Lake is sourced from the Lost River (also known as Bitter Creek). The lake is seasonally dessicated, as in this July 2016 photo. Salt deposits make up much of the lake bed sediments.

of Pleistocene alluvial sediments built up above the floodplain. The easternmost part of the refuge includes bluffs of Permian Seven Rivers Formation gypsum and red mudstone.

As the Pecos River enters the Roswell Artesian Basin, it leaves a narrow valley and begins meandering freely on a broad floodplain. In the valley, the river flows directly atop Permian Seven Rivers rocks. Southward, in the main refuge tract, flood-plain alluvium has buried the Permian redbeds to a depth of as much as 100 feet near the eastern bluffs.

Refuge surface waters are extremely saline due to dissolution of underlying Permian rocks, and hydrologic conditions are similar to those at nearby Bottomless Lakes State Park. Groundwater originating in the Sacramento Mountains to the west flows down-gradient beneath the Pecos Slope until it encounters the Seven Rivers beds. There, hydraulic pressure forces water to the surface through the gypsum bedrock into springs and sinkholes. Before depletion by widespread groundwater pumping for irrigation during the mid-20th century, this artesian groundwater rose to the surface in numerous springs and lakes. Chloride concentrations measured in the refuge's springs, sinkholes, and Bitter Lake itself—located in a restricted part of the refuge, but visible from the bike trail—range from 1,100 to 3,500 parts per million (ppm). For comparison, the EPA secondary drinking-water standard for total dissolved solids is 500 ppm. Nevertheless, these saline soils and waters provide habitat for a number of threatened and endangered species of fish, snails, and other small invertebrates.

Geologic map of the Bitter Lake National Wildlife Refuge.

Geologic History

During Middle Permian time (about 270 to 260 million years ago), the area was part of the Northwest Shelf just north of the Delaware Basin. Extensive lagoons and intertidal mud flats that were repeatedly flooded and dried, resulting in the deposition of muds and precipitation of gypsum and other salts. Over time, these deposits were lithified into the mudstone and gypsum beds that now comprise the Seven Rivers Formation. The majority of the modern landscape at Bitter Lake has developed in the last 300,000 years or so. Initially, the older, higher Orchard Park alluvial terrace (upon which the majority of Roswell is built) was deposited primarily by rivers flowing from the west. The younger Lakewood terrace was then deposited in three phases during the Wisconsinan, the most recent glaciation, resulting in three distinct terrace levels above the floodplain. The Lakewood terrace is inset beneath the Orchard Park and was deposited entirely by the Pecos River.

The oldest sediments that comprise the Pecos floodplain were deposited during the closing of the last glacial period roughly 25,000 to 12,000 years ago. These reveal that the Pecos River was a braided stream with many channels at this time. As the arid, early Holocene (the most recent geological epoch) began about 9,000 years ago, water and sediment discharge were greatly reduced, and the river began to meander. Meander size and amplitude has grown throughout the rest of the Holocene, and the river has gradually shifted from farther west to its present locality near the eastern bluffs.

Geologic Features

From the Visitor Center on the edge of the Orchard Park terrace, approximately 25 feet above the floodplain, the loop road drops down onto the uppermost Lakewood terrace level. The Dragonfly Trail drops off the upper Lakewood terrace onto the intermediate terrace level, before reaching the small pier overlooking the flooded area on the edge of the floodplain. The Goose Hill overlook, 2.8 miles from the Visitor Center, sits on the Orchard Park terrace and offers another view of the floodplain and river about a mile distant. The Desert Upland Trail provides a short walk on the lowest Lakewood Terrace level a mile or so further south.

From here, the tour road heads east across an impoundment levee above the backswamp

Satellite image of the wildlife refuge area showing meander bends on the Pecos River with numerous large sinkholes in the Seven Rivers Formation outcrop area to the east.

to the Oxbow Trail, a two-mile hike circumnavigating an oxbow lake formed by a meander cutoff of the Pecos River. The trail provides visitors an excellent opportunity to explore landforms created by a meandering river (cutbanks and point bars) while enjoying bird watching along the deep-water lake, often surrounded by endangered Pecos sunflowers.

River meanders form and grow laterally due primarily to erosion of the outside bend of a river bank (called the cutbank) followed by the deposition of that material further downstream on the inside of a meander bend (called the point bar). This results in the lateral migration of that portion of the river channel toward the cutbank (see diagram). As meanders increase in size with time, the overall river length increases and the effective river slope decreases. Eventually meanders can become so convoluted that, during a flood, a river may cut across the neck of the meander, restoring a shorter channel and increasing both the slope and velocity of the river. After the flood, the abandoned channel of the meander becomes an isolated oxbow lake, which eventually fills with clay.

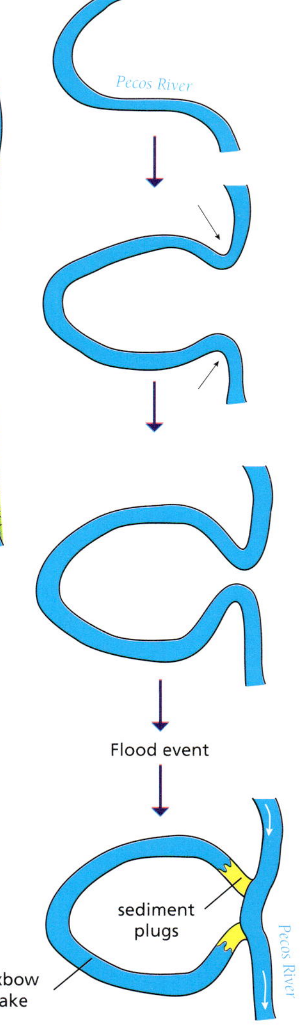

Left: **Model of meander formation by lateral meander migration caused by helical flow.** *Right:* **Evolution of formation of an oxbow lake via meander cutoff.**

Between mile 4.8 and 5, another meander of the Pecos River can be seen about 100 feet to the east of the road. This meander, which developed during the 19th and early 20th centuries, was subject to a man-made meander cutoff across its neck in the 1950s. Several cutoffs and channelized reaches were engineered on the Pecos during this period in an effort to speed up delivery of water downstream to Texas as required under the Pecos River Compact of 1949. River ecosystems were greatly degraded as a result. However, habitat restoration on 6 river miles through the refuge, including this 1.5-mile meander, was undertaken from 2007 to 2009. Restoration included excavating a small channel through the sediment plug that filled the old meander, followed by damming of the cutoff at the top of the old meander, causing the river to reoccupy the old channel. Engineering the river back to its natural state has greatly improved the aquatic habitat. In meanders, quiet water pools develop during low flow, ideal for protecting fish spawned during high-water events primarily caused by upstream releases from Santa Rosa Reservoir. These sheltered environments did not exist in the straight cutoff channel.

— *David J. McCraw and Lewis A. Land*

The Ink Pot sinkhole is developed in Permian redbeds just to the north of the Salt Creek valley in the refuge's Salt Creek Wilderness.

From the south side of Roswell, take US 380 (Second Street) east about 3 miles to Red Bridge Road. Follow Red Bridge Road north to Pine Lodge Road and travel east to the refuge entrance gate. From the north side of Roswell, take US 285 (Main Street) north to Pine Lodge Road. Take Pine Lodge east for about 6 miles to the refuge entrance.

The highly-popular Dragonfly Festival occurs each year in early September and includes free tours to sinkholes and other restricted parts of the refuge (by reservation only).

The Salt Creek Wilderness is located approximately 7.5 miles north of the US 70/285 route divide north of Roswell on US 70. The parking lot is on the north side of the road just west of the Pecos River. From the entrance, an old road heads due north and provides hiking access to rock outcrops and sinkholes. For more information:

Bitter Lake National Wildlife Refuge
4200 East Pine Lodge Road
Roswell, New Mexico 88201
(575) 622-6755
www.fws.gov/refuge/bitter_lake

Mescalero Sands North Dune Off-Highway Vehicle Area

BUREAU OF LAND MANAGEMENT

The Mescalero Sands North Dune Off-Highway Vehicle (OHV) Area lies
in east-central Chaves County. The North Dune area is but a small part
of the eolian (wind-deposited) Mescalero Sands, which cover much
of the Mescalero Plain. Named for Mescalero Apache hunters who once
roamed the area, the sands are covered mostly by shinnery
oaks and Torrey mesquite, but host some large cottonwood
trees as well. The Mescalero Sands are interlayered quartz-rich
sand deposits separated by buried soils. This interlayering
records dramatic climatic changes during the past 100,000+
years. The soils developed during periods of landscape
stability under wetter, cooler conditions, while the sand layers
indicate episodes of aridity and sediment instability. Also
clearly visible to the east, although not present in the park, is
the Caprock Escarpment (Mescalero Ridge, the eroded edge of
the High Plains)—see introduction to this section of the book.

Regional Geologic Setting

The Mescalero Plain is largely underlain by Permian
mudstone and gypsum to the south and west and by Triassic
shale and sandstone in the north and east. The blanketing
Mescalero Sands are just one of several Quaternary eolian
sands sheets found throughout the Great Plains from the
Nebraska Sand Hills in the north to the Monahans dunes of
west Texas to the south. Unlike most sand sheets that trend
west to east, the Mescalero Sands run north-south because
they were piled up against the Mescalero Ridge, to the
immediate east. The sands eventually turn to the southeast,
south of Hobbs, and merge with the Monahans Sandhills.
Sands within the Mescalero and Monahans dunes were
scoured from the Pecos River and its floodplain by westerly
winds beginning in the late Pleistocene.

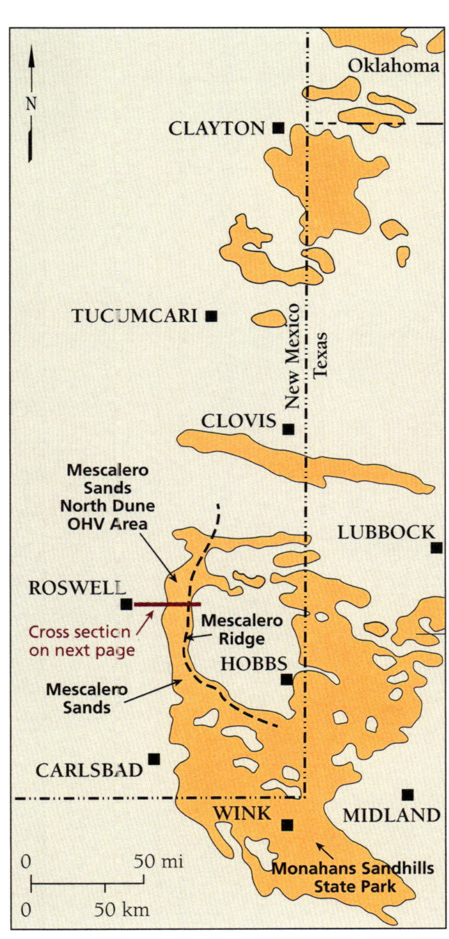

High Plains sand sheets of eastern New Mexico.

Geologic History

The well-developed Mescalero soil formed on Permian and Triassic
beds exposed on the Mescalero Plain during much of the early to
middle Pleistocene, a time of landscape stability. Sand and gravel shed
from the Sacramento Mountains to the west were carried by streams
and deposited in the Roswell Artesian Basin in a broad series of alluvial

Panoramic view of the
Mescalero Sands North Dune area
adjacent to Mescalero Ridge or
Caprock Escarpment.

fans. There is no evidence that the Pecos River flowed through the
basin until the middle to late Pleistocene, sometime between 600,000
to 300,000 years ago.

Roughly 100,000 to 70,000 years ago, a dry climate began to allow
strong westerly winds to erode sand from the Pecos River and deposit
it on the Mescalero soil. During a subsequent glacial period about
60,000 to 13,000 years ago, increased precipitation caused another soil
to develop upon the eolian sand, which was stabilized by a sagebrush
grassland similar to that of modern-day Idaho. Springs were plentiful,
high groundwater levels filled playa lakes, and spring and cienega
(alkaline to freshwater wetlands) deposits formed. This was followed
by dramatic drying 12,000 to 9,000 years ago, when winds again
began to erode the landscape and a second eolian sand was deposited.
Following this eolian episode, the sand again became relatively immo-
bile as the shinnery oak community was established and another soil
developed. The present-day landscape was established in the 1880s,

Geologic cross section through
gently eastward-inclined sedimentary
rocks that are capped by the
Mescalero Sands.

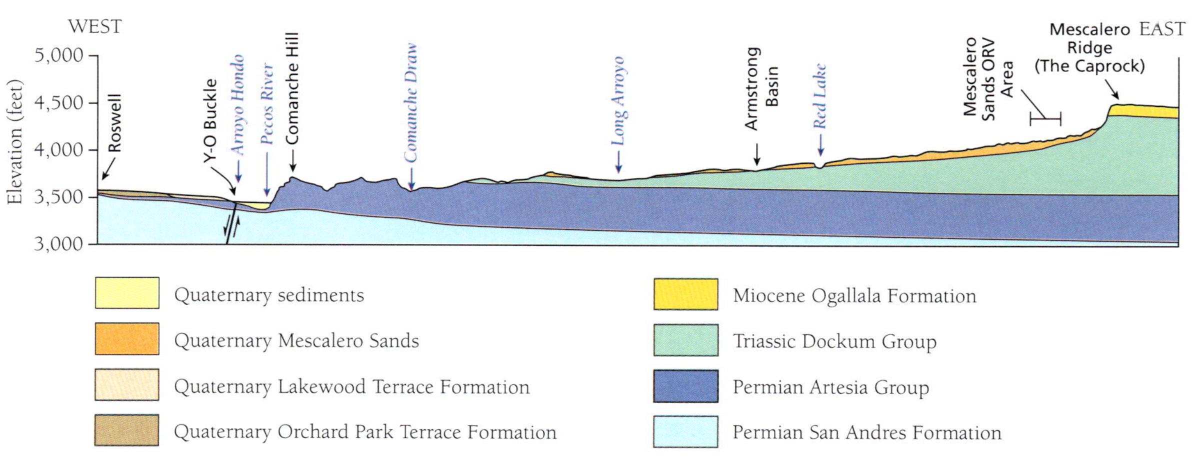

when large numbers of grazing livestock were introduced, vegetative cover was reduced or removed, and the Mescalero Sands were once again mobilized by the wind to form local dunes.

Geologic Features

Large areas of the Mescalero Sands are now coppice dunes (isolated mounds stabilized by vegetation), which form when wind speed is lowered by clumps of vegetation and sand is deposited in a circular mound around the shrub. They do not occur in the North Dune area, however, where curved parabolic dunes have instead formed perpendicular to the prevailing wind direction, with their tips anchored by vegetation. Sand in parabolic dunes is scoured from depressions termed blowouts immediately upwind of the dunes.

Blowouts often contain yardangs (unvegetated, wind-sculpted, sandblasted features shaped into flags and horns). These are cut into the underlying, reddish, buried soil, primarily located in the eastern North Dune Area. The older Pleistocene sand beneath this soil often displays churned and burrowed textures created by plants and animals. This includes the Pleistocene cicada nymphs, which produced burrows with crescent-shaped backfills.

—David J. McCraw

If You Plan to Visit

Although NM 249, US 380, and US 82 all traverse the Mescalero Sands, the only accessible active dunes are in the North Dune area. This area has toilet facilities and covered picnic tables, but no water.

Partially stabilized sand dunes.

As this is primarily a designated recreational area for off-road vehicles, weekday visitation is highly recommended if one wants to explore the dunes on foot. From Roswell, travel east 45 miles on US 380 and turn south at the park sign. From Tatum, travel west 27 miles on US 380 and turn south at the park sign. For more information:

Bureau of Land Management
Roswell Field Office
2909 W. Second Street
Roswell, NM 88201-2019
(575) 627-0272
www.blm.gov/nm/st/en/prog/recreation/roswell/mescalero_sands.html

PART 6

THE SOUTHERN GREAT PLAINS

The Great Plains province is a vast, geologically stable region in the western part of the central United States. This region stretches north to south from the Pleistocene glacial-till deposits near the Canadian border through the Edwards Plateau of Texas, and east-west from the tall-grass prairie of the central lowlands to the Rockies, encompassing about a half-million square miles. In southeastern New Mexico, the Great Plains consists of two subdivisions: the Pecos Valley and the Southern High Plains, also termed the Llano Estacado.

A broad, semiarid plateau along New Mexico's eastern edge that spans the border with Texas, the Southern High Plains attain an elevation of as much as 4,600 feet in the area east of Fort Sumner and slope gently southeast at 10–15 feet per mile. The Southern High Plains are underlain by the middle Miocene–lower Pliocene Ogallala Formation, a major aquifer in the region.

The Pecos Valley portion of the Great Plains province was carved by the Pecos River and its tributaries following the cessation of deposition of the Ogallala Formation about 5 million years ago. The Pecos River heads in the Sangre de Cristo Mountains east of Santa Fe and flows south to where it joins the Rio Grande in Texas near Del Rio. The evolution of the Pecos River in southeastern New Mexico has been strongly influenced by subsidence of sinkholes and basins caused by dissolution of gypsum in underlying Permian strata.

Geologic History

The Great Plains constitutes the southwestern part of the North American craton, the geologically stable core of our continent. Beneath the Great Plains, the crust is thicker (about 25–30 miles thick) than it is beneath the more easily deformed periphery of North America. The Southern Great Plains were moderately deformed by a continental collision event along the eastern and southern margin of the craton about 300 million years ago, during the formation of the supercontinent Pangea. That orogeny gave rise to the Ancestral Rocky Mountains during the Pennsylvanian and early Permian, but since then the region

OPPOSITE: **Remains of the mission church at Gran Quivira (background) and foundations of dwellings and storage buildings constructed of San Andres Formation limestone.**

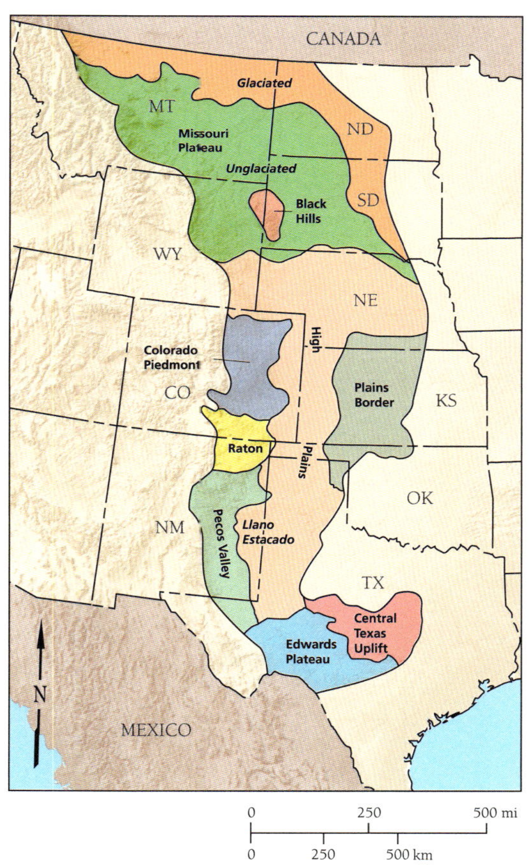

The Great Plains and its subdivisions, which in southeastern New Mexico include the Southern High Plains (or Llano Estacado) and the Pecos Valley.

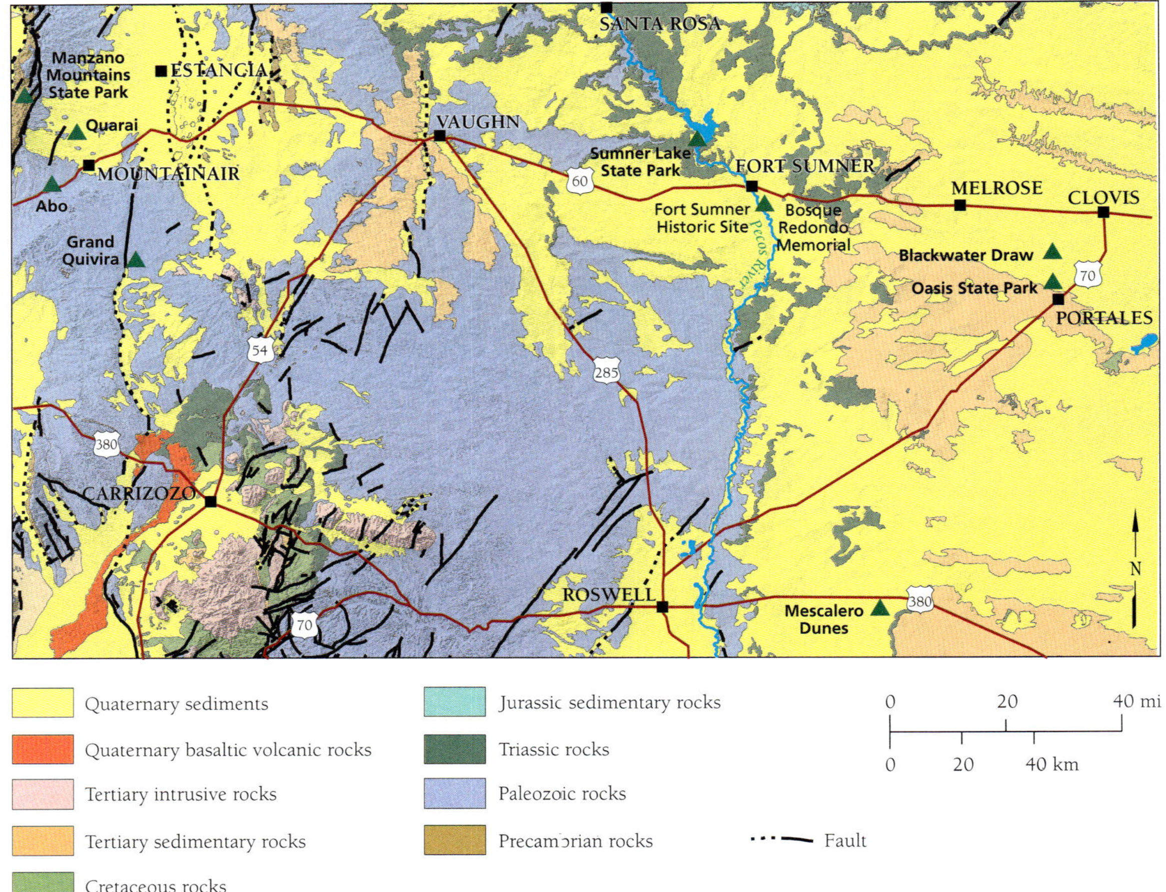

Geologic map of the Great Plains of east-central New Mexico.

has experienced no significant mountain-building events. That is not to say the region has not been subsequently uplifted—it has been, but as a vast, coherent block, not as individual mountain ranges. The older Paleozoic rocks in the Great Plains region are still preserved beneath a relatively thin cover of Mesozoic and Cenozoic strata but, at least in New Mexico, they are exposed primarily along the region's western margin.

The Southern Great Plains were at or near sea level throughout the Mesozoic. Even during the Laramide orogeny (75–45 million years ago), when the modern Rockies began to rise to the west and northwest, the Southern High Plains region remained not far from sea level. It was not until about 30 million years ago that the region began to be significantly uplifted, about the same time as the culmination of massive Oligocene volcanism in the San Juan Mountains of Colorado, in the Mogollon–Datil and Sierra Blanca fields of New Mexico, and in the Sierra Madre Occidental of Mexico. This huge magmatic event caused density changes and upwelling in the mantle, leading to regional uplift and nearly mile-deep erosion on the Southern High Plains prior to deposition of the Ogallala Formation.

The Ogallala Formation began to be deposited on the Southern Great Plains about 12 million years ago. Sediments of the Ogallala Formation in southeastern New Mexico were deposited by rivers that drained the western flanks of the Sangre de Cristo, Sandia, Manzano, and Sacramento mountains, as well as high-standing Oligocene igneous intrusions in Sierra Blanca–Capitan area. The cessation of erosion and the beginning of Ogallala deposition was probably due to increased

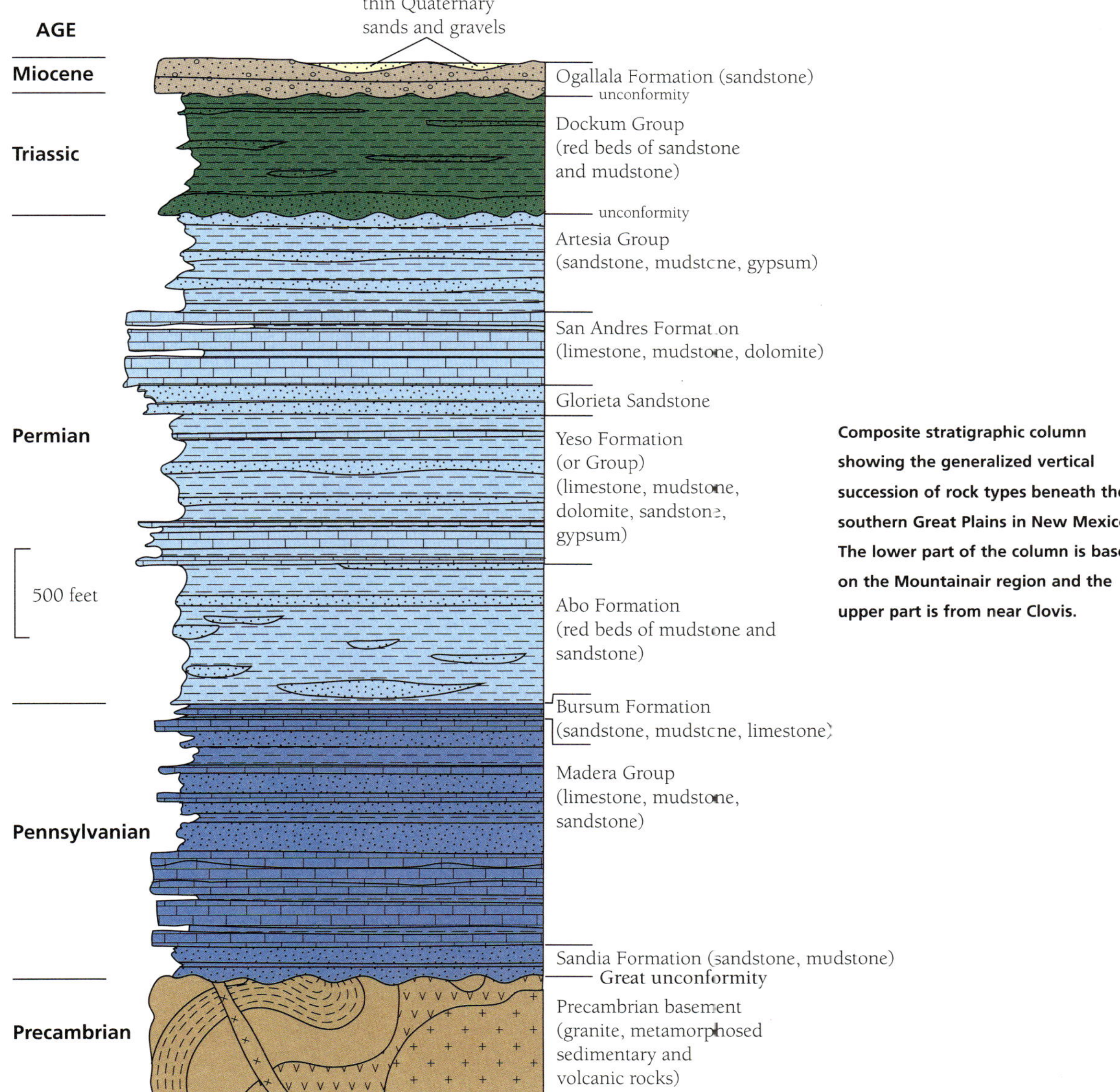

Composite stratigraphic column showing the generalized vertical succession of rock types beneath the southern Great Plains in New Mexico. The lower part of the column is based on the Mountainair region and the upper part is from near Clovis.

Massive gray calcrete (caliche), representing numerous stacked calcareous soils, forms a caprock on buff-colored Ogallala Formation exposed along the Mescalero escarpment near NM 249 east of Hagerman.

sediment supply from rising mountain ranges along the flanks of the Rio Grande rift and to the effects of increased aridity. Aggrading Ogallala sediments eventually buried the hills, valleys, and ridges that had been carved into older strata (mostly Upper Triassic and Lower Cretaceous beds) during the previous erosional episode. In places, these valleys were as much as a few hundred feet deep. Deposition of eolian sand and silt by prevailing southwesterly winds was a major process affecting the more southerly parts of the Ogallala Formation.

As much as 500 feet of sand, silt, mud, and gravel had accumulated by the time deposition of the Ogallala Formation ceased about 5 million years ago. The end of Ogallala deposition was probably a response to an increase in monsoonal precipitation during the latest Miocene, which increased the erosive power of streams in the region. The Southern High Plains once again began to erode, giving rise to a regime of valley-cutting and scarp retreat that continues to the present day. The broad, now-abandoned top of the Ogallala Formation began to accumulate calcareous soils. These soils form a dense caprock caliche, commonly tens of feet thick.

The evolution of the Pecos Valley is complex and not yet fully understood. Possibly during deposition of the Ogallala Formation, a north-flowing precursor to the Pecos River was present in southeastern New Mexico. These poorly dated river deposits are part of the Gatuña Formation. Preserved mostly in sinkholes and solution-subsidence basins south of Artesia, these deposits contain sand and gravel derived from at least as far south as the Big Bend area of Texas.

During Ogallala deposition, the early, upper part of the Pecos River (with headwaters in the Sangre de Cristo Mountains, similar to today) drained eastward across the High Plains region toward Lubbock.

Eastward drainage continued after the end of Ogallala deposition, during which time the river began to cut the Portales Valley. The Portales Valley becomes shallower to the east, suggesting that regional uplift and eastward tilting of the High Plains continued during valley cutting. During this time, the upper Pecos River formed the headwaters of the Brazos River of Texas.

The south-flowing, lower Pecos River probably began as a local tributary to the Rio Grande near Del Rio, Texas. It encroached northward by headward erosion and solution collapse caused by dissolution of gypsum in the underlying Permian strata. The lower Pecos River eventually captured the Upper Pecos drainage near Fort Sumner, at which time the Portales Valley was abandoned and the upper Pecos River became a tributary of the Rio Grande, as it is today. The timing of this capture event is poorly constrained, occurring sometime during the Pleistocene.

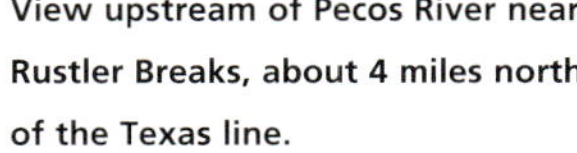

View upstream of Pecos River near Rustler Breaks, about 4 miles north of the Texas line.

Water Resources

On the High Plains, water has long been critical for life and livelihood. Both farming and ranching, two mainstay occupations in the region, require substantial water. In pioneer times, rainfall sufficed for small-scale operations, but following the "dust-bowl days," people realized that natural rainfall is too sparse and unpredictable for a stable existence. In addition, by the mid-20th century, large-scale farming and ranching operations became an economic necessity that required exploitation of water sources other than just rainfall.

Satellite image (Landsat) of the Portales–Clovis area of New Mexico and adjacent Texas showing a remarkable number of green center-pivot irrigation circles that are supplied with water from the High Plains aquifer.

351

352

Simplified U.S. Geological Survey map of the High Plains aquifer showing major drawdown of the surface of the water table in its central and southern parts.

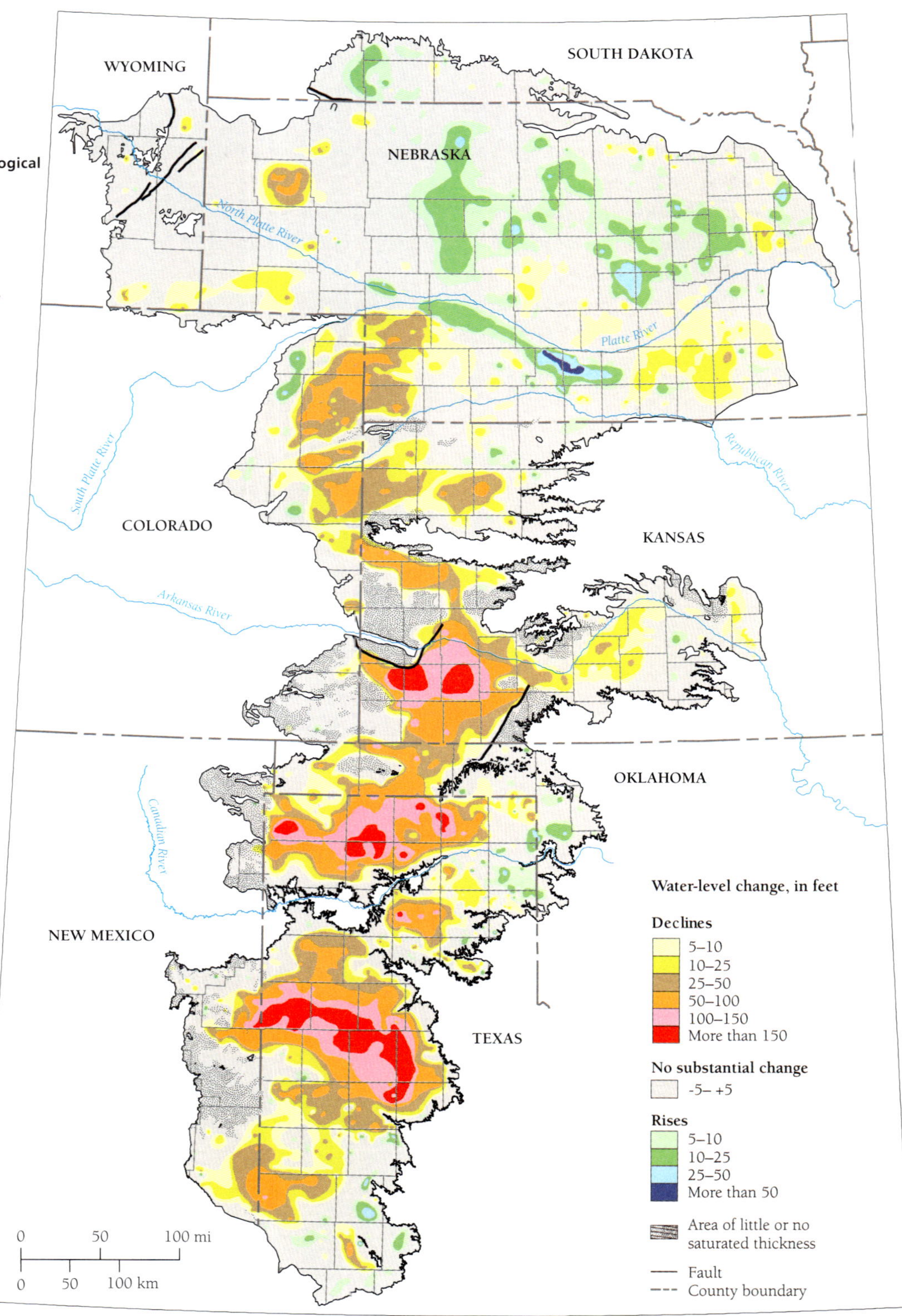

Fortunately, the High Plains sits on one of the greatest aquifers in the world. The shallow High Plains aquifer (mostly within the Ogallala Formation) underlies parts of eight states, and by the end of the 20th century it supplied nearly one-third of all agricultural groundwater withdrawals in the United States. Groundwater-based farming almost completely supplanted earlier rainfall-dependent farming, and thousands of wells supplied center-pivot irrigators with abundant water. The good news was short lived, however. The High Plains aquifer, although huge in area and volume, has very little recharge in most areas, especially in New Mexico and adjacent west Texas. The Pecos River long ago cut to a level below the Ogallala Formation and thus cut off recharge from the Sacramento and Sangre de Cristo mountains to the west. Thus, only the roughly 18 inches of annual precipitation (about half of the average for the United States as a whole) adds water to the aquifer in this region. Because such precipitation was inadequate for agriculture in the first place, it clearly cannot significantly restore a heavily used aquifer.

The future of the High Plains aquifer, at least in New Mexico and west Texas, is not promising. Conservation measures such as Low Energy Precision Application (LEPA) irrigators and laser-leveling of fields have slowed, but not reversed, water-level declines. Many agricultural areas have been abandoned, especially in New Mexico where more than 80 percent of the land area that once could sustain commercial irrigation can no longer be used for that purpose. Many ranches and small towns now lie abandoned—a sad testimony to unsustainable water use.

A growing savior in this region is a new brand of sustainable "farming"—the harvesting of renewable energy, both wind and solar. Neither requires much water, and now many areas, especially along the High Plains escarpments, are studded with wind turbines and huge arrays of photovoltaic panels. The availability of locally-sourced energy may allow a future trend of pumping and treatment of groundwater from deeper

Now abandoned Taiban Presbyterian Church on the edge of water-starved High Plains. The town is famous as the site of Pat Garrett's capture of Billy the Kid in 1880.

**Modern wind turbines on the
High Plains near Dora, New Mexico.**

aquifers. The fact that this deep groundwater will be less plentiful, more expensive to produce, and more saline will bring its own share of problems, as may our changing climate. Only time will tell how these issues will be resolved.

—*Steven M. Cather and Peter A. Scholle*

Manzano Mountains State Park
NEW MEXICO STATE PARKS

Manzano Mountains State Park is located in Torrance County on the eastern side of the Manzano Mountains, near the village of Manzano. The park features camp sites, trails, and a Visitor Center. Bordered on the west by the Cibola National Forest, it is a nice place to get away from the summer heat and enjoy the fresh air, shady forests, and hiking. The park is an ideal base camp from which to explore surrounding areas, whether it be the Salinas Pueblo Mission sites to the south or the spectacular autumnal colors of the bigtooth maple trees in Fourth of July Canyon to the north.

Geography

The Manzano Mountains State Park, at an elevation of about 7,300 feet, encompasses woodlands of piñon, juniper, and ponderosa near the transition to lower-elevation grasslands to the east. Manzano Peak rises to an elevation of 10,098 feet approximately 5 miles west of the park. East of the park, the topographically-closed Estancia basin lies within the transition to the Great Plains physiographic province. The floor of the Estancia Basin was occupied by the large perennial Lake Estancia during the last ice age, about 20,000 years ago. Runoff from the Manzano Mountains was a major contributor to this lake.

The Rock Record

The bedrock underlying Manzano Mountains State Park consists of red mudstones and sandstones of the lower Permian Abo Formation that were deposited in river systems that flowed south from the Ancestral Rocky Mountains about 290 million years ago.

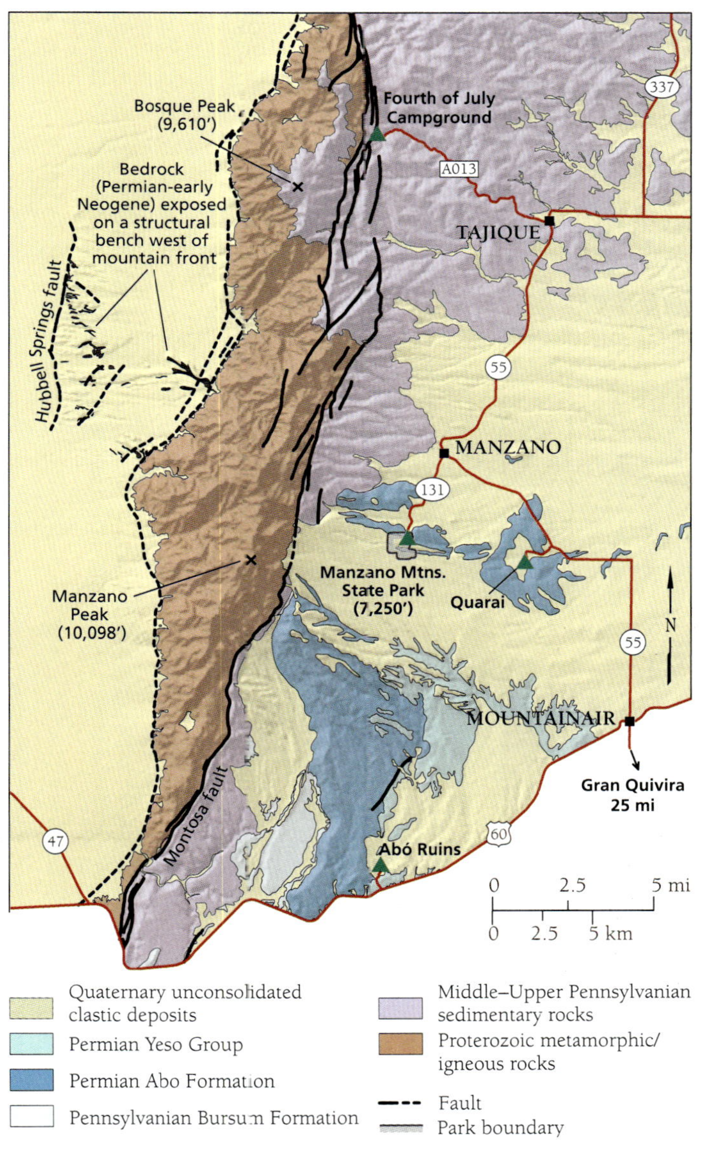

Geologic map of the Manzano Mountains area in central New Mexico, including Salinas Pueblo Missions sites and Fourth of July campground.

The rocks are red because of the presence of iron oxide. The formation is well known geologically for its fossil tracks left by insects and amphibians, skeletal remains of sail-back synapsids ("pelycosaurs"), and as a host-rock for copper mineralization. Thin sandstone beds within the formation have been extensively utilized for building material, both as wall-rock and flagstones. The formation is a little over 1,000 feet thick at Abo Pass to the southwest. Pennsylvanian marine limestones, which underlie the Abo Formation, are present nearby, including along the road to the Cibola National Forest Fourth of July Campground. These limestones contain brachiopods, crinoids, and bryozoans, and provide "hunting" pleasure for naturalists and children alike.

The state park is surfaced largely by unconsolidated gravels derived from the Manzano Mountains. These deposits form an alluvial cover over the Abo Formation and consist mostly of debris eroded from metamorphic rocks (gneiss, quartzite, and schist) in the Precambrian core of the mountains. The gravels are relatively thin in the park, generally about ten feet thick or less. However, they grade into thicker deposits to the east beneath the floor of the Estancia Basin, where they comprise an important groundwater aquifer.

Geologic History and Geologic Features

Crustal shortening beginning in the Late Cretaceous resulted in a mountain-building episode (Laramide orogeny) that helped to form the present-day Rocky Mountains. In the vicinity of Manzano Mountains State Park, Laramide tectonic forces resulted in faulting and uplift of the Earth's crust. The main, roughly north-south zone along which this faulting occurred (the Montosa fault) lies about 3.5 miles to the west of the park. West of the Montosa fault, Precambrian crystalline rocks are exposed. Paleozoic sedimentary rocks, including the Abo Formation that underlies the park, crop out east of the fault.

By Late Tertiary time, Laramide compressional stresses had ended and the Earth's crust began to experience crustal

Autumnal colors near Fourth of July campground.

stretching, which resulted in opening of the Rio Grande rift to the west. The present-day Manzano Mountains are an eastward-tilted fault block along the eastern flank of the Rio Grande rift. Subsequent weathering, erosion and transport of debris from the Manzano Mountains resulted in deposition of the gravels that cover much of the park today.

More recently, east-flowing drainages have cut through the thin veneer of gravel, exposing the underlying Abo Formation. Features to look for while examining the Abo include ripple marks, cross bedding, and mud cracks. Also, be sure to look for fossil tracks of creatures that wandered across this landscape nearly 300 million years ago!

—Bruce D. Allen

If You Plan To Visit

The turnoff to Manzano Mountains State Park is located on NM 55, about 13 miles northwest of Mountainair, NM, along the eastern side of the Manzano Mountains. Take NM 131 at its junction with NM 55 in the historic village of Manzano, and proceed 3 miles southwest to the park entrance. Road signs indicate the route to the park. Camp sites range from primitive to fully developed. Fees vary—visit the park's website.

Manzano Mountains State Park
Mile Marker 3, Hwy 131
Mountainair, NM 87036
(505) 469-7608
www.emnrd.state.nm.us/spd/manzanomountainsstatepark.html

357

Roadcut exposure of Pennsylvanian-age limestone and shale on the road to the Fourth of July campground.

An exposure of Abo Formation redbeds within Manzano State Park.

Salinas Pueblo Missions National Monument

The Salinas Pueblo Missions National Monument (SPMNM) is located in the westernmost part of the Great Plains province, in Torrance County near Mountainair. The monument consists of three small and widely separated units—Abó, Gran Quivira, and Quarai.

When the Spanish arrived in this area in the late 1500s they found multiple inhabited pueblos, and they referred to this area as the Salinas District. However, because of drought, famine, disease, and Apache raiding, by the late 1600s the entire Salinas District was depopulated of both Native Americans and Spaniards. What remains today are reminders of this earliest contact between Pueblo Indians and Spanish colonists: the ruins of three mission churches and a partially excavated pueblo that Juan de Oñate called Las Humanas or, as it is now known, Gran Quivira Pueblo.

The headquarters of the SPMNM is located in Mountainair, which can be reached by taking I–25 south from Albuquerque to Belen, then NM 47 southeast to US 60, then east 21 miles to Mountainair. The three park units are discussed in separate sections with directions given from Mountainair.

OPPOSITE: Abó dwelling and mission walls constructed of local, red Permian sandstone blocks.

Abó Unit

The Abó Unit of the SPMNM is about 9 miles west of the town of Mountainair, New Mexico. It encompasses approximately 270 acres and was made a National Historic Landmark in 1962. Although there are many unexcavated pueblo ruins at this park, the most prominent features are the remains of the rock-walled Mission of San Gregorio de Abó, initially constructed from 1622 to 1628 and expanded from 1640 to 1658. There is a visitor station, a 0.25-mile trail through the mission ruins, and a 0.5-mile trail around the unexcavated pueblo ruins.

Geography and Geologic Province

The Abó Unit is on the eastern slope of the Manzano Mountains, a fault-block mountain range that forms the transition between the Rio Grande rift and the Great Plains province to the east. The core of the Manzano Mountains is made up of Proterozoic (late Precambrian) rocks, many of which are metamorphic. These rocks are capped by upper Paleozoic sedimentary rocks of Pennsylvanian and early Permian age, about 320–290 million years old. The sedimentary rock layers are

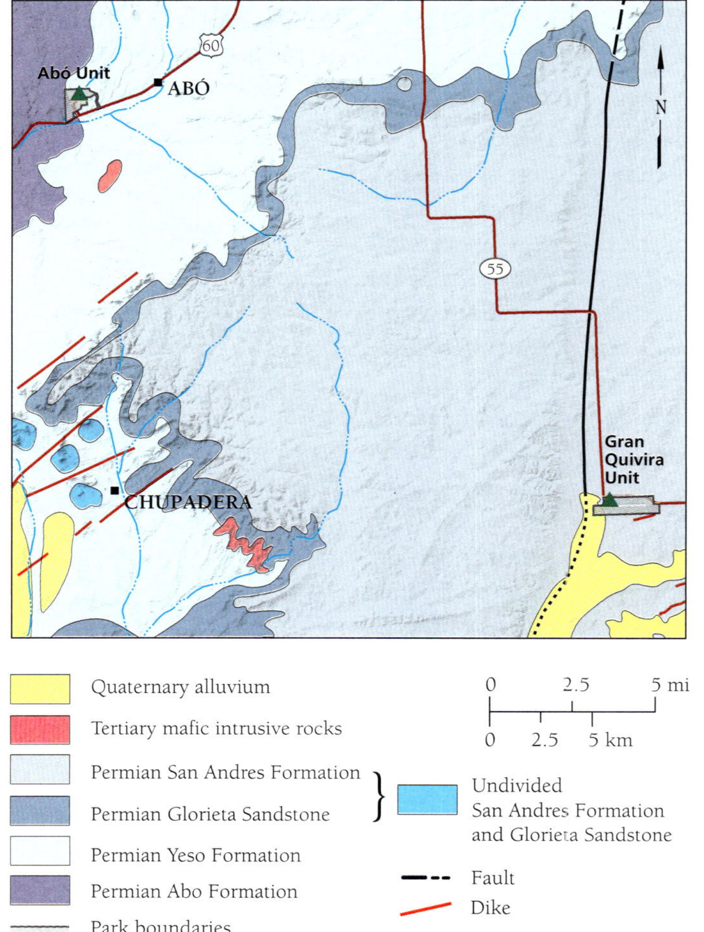

Geologic map including the areas of the Salinas Pueblo Mission National Monument. The geology of the Quarai unit is comparable to that at the Abó unit and is shown on the geologic map in the preceding chapter on the Manzano Mountains State Park.

tilted to the east, so that the farther east you are from the crest of the Manzano Mountains, the younger the rocks are. Thus, at the location of the Abó Unit, the bedrock is of early Permian age (about 290 million years old), whereas older, Pennsylvanian rocks (about 300–320 million years old) are exposed to the west.

The Rock Record

The reddish bedrock exposed in the Abó Unit belongs to the lower Permian Abo Formation and overlying Yeso Group. Locally, the Abo Formation is almost 1,000 feet thick, but only its uppermost layers are exposed at the Abó Unit. These are layers of coarse-grained sandstone with crossbeds and ripple marks interbedded with layers of brick-red to chocolate-brown mudstone. These beds contain an extensive fossil record of plants, as well as footprints, and bones of early amphibians and reptiles.

The lower part of the Yeso Group is almost entirely orange-colored sandstone, except for a few thin layers of dolomite. The sandstone is fine grained and has many ripple marks. Some of the sandstone layers are rich in gypsum. Both fossil footprints and foliage impressions of conifers and seed ferns are present in the Yeso Group in the Abó Unit.

Geologic History

During the Pennsylvanian, multiple continents came together to form a single supercontinent called Pangea, which remained amalgamated until the Late Triassic. The collision of continents that created Pangea produced a series of geologic uplifts across much of the southern and western United States that geologists refer to as the Ancestral Rocky Mountains.

During early Permian time, the Abo Formation was deposited by rivers that drained the ancestral Rocky Mountains in northern and central New Mexico. These rivers built a vast floodplain across much of central New Mexico as they flowed southward to the seashore near the present-day latitude of Las Cruces. Located near the early Permian equator, this floodplain was a hot and humid place under a monsoonal climate. Thus, each year months of rain were followed by months of dry weather, not unlike present-day New Mexico. The seasonally arid Abo floodplain was shrouded in a dense vegetative cover of conifers and seed ferns. Scorpions, beetles, and other arthropods lived in the underbrush, hunted by small amphibians and reptiles. Larger reptiles, notably the sail-backed *Dimetrodon*, were the top predators.

After Abo deposition ended, the climate dried, and a vast desert appeared across much of the Four Corners and northern New Mexico. In central New Mexico, where the Abó Unit is located, that desert met an arid coastal plain of the Permian Basin seaway of west Texas and southeastern New Mexico. Deposition of the lower Yeso Group took place on that coastal plain as windblown sand sheets and small dunes, on tidal flats (gypsum-rich sandstones) and in restricted marine embayments (dolomites).

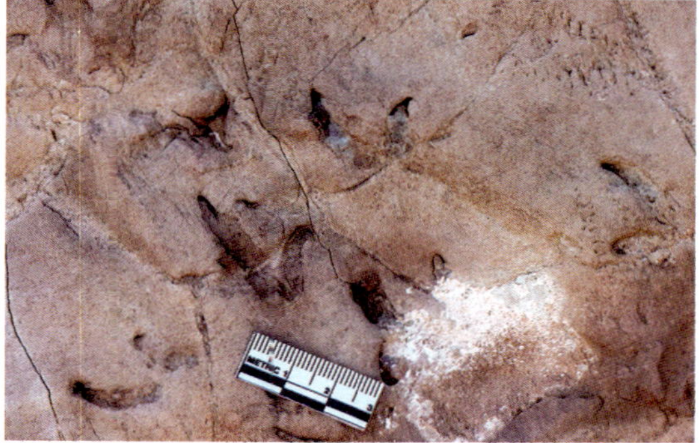

Reptile tracks in sandstone bed near the base of the Yeso Group. These tracks are called *Dromopus*, and were made by a very primitive reptile with slender, curved and clawed toes.

Geologic Features

Red and orange sandstones are very visible at the entrance to the Abó Unit on US 60. They are also well exposed in the drainage of Cañon de Espinoso, on the drive into the Abó Unit. Most of these

Fluvial deposits of the Abo Formation including a stream-channel margin in the center and center left of the photo.

sandstones belong to the lower Yeso Group. Among the sandstones are a few thin layers of dolomite, and some of the sandstones have a high content of gypsum.

Beneath them are maroon to light-red rocks that include much mudstone—the beds of the Abo Formation. These beds are at the top of what geologists call the "type section" of the Abo Formation. A type section is the rock column that geologists consider as "typical" of the formation. U.S. Geological Survey geologist Willis T. Lee (1864–1926) coined the name Abo Formation in 1909 after Abó Canyon (albeit without using the accent mark over the letter "o") and designated the rock column directly south of the Abó Unit as the type section of the formation.

The pueblo at Abó Ruins was built primarily from local red sandstone slabs of the lower Permian Abo Formation and Yeso Group.

If You Plan To Visit

To reach the Abó Unit, drive about 9 miles southwest of Mountainair on US 60; turn right (north) on NM 513 and go 0.5 mile to the park. For more information: **www.nps.gov/sapu/learn/historyculture/abo.htm**

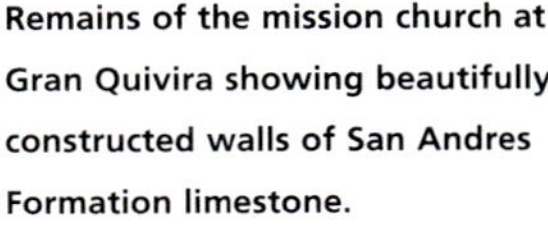

Remains of the mission church at Gran Quivira showing beautifully constructed walls of San Andres Formation limestone.

Gran Quivira structures on top of a hill with unexposed San Andres Formation.

Gran Quivira

The Gran Quivira Unit is located about 26 miles southeast of the town of Mountainair. At 611 acres, it is the largest of the three units (Abó, Gran Quivira, and Quarai) of the SPMNM.

Prior to Spanish contact during the late 1500s, Gran Quivira (Cueloze as it may have been called by the pre-Spanish Jumano people or Las Humanas, its earliest Spanish name) was an enormous community with multiple pueblos and kivas. Part of this city was Gran Quivira Pueblo, a 226-room structure from the Pueblo IV period (about A.D. 1300–1600) that is the largest and only fully excavated pueblo at the unit. The remains of two relatively short-lived mission churches dating from 1630 to about 1672, San Isidro and San Buenaventura, can be seen here, and exhibits at the Visitor Center provide a good overview of the history of the site.

Geography and Geologic Province

Gran Quivira sits on top of Chupadera Mesa, a large tableland near the western edge of the Great Plains where it adjoins the Rio Grande rift. Chupadera Mesa extends from near Mountainair in Torrance County on the north to near Carrizozo in Lincoln County on the south, a distance of about 55 miles. The mesa is developed in lower Permian rocks (approximately 280 million years old). To the west of Gran Quivira, the Manzano Mountains form the skyline.

The Rock Record

Between the Manzano Mountains and Chupadera Mesa lies dissected topography that exposes Pennsylvanian and Permian sedimentary rocks. Pennsylvanian rocks are only exposed on the crest of the mountain range or in nearby canyons, well west of Mountainair and Abó.

Aerial view of Gran Quivira showing churches, residence structures, and several circular kivas. The park Visitor Center is at the far left.

However, along and southeast of US 60, red rocks of the lower Permian Abo Formation and overlying Yeso Group are well exposed in arroyos and along cuestas and the slopes of mesas. More resistant rocks, the Glorieta Sandstone and San Andres Formation, overlie the red rocks and form the caprock of Chupadera Mesa. The top of the mesa is locally cut by geologically young (Tertiary) faults and dikes.

The Yeso Group along the western side of Chupadera Mesa is about 600 feet thick and consists mostly of slope-forming, salmon-colored siltstones and claystones with a few prominent layers of dolomite and gypsum capped by red sandstone. The Glorieta Sandstone is nearly 250 feet thick and is mostly tan and yellowish-orange, quartz-rich sandstone that displays crossbeds and ripple marks. The San Andres Formation is about 280 feet thick and is mostly gray beds of limestone. Most of these limestones were originally muddy, and most do not have obvious fossil fragments.

Essentially all of the construction at Gran Quivira was done using slabs of limestone from the San Andres Formation. Unfortunately, the San Andres rocks underlying the hill on which Gran Quivira is located also contain layers and nodules of gypsum, which is very soluble in water. Thus, over the centuries, gypsum removal by groundwater has led to cracking and even collapse of some of the massive stone structures at the site.

Geologic History

As discussed under the Abó Unit section, during the early Permian, all of the world's continents were united in a single supercontinent called Pangea. What is now New Mexico was along the western shoreline of Pangea nearly at the equator. The Chupadera Mesa area, northwest of an interior seaway, was an arid coastal plain during deposition of the Yeso Group sediments. As sea levels fluctuated, sand dunes formed during times of low sea levels, whereas tidal flats and restricted marine embayments formed during high sea levels. Reddish, gypsum-rich siltstones represent silt from deserts that was blown by the wind into shallow brine pools on tidal flats, where gypsum was being precipitated.

The Glorieta Sandstone was deposited as the climate continued to become more arid, creating a vast desert that extended from the Grand Canyon region of Arizona to central New Mexico.

During deposition of the San Andres Formation, sea levels were generally high, and the seas expanded to the west across most of central New Mexico and into eastern Arizona. Limestones of the San Andres Formation record shallow, mainly marine deposition in tropical waters. Locally, fossils of the inhabitants of that seaway, mostly brachiopods, bivalves (clams), and nautiloid cephalopods, can be found in the San Andres Formation.

Geologic Features

The top of Chupadera Mesa is a rolling plain mantled by soil covered with grassland. Therefore, little bedrock can be seen at Gran Quivira. The best places to see the Permian strata that underlie the mesa are along the highway from Mountainair to Gran Quivira, or southwest of Mountainair on US 60, and looking east toward Chupadera Mesa.

San Andres limestone blocks used to construct dwellings and kivas at Gran Quivira.

The Gran Quivira Pueblo was built of irregular blocks of limestone from the lower Permian San Andres Formation. In some of these limestone blocks, fossils of marine organisms, especially brachiopods, can be seen.

If You Plan To Visit

To reach the Gran Quivira Unit, drive 26 miles south of the town of Mountainair on NM 55 (making numerous right-angled turns en route). There is a Visitor Center near the parking lot. A 0.5-mile trail leads through partially excavated pueblo ruins and the ruins of the mission church. For more information:

www.nps.gov/sapu/learn/historyculture/gran-quivira.htm

Quarai Unit

The Quarai Unit of SPMNM is located near Punta de Agua, about eight miles north of Mountainair. At 97 acres, Quarai is the smallest of the three units of Salinas Pueblo Missions National Monument.

Pueblo Quarai (original name was Cuarac) was a very large pueblo when the Spanish first arrived in 1598. This was mainly due to the presence of a year round water source flowing from springs along nearby Zapato Creek (Cañon Zapato)—that canyon also provided relatively accessible rock for building construction. The Quarai Mission and Convento were setup at Quarai in 1626 by Fray Juan Gutierrez de la Chica. Construction on Nuestra Señora de la Purísima Concepción de Cuarac started in 1627 and continued to 1632. As at the other two mission sites, a combination of disease, drought, famine, and Apache raiding led to its abandonment in 1678.

Geography and Geologic Province

Like the Abó Unit, Quarai is on the eastern slope of the Manzano Mountains. The Manzano Mountains consist of a Proterozoic core of mostly metamorphic rocks capped by limestones of Pennsylvanian age. These Pennsylvanian rocks are also exposed in canyons and ridges near

El Misión Nuestra Señora de la Purísima Concepción de Cuarac Church at Quarai.

the mountain crest. However, farther east, at Quarai, the bedrock is younger—of early Permian age.

Quarai is located north of Chupadera Mesa on the western edge of the Great Plains physiographic province, near its transition with the Rio Grande rift.

Rock Record

Bedrock at and around the Quarai Unit, as at the Abó unit, consists of red sandstones and mudstones of the lower Permian Abo Formation, overlain by orange sandstones of the lower Yeso Group. In this area, the Abo Formation is nearly 1,000 feet thick and is mostly slope-forming mudstone and ledgy beds of sandstone that contain crossbeds and ripple marks. These Abo sandstones are darker colored, coarser grained and more feldspar-rich than are sandstones of the overlying Yeso Group.

The lower Yeso Group near Quarai consists mostly of orange beds of fine-grained sandstone, typically with ripple marks. A few thin beds of dolomite are also present, and many of the sandstones are rich in gypsum.

At and around Quarai, Permian bedrock is not well exposed. Instead, it is largely covered by Quaternary alluvial gravel derived from the Manzano Mountains. The clasts of this gravel range in size from pebbles to boulders and are made up of Precambrian rocks from the mountain core (mostly gneiss, schist, and quartzite) as well as Pennsylvanian limestone and Permian sandstone. The valley of Cañon Zapato is full of finer grained stream-channel alluvium of Quaternary age.

Geologic History

The geologic history of the bedrock at the Quarai Unit is the same as at the Abó Unit—river floodplains of the Abo Formation followed by the arid coastal plain of early Yeso deposition.

Geologic Features

The rimrock and canyon walls of Cañon Zapato expose beds of lower Permian sandstone at Quarai. The pueblo walls are built from these sandstones.

—Spencer G. Lucas

Aerial view of Quarai including (at upper center) El Misión and Convento with its unusual central square kiva.

El Misión Nuestra Señora de la Purísima
Concepción de Quarai Church.

If You Plan To Visit

To reach the Quarai Unit, drive 8 miles north of Mountainair on NM 55, then turn west at the signpost and drive 1 mile. There is a Visitor Center and a 0.5-mile trail through the ruins and stands of old cottonwood trees. For more information:

Salinas Pueblo Missions
National Monument
PO Box 517
Mountainair, NM 87036-0517
(505) 847-2585
**www.nps.gov/sapu/learn/historyculture/
quarai.htm**

Sumner Lake, Bosque Redondo, and Fort Sumner

Located in the Pecos River Valley, 13 miles northwest of Fort Sumner, Sumner Lake (known until 1974 as Alamogordo Reservoir) offers a variety of pursuits to the outdoor enthusiast, including camping, fishing, hiking, boating, birding, and horseback riding. Formed by damming of the Pecos River by the U.S. Bureau of Reclamation in 1936–1937, Sumner Lake is a Y-shaped reservoir with arms extending up the valleys of the Pecos River to the northwest and Alamogordo Creek to the north. At an elevation of 4,300 feet, the lake surface encompasses about 4,500 acres when at full capacity. The area around Sumner Lake displays good exposures of the Upper Triassic Dockum Group and Pleistocene-age topographic benches.

Regional Setting

Sumner Lake is located in the Pecos valley portion of the vast Great Plains province, between the High Plains to the east and the Southern Rocky Mountains and Rio Grande rift to the west. The headwaters of

Sumner Lake with shoreline exposures of sandstones of the Dockum Group at sunrise.

the Pecos River lie to the northwest in the Sangre de Cristo Mountains, east of Santa Fe. About 6 miles wide near Sumner Lake, the modern valley of the Pecos River exposes red beds of the Dockum Group and younger alluvial deposits.

The Rock Record

The red beds around Sumner Lake were deposited in a very large basin (the Chinle–Dockum basin), which formed in southwestern North America during the Late Triassic, about 230–205 million years ago. This basin stretched from west Texas across parts of New Mexico, Colorado, Arizona, and Utah to central Nevada. About 700 feet thick in east-central New Mexico, the basin-fill consists mostly of red-hued sandstones and mudstones. These beds are termed the Dockum Group, although some workers have applied the Arizona term Chinle to these beds.

The broad, nearly flat upland surfaces that lie about 100–300 feet above Sumner Lake to the east and west formed at a time when the floor of the Pecos valley was higher than it is today. An example of such a surface is traversed by US 84 east of the state park. These broad, gently sloping, low-relief surfaces are typically underlain by 10–200 feet of sand and gravel of the ancestral Pecos River and its tributaries. These surfaces are commonly capped by thick accumulations of soil-formed calcium carbonate (caliche) that are indicative of landscape stability in a semiarid to arid climate.

Geologic map of the Sumner Lake-Bosque Redondo area.

Geologic History

During Late Triassic time, a broad region of southwestern North America began to subside behind a developing subduction system and volcanic arc along the west coast. This subsidence produced an immense basin, termed the Chinle–Dockum basin. Rivers within the basin generally flowed west or northwest toward an ancient ocean in Nevada. Lying only a few degrees north of the equator, rivers in what is now east-central New Mexico deposited oxidized sediments in a tropical setting. These sediments became reddened by development of iron oxide cements during burial, giving rise to the characteristic coloration of the Dockum Group. After Dockum deposition ceased about 205 million years ago, episodic accumulation of marine and continental deposits continued through parts of Jurassic, Cretaceous, and early Cenozoic time, burying the Dockum Group to a depth of nearly a mile.

Regional uplift began about 30 million years ago in response to changes in mantle density caused by regional volcanism. Uplift led to thousands of feet of erosion that stripped off most of the strata overlying the Dockum Group. Erosion continued until about 12 million years ago, when the Ogallala Formation began to be deposited by east-flowing rivers that drained the Southern Rocky Mountains. Ogallala sediments continued to accumulate until about 5 million years ago. Now mostly restricted to the High Plains to the east, the Ogallala Formation formerly extended westward beyond the Sumner Lake area. The top of the Ogallala Formation projects to an elevation nearly 1,000 feet above Sumner Lake. This gives an estimate of the depth of erosion by the Pecos River since 5 million years ago.

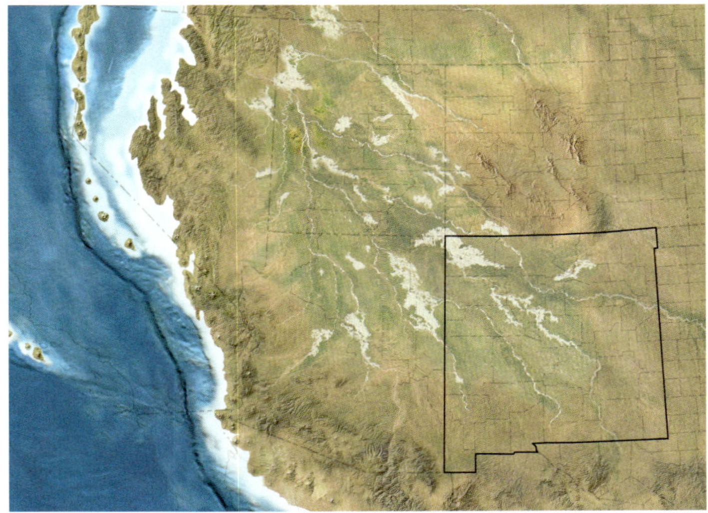

Paleogeographic map of the southwestern United States showing the vast Chinle–Dockum river basin that drained northwest to the sea. 205–230 million years ago.

The post-Ogallala cutting of the Pecos Valley was interrupted by an episode of valley widening, sediment accumulation, and soil development, probably during the early Pleistocene. This episode is represented by the broad topographic benches east and west of Sumner Lake, a few hundred feet above the lake. The Pecos River has subsequently incised 100–300 feet beneath these surfaces.

Geologic Features

The Dockum Group is exposed in many places around Sumner Lake. Especially good are the exposures at the dam spillway, where the basal sandstone of the Dockum Group (the Santa Rosa Sandstone) is overlain by interbedded mudstones and thin sandstones (the Tecovas Formation). Lenticular, sand-filled fluvial river channels surrounded by finer-grained floodplain deposits are clearly visible, especially in the Santa Rosa section.

The Pleistocene topographic benches are easily viewed along US 84 east of Sumner Lake, and southwest of the lake on Buchanan Mesa along US 60. The alluvial deposits underlying the benches can be seen along the road to Sumner Lake State Park, about 4 miles west of the turn-off from US 84.

Dockum Group exposures at the spillway of the Sumner Lake dam. The lower two-thirds of the cliff is the Santa Rosa Sandstone; the upper third is the Tecovas Formation.

Bosque Redondo Memorial and Fort Sumner Historic Site

Located about 5 miles southeast of the modern town of Fort Sumner, the old Fort Sumner was the headquarters of the large (million acre) Bosque Redondo Indian Reservation. Diné (Navajo) and the N'de (Mescalero Apache) men, women, and children were forcibly rounded up and marched to the area starting in 1863 where they lived in squalor on those lands, with widespread hardship and starvation, until 1868. These two sites, directly adjacent to each other, detail that shameful history and honor the victims with an interpretive trail and a new Visitor Center of striking design by Navajo architect David Sloan.

Near the northeast corner of the Fort Sumner Historic Site lies the grave of William H. Bonney, a.k.a. Billy the Kid. Three months after shooting his way out of jail in Lincoln, New Mexico, Billy the Kid was tracked down and killed at old Fort Sumner by Sheriff Pat Garrett on July 14, 1881.

Although there is little of geologic interest at Bosque Redondo, it allows access to the floodplain, channel and terraces of the Pecos River and provides views of nearby irrigation canals that show the importance of Pecos River waters to the agricultural productivity of the region.

—*Steven M. Cather*

Bosque Redondo Visitor Center.

If You Plan to Visit

Sumner Lake State Park is located 13 miles northwest of Fort Sumner. It can be reached by traveling west about 7 miles from US 84 on NM 203. For more information:

Park Superintendent
32 Lakeview Lane
Sumner Lake, NM 88119
(575) 355-2541
www.emnrd.state.nm.us/SPD/sumnerlakestatepark.html

Memorial to the Long Walk of the Navajo, a deportation (1863-68) in which Navajo and some Mescalero Apache families were forced to walk from their land in present-day Arizona to resettlement in eastern New Mexico.

Oasis State Park
NEW MEXICO HISTORIC SITES

The lake at Oasis State Park.

Nestled among sand hills in a broad, shallow valley on the High Plains about 5 miles north of Portales, Oasis State Park is a popular weekend destination. The park hosts a man-made lake, numerous campsites, and a Visitor Center, as well as opportunities for fishing, hiking, and birding. The area around the park provides evidence of a surprisingly complex geologic history, one that includes regional uplift, deposition of a widespread apron of alluvium (the Ogallala Formation) derived from the Rocky Mountains to the west, subsequent cutting of the Portales Valley by the early Pecos River, and the presence, during the Late Pleistocene, of abundant springs and ponds that drew early humans to the area.

Regional Setting

The High Plains extend from South Dakota to southeastern New Mexico and west Texas, a vast region underlain mostly by the middle Miocene to lower Pliocene Ogallala Formation. The Ogallala Formation, typically several hundred feet thick, consists of sand, gravel, and mud that were derived mostly from erosion of the Southern Rocky Mountains to the west and deposited by rivers and winds beginning about 12 million years ago. As noted in the introduction to Part 6, the Ogallala Formation forms a major part of America's most important groundwater aquifer, supporting agriculture in eight mid-continent states. In east-central New Mexico, the Ogallala Formation buried an eroded landscape atop the Upper Triassic Dockum Group and, in some areas, Lower Cretaceous marine rocks. The topography beneath the Ogallala Formation consists of buried ridges, hills, and valleys. These paleovalleys sloped eastward at about 15 feet per mile and were as much as a few hundred feet deep. One of these paleovalleys lies approximately beneath the modern Portales Valley.

The upper surface of the Ogallala Formation has remarkably low relief, and traveling across it has been compared to sailing the ocean. It slopes gently southeastward and is topped in many places by a caprock consisting mostly of calcium carbonate (caliche) that accumulated in soils after deposition of the Ogallala Formation ceased about 5 million years ago.

The Rock Record

The Ogallala Formation and underlying Mesozoic rocks are poorly exposed at Oasis State Park. Interested readers can view good outcrops of these rocks in the Taiban Mesa area between Fort Sumner and Melrose, 40 miles to the northwest. Most of the landscape near Portales is veneered by thin, post-Ogallala (Pliocene and Quaternary) sediments that obscure the older strata. The nearby Blackwater Draw quarry, however, provides an excellent view of uppermost Ogallala and post-Ogallala deposits.

The Portales Valley is a broad, shallow, east-southeast-trending valley that was cut into the Ogallala Formation after cessation of Ogallala deposition. Near Portales, the valley floor is about 10 miles wide and lies about 260 feet lower than the town of Clovis, which

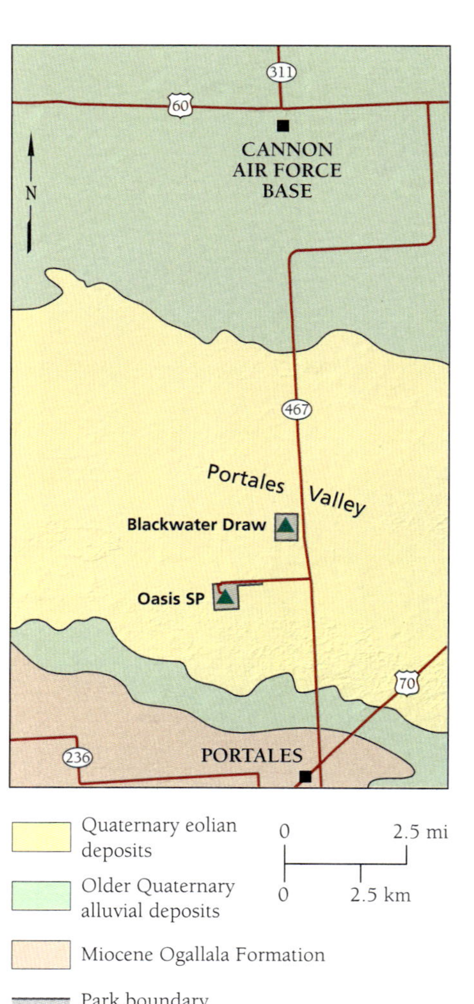

Geologic map of the Portales Valley.

sits on its north rim. The floor of the Portales Valley slopes eastward less steeply than the top of the Ogallala Formation into which it was incised. Consequently, the valley becomes shallower to the east. Near Lubbock, Texas, the valley is almost imperceptibly shallow. The floor of the Portales Valley is underlain by alluvium as much as 120 feet thick, the unexposed, basal part of which contains gravels of the ancestral Pecos River. The age of these gravels is poorly known, but is probably Pliocene and/or Pleistocene.

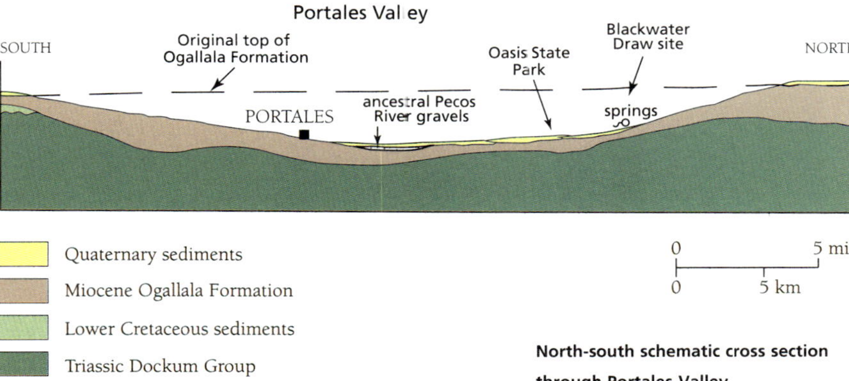

North-south schematic cross section through Portales Valley.

Geologic History

During the Oligocene to early Miocene (about 30–15 million years ago), the southern High Plains and adjoining regions were broadly uplifted due to mantle-density changes caused by major volcanism to the west and southwest. Uplift caused profound erosion (nearly cne mile deep), exposing Upper Triassic and Lower Cretaceous strata in the region near Portales.

About 12 million years ago, deposition of the Ogallala Formation began in response to rising rift-related mountains to the west, burying the older erosional topography. The relatively dry climate fostered deposition of Ogallala sediments via ephemeral stream and wind transport until about 5 million years ago, when increased monsoonal precipitation ended Ogallala deposition, incision began, and the cap-rock caliche started to develop on the abandoned depositional surface.

Since the end of Ogallala deposition, the southern High Plains has experienced mostly erosion punctuated by minor episodes of deposition. During this time, the east-flowing upper Pecos River, which served as the headwaters of the Brazos River of Texas, cut the Portales Valley. The valley is deeper to the west, implying that slight eastward tilting of the High Plains was occurring during valley incision.

The upper Pecos River was subsequently captured near Fort Sumner by the south-flowing lower Pecos River, whereupon the upper Pecos became a tributary to the Rio Grande and the Portales Valley was abandoned to local stream and wind processes. During the late Pleistocene (about 13,000 years ago), increased rainfall and associated high water tables gave rise to springs and ponds along the north margin of the

Vegetatively stabilized eolian dunes near Oasis State Park.

Portales Valley. These attracted a wide variety of large mammals, which in turn drew hunters of the Clovis and Folsom cultures to the area (see Blackwater Draw site).

Geologic Features

The sand dunes around Oasis State Park are perhaps the most interesting geologic features in the area. The park lies in a swath of eolian dunes about 50 miles long and 3–12 miles wide that extends east-west along the northern margin of the Portales Valley. Most of the dunes have crescentic, barchan shapes. Dunes advance as wind sweeps sand up the windward face of the dune to the crest, whereupon grains cascade down and are deposited on the steep leeward face of the dune, forming steep crossbeds that incline generally in the down-wind direction. Dunes are as much as 40 feet high and are now mostly stabilized by vegetation, indicating wetter conditions today than when the dunes first formed. Some of the smaller dunes are still active, in part due to destruction of stabilizing vegetation in areas of heavy foot traffic.

—*Steven M. Cather*

If You Plan to Visit

Oasis State Park is 1.8 miles west of NM 467, about 5 miles north of Portales. For more information:

Oasis State Park
1891 Oasis Road
Portales, NM 88130
(575) 356-5331
www.emurad.state.nm.us/SPD/oasisstatepark.html

Blackwater Draw National Historic Landmark

Blackwater Draw Locality 1 (a.k.a. the Clovis site) is an archaeological park located on the north side of the Blackwater Draw in Roosevelt County. This landmark is a collection of well-preserved archaeological sites situated around the remnants of a Pleistocene-age lake. The area is renowned for its remarkable preservation of paleontological and cultural remains associated with the earliest defined culture in North America, which is about 13,000 years old. The Blackwater Draw

A southwest-facing overview toward the Interpretive Center at Blackwater Draw National Historic Landmark.

National Historic Landmark lies between the cities of Clovis and Portales and covers approximately one square mile on the north side of the Portales Valley, an ancient river valley carved into the Llano Estacado by the ancestral Pecos River.

The site is dominated by three large quarry pits and associated sediment piles generated from the decades of gravel mining that heavily impacted the site. The South Pit contains one of the few visible exposures of the Miocene Ogallala Formation in the region. Ogallala gravels exposed in the north wall of the pit record river systems that drained the Rocky Mountains to the northwest, prior to the cutting of the Portales Valley. A fossil elephant-like gomphothere recovered from the South Pit indicates the Ogallala gravels there are several million years old. The Ogallala Formation at the Clovis site is overlain by lake, spring, stream, and wind-deposited sediments, ranging in age from about 13,000 years to the present, which accumulated along the north margin of the Portales Valley.

Prehistorically, the site was one of many small spring-fed lake basins that covered the Southern High Plains. Stratigraphic evidence tells us that this former wetland began to dry up about 8,500 years ago and, although there have been periods of respite, the area has remained semi-arid through the present day. The region held abundant shallow groundwater well into the 20th century, but recent industrial agricultural practices have led to substantial depletion, lowering the local water table by as much as several hundred feet in some areas. Until 1974, the pits regularly filled with water from seeps, creating conditions for dredge mining from barges. More recently, a combination of local conservation efforts and natural reclamation has reestablished a native plant community now vital to wildlife.

Overview and Significance

The Clovis site was discovered by local artifact collectors in the late 1920s. The site contained not only many spear points, but many large, well-preserved mammal bones. Scientific research began in earnest in 1932 and continues through the present day under the auspices of Eastern New Mexico University. During the first few field seasons of the 1930s, a human presence beginning at the end of the Pleistocene was identified with the discovery of tools made from bone, ivory, and stone in association the remains of extinct animals. These finds demonstrated for the first time, beyond any doubt, that humans were hunting and butchering mammoth

An iconic Clovis point from a mammoth kill at Blackwater Draw Locality 1.

and other now-extinct animal species in the Americas. Subsequent excavations over the following decades showed that mammoth and bison hunting was practiced regularly around the pond margin and that the Clovis locality was a seasonally utilized hunting and butchering site. Other animals discovered in human context include (but are not limited to) camelops, dire wolf, short-face bear, horse, giant sloth, deer, elk, turtle, and pronghorn.

This north-facing profile provides a rare opportunity to examine the underlying sediments of the Llano Estacado.

Geomorphology and Unique Preservation

The Llano Estacado, on which the Clovis site is located, is the southern expression of the High Plains in the region south of the Canadian River valley. This high plateau, and particularly the Portales Valley, has aggraded mostly by wind-borne sediment deposition over the past several million years, covering and preserving a remarkable geologic record of a changing environment. The Southern High Plains are extremely flat and have little erosion by flowing water, creating conditions nearly perfect for the preservation of delicate materials, including bone and ivory tools. This level of landform stability is almost unknown elsewhere in archaeological contexts. In some areas, the cultural deposits lie more than 15 feet below the surface and may not have come to the attention of the scientific community were it not for the massive blowouts by wind erosion during the Dust Bowl conditions of the 1930s and again in the 1950s.

When an animal dies or humans leave natural detritus behind, most environments consume the animal tissues rather quickly, including bone and ivory, leaving only the most durable materials behind

Stratigraphic profile in the northern pit during excavations circa 1963 showing the depth of deposits to the Clovis cultural materials; shovel for scale. Some portions of the site with the earliest human occupations dating back over 13,000 years are blanketed by 15 feet of sediment.

(generally stone) for archaeologists to study. However, in the Blackwater Draw region, the unique combination of shallow, often anaerobic groundwater with high calcium content, land-form stability, and continuous aggradation encapsulated whole animal skeletons, pollen, charcoal, diatoms, and the tools left behind by the first human inhabitants.

While the archaeological site serves as a tourist stop for the archaeo-logically inclined, the primary function of the Blackwater Draw site is a research facility for Quaternary studies of all types. Eastern New Mexico University curates an enormous collection of artifacts, faunal specimens, and geologic information from decades of work in the area, offering many possibilities for students and professionals in the archaeological, paleontological, and environmental communities.

—*George Crawford*

If You Plan to Visit

To reach Blackwater Draw National Historic Landmark take US 70 to the northeastern outskirts of Portales; then turn onto NM 467 and go north for about six miles. The archaeological site is on the west side of the road and has a Visitor Center and a unique inter-pretive display of in situ cultural materials. There is a 1-mile nature trail that traces the margin of the former lake with signs explaining some of the unique cultural aspects of the site. This exceptional property is managed by Eastern New Mexico University, which also has an associated museum dedicated to the Clovis site. The museum is accessible year-round at its new location in Lea Hall, on the university campus in Portales. The National Landmark is open seasonally from April through October; please call for hours of operation. For more information:

Blackwater Draw Museum
Eastern New Mexico University
1500 S. Ave K, Lea Hall, Rm. 163
Portales, NM 88130
(575) 562-2202
my.enmu.edu/web/blackwater-draw/home

Late Paleoindian age *Bison antiquus* from the South Bank kill area, visible in the Interpretive Center.

agate—A translucent variety of cryptocrystalline quartz that has a waxy luster and a wide variety of colors.

alluvial fan—A fan-shaped mass of sediment deposited by a stream at the mouth of a canyon in arid or semiarid regions.

alluvium—Unconsolidated mud, sand, and/or gravel deposited by rivers, streams, or sheetwash.

alunite—A mineral that is generally a hydrothermal-alteration product of feldspar-rich igneous rocks.

amalgamate—To create a craton by joining together two different geologic terranes.

ammonite—A marine mollusk (cephalopod) characterized by a coiled shell with sutures having finely divided lobes and saddles.

amygdules—Voids in volcanic rock formed by gas bubbles and later filled by other minerals.

andesite—A dark, finely crystalline volcanic rock composed principally of plagioclase feldspar and one or more mafic (Fe- and Mg-bearing) constituents.

angular unconformity—The surface between two groups of rocks of different ages whose bedding planes are not parallel or in which the older, underlying rocks dip at a different angle (usually steeper) than the younger, overlying strata.

anhydrite—A mineral consisting of anhydrous calcium sulfate ($CaSO_4$).

anticline—An arch-like fold in which strata dip in opposite directions from a common ridge or axis; the core of an anticline contains the older strata.

apron—An extensive blanket-like deposit of alluvial, glacial, eolian, marine, or other unconsolidated material derived from an identifiable source, and deposited at the base of a mountain, in front of a glacier, etc.

aragonite—The orthorhombic version of calcium carbonate ($CaCO_3$).

arthropod—Any one of a group of solitary marine, freshwater, and aerial invertebrates belonging to the phylum Arthropoda, characterized chiefly by jointed appendages and segmented bodies.

aquifer—A body of rock or sediment capable of storing and transmitting water.

ash (volcanic)—Fine-grained uncemented volcanic rock and glass ejected during eruptions.

ash flow—A superheated mixture of volcanic ash and gas that forms typically as a result of the explosive eruption of viscous lava that flows across the Earth's surface.

back reef—The landward side of a reef. In some places, as on a platform-edge reef tract, "back reef" refers to the side of the reef away from the open sea, even though no land may be nearby.

basalt—A dark-colored, finely crystalline volcanic rock, rich in iron and magnesium, which typically erupts from fissures and low-relief volcanoes.

basement—The crust of the Earth below sedimentary deposits, extending downward to the Mohorovicic discontinuity.

basin—A large depression or area of subsidence in which strata incline inward.

batholith—A large igneous intrusion having an aerial extent of at least 40 mi^2 and extending deep into the Earth's crust.

biota—All living organisms of an area; the flora and fauna considered as a unit.

bluff—Any cliff with a steep broad face.

bombs—Volcanic fragments, from the size of an apple and upward, ejected from a volcano.

brachiopod—A bivalved marine invertebrate, commonly found as fossils, mainly in Paleozoic rocks, but still existing today in cold marine waters.

brackish water—A term for water with an intermediate salinity between average seawater and fresh water.

breccia—A coarse-grained clastic deposit formed of broken angular fragments of rock.

bryozoan—A small, colonial aquatic invertebrate that resembles either twigs or has moss-like patterns.

calcareous—Said of a substance that contains calcium carbonate (calcite).

calcite—A trigonal mineral composed of calcium carbonate ($CaCO_3$). The primary mineral of limestones.

caldera—A large volcanic crater, often 10–20 miles in diameter, that forms after a gas-charged viscous magma body erupts explosively. Following the eruption, the shallow roof of the chamber collapses, forming a large, circular depression.

caliche—Gravel, sand, silt, etc., cemented by calcium carbonate. Forms at the surface near the base of the upper soil layers.

cephalopod—A marine mollusk. Fossil forms have a chambered calcareous shell. Modern forms include nautiloids, squid and octopus.

chalcedony—A translucent variety of fibrous quartz having a waxy luster. May be white, blue, brown, gray, or black.

chert—See agate.

cinder—Small rock fragments blown from an erupting volcano.

cinder cone—A steep hill composed of pyroclastic debris (cinders) built around a volcanic vent.

clast—An individual constituent, grain, or fragment of a sediment or rock, produced by the mechanical or chemical disintegration of a larger rock mass.

clastic—The general term used for rock or sediment composed of broken fragments of pre-existing rocks that have been transported some distance.

conglomerate—A coarse-grained, poorly sorted clastic rock formed principally of rounded fragments of rock larger than 2 mm in size.

corundum—An extremely tough mineral (Al_2O_3) with a hardness of 9 on the Mohs scale, often used as an abrasive.

crater—A steep-walled depression on top of a volcanic cone above the pipe or vent that feeds the volcano.

craton—A part of the Earth's crust that has attained stability and has been little deformed for a prolonged period.

crinoid—An echinoderm with numerous radiating arms, typically attached to the sea floor. The segmented stems of Paleozoic crinoids are especially abundant as fossils in Mississippian and Pennsylvanian rocks.

crust—The outermost layer of the Earth that is composed of sedimentary, igneous, and metamorphic rocks.

crystallization—To form a solid from a liquid.

dacite—A volcanic rock consisting of plagioclase and orthoclase feldspars, quartz, and mafic minerals such as pyroxene.

debris flow—A moving mass of rock fragments, soil, and mud, more than half of the particles being larger than sand size. Slow debris flows may move less that 3 feet per year; rapid ones can reach 100 mi per hour.

deformation—The process of changing shape of a rock through pressure and temperature.

detritus—A collective term for loose rock or mineral material that is mechanically eroded from pre-existing rocks and transported some distance.

diagenesis/diagenetic—All changes, both physical and chemical, that occur to sediments after deposition.

dike—A tabular intrusive body of rock that cuts across adjacent rocks.

diopside—A magnesium and calcium pyroxene mineral ($MgCaSi_2O_6$).

diorite—An intrusive, coarsely crystalline igneous rock that is equivalent to volcanic andesite.

dip—The inclination of strata over an area expressed in degrees from horizontal.

dissolution—The process of dissolving minerals in water.

dolomite—A term used for both a mineral and a sedimentary rock with the composition $CaMg(CO_3)_2$.

dome—Two anticlines that intersect each other. This forms distinct, rounded spherical protrusions on the Earth's surface, like a dome on a building.

embayment—A recess in a coastline, like a bay.

endemic—Said of an organism or group of organisms that is restricted to a particular region or environment.

eolian—Deposited by wind.

epidote—A secondary mineral ($Ca_2Al_2(Fe,Al)(SiO_4)(Si_2O_7)O(OH)$) formed during metamorphism.

escarpment—A long, more or less continuous cliff or relatively steep slope facing in one general direction, breaking the continuity of the land by separating two level or gently sloping surfaces, and produced by erosion or faulting.

evaporites—A mineral deposit formed by evaporation of a body of water.

extensional fault—A break (with displacement) formed by stretching the Earth's crust.

extrusive—Refers to igneous rocks that erupted and quickly cooled at the Earth's surface.

facies—A body of rock that has common composition, appearance, biologic content and/or mode of formation.

fault—A fracture or break along which displacement has occurred.

feldspar—The most common group of rock–forming minerals with the general formula $(K,Na)AlSi_3O_8$–$CaAl_2Si_2O_8$. They make up more than 50% of the Earth's crust. Since they easily break down during physical and chemical weathering, feldspars are less abundant than quartz in terrigenous sediments.

felsic—The term applied to igneous rocks with an abundance of light-colored minerals, typically quartz, feldspar, and muscovite.

fissure vent—An elongated crack in the earth's surface from which lava and/or pyroclastic material has erupted.

fluorite—A cubic mineral (CaF_2) that occurs in many different colors (often blue or purple), has a hardness of 4 (Moh's scale) and is commonly associated with ore deposits.

fluvial—A sedimentary deposit transported and laid down by a stream.

fore reef—The seaward side of a reef that consists of a slope covered with coarse and fine reef debris.

formation—A mappable rock unit, consisting of one or more types of rocks deposited essentially without interruption and distinctive from rock units above and below.

fusulinid foraminifera—An extinct protist single-celled organism that formed rice-shaped calcium carbonate shell.

gastropod—A mollusk characterized by a distinct head with eyes and tentacles, and, in most cases, a single, unchambered shell of aragonite ($CaCO_3$). Modern-day gastropods include snails and abalone.

geochronology—The science of dating and determining the time sequence of events in the history of the Earth.

gneiss—A foliated metamorphic rock, formed by regional metamorphism of igneous rock (like granite), in which mineral grains are aligned in distinct bands or lenses.

graben—A downthrown fault block bounded on either side by normal faults.

granite—A light-colored igneous rock composed mainly of quartz and feldspars. Granites have large crystals formed from the slow cooling of a magma chamber.

granodiorite—A group of coarsely crystalline plutonic rocks intermediate in composition between quartz diorite and quartz monzonite, containing quartz, plagioclase, potassium feldspar, and biotite, hornblende, or, more rarely, pyroxene, as the mafic components.

greenhouse—Times in the geologic past when the Earth's climate was hotter and, thus, are times of little or no global glaciation.

group—A formal rock unit combining two or more formations.

gypsum—A mineral that is composed of hydrated calcium sulfate ($CaSO_4$ $2H_2O$)

halite—A cubic mineral ($NaCl$); also known as native salt or table salt.

hoodoos—The informal term for erosional landforms that resemble spires, pillars, and columns. They form through the differential weathering of horizontal sediments, often in arid landscapes.

hornblende—The most common mineral of the amphibole group $(Ca,Na)_{2-3}(Mg,Fe^{+2}, Fe^{+3},Al)_5(OH)_2$ $[(Si,Al)_8O_{22}]$ with a variable composition. It is typically black, dark green, or brown and is a common metamorphic mineral in gneiss and schist.

hydromagmatic—A general term to include all processes, subsurface or surface, involving interaction of magma or magmatic heat with meteoric or connate water in the Earth.

hypersaline—Water that has salinity much higher than seawater.

icehouse—Times in the geologic past when the Earth's climate was colder producing large ice-sheets. Possibly caused by lower CO_2 and other greenhouse gases in the atmosphere.

igneous rock—A rock formed from molten material. Igneous rocks can crystallize at or near the surface or deep beneath the surface.

ignimbrite—The term for a volcanic deposit that results from a pyroclastic flow; includes ash-flow tuffs and welded tuffs.

incision—The process whereby a downward-eroding stream deepens its channel or produces a narrow, steep-walled valley.

intrusive—Referring to igneous rocks that are emplaced and cool slowly at depth beneath the Earth's surface (for example, a granite pluton).

karst—A type of typography that generally forms in limestone terrains and is characterized by subsurface dissolution features (caves and sinkholes) and an absence of surface streams.

lahar—A destructive mudflow that occurs on the slope of an erupting volcano.

lava tube—A subterranean hollow tunnel or cave in lava, which forms in a variety of ways as molten lava escapes from an eruptive vent.

lazulite—An azure-blue to violet-blue or bluish-green mineral $MgAl_2(PO_4)_2(OH)_2$.

leeward—Said of the side or slope sheltered or protected from the wind.

limestone—A sedimentary rock composed chiefly of calcium carbonate (calcite). Limestone can form by either organic or inorganic means.

liquefaction—Catastrophic failure of loose, cohesionless material that is saturated with water, usually occurring after an earthquake. Can occur at the Earth's surface or below the ocean.

lithic—A rock deposit that contains abundant fragments of previously formed rocks.

maar—A low-relief volcanic crater formed from a shallow steam and magma explosion.

mafic—The term applied to igneous rocks composed chiefly of dark colored ferromagnesian minerals (rich in iron and magnesium).

magmatism—The development and movement of magma, and its solidification to igneous rock.

magnetite—A black, cubic, magnetic mineral $(Fe^{2+}Fe^{3+})Fe^{3+}O_4$. Common in all types of rocks.

malpais—Spanish for "bad lands," usually applied to the rough-surfaced areas covered by basalt.

mantle—The layer in the Earth's interior between the core (center) and the crust.

metamorphic rock—A rock that has formed from pre-existing rocks at depth, without complete melting, in response to changes in heat, pressure, stress, and the chemical environment.

mollusk—An invertebrate belonging to the phylum Mollusca. Among the classes included in the mollusks are the gastropods, bivalves, and cephalopods.

monocline—A single bend in otherwise uniformily dipping or horizontal strata.

monzodiorite—A intrusive igneous rock with less than 5% quartz and 65 to 90% plagioclase and albite.

mudstone—A hardened sedimentary deposit made up of particles of clay and silt.

normal fault—A type of fault where the footwall moves upward with respect to the hanging wall.

olivine—An olive-green orthorhombic mineral $(Mg,Fe)_2SiO_4$. Olivine is a common rock-forming mineral of basic, ultrabasic, and low-silica igneous rocks (gabbro, basalt, peridotite, dunite). It crystallizes early from magmas at high temperatures.

ooids—A spherical to ellipsoidal grain (0.25 to 2.00 mm in diameter) with a nucleus covered by one or more precipitated concentric coatings that have radial and/or concentric orientation of constituent crystals. When part of a rock, it is termed oolitic.

orogeny—Mountain building episode, generally involving large-scale faulting, magma intrusions, uplift, volcanic activity, and deformation.

pahoehoe—The term for basaltic lava flows that are characterized by a smooth, billowy or ropey surface.

pelycosaurs—An informal group of Late Paleozoic synapsids (mammel-like reptiles). Dimetrodon is a common member of this group in New Mexico.

peridot—Gem-quality olivine.

pinnacle—A tall slender tapering tower or spire-shaped pillar of rock.

pisoid—A spherical to ellipsoidal grain that has concentric laminations. Similar to an ooid, but larger (>2 mm in diameter). When part of a rock, it is termed pisolitic.

phenocryst—A large crystal of a mineral in a porphyritic rock.

piedmont—A flat to gently sloping surface formed at the base of a mountain, like a thick alluvial aprons.

placer—Mineral deposit formed by mechanical concentration, usually by water, of heavy mineral particles, such as gold, from weathered debris.

playa—A dry, flat, vegetation-free area at the lowest part of an undrained desert basin, typically filled with evaporite deposits and sometimes occupied by ephemeral lakes.

pluton—A deep-seated igneous intrusion.

porosity—The percentage of open space between grains or in fractures in rock or sediment.

porphyry—An igneous rock containing two distinctly different sizes of crystals—typically large crystals (phenocrysts) in a fine-crystalline groundmass. Represents two different modes of cooling.

potash—An industry term for a group of potassium-bearing salts that includes potassium chloride and may also refer to potassium sulfate, nitrate, and oxide.

pyrite—A brassy yellow, metallic cubic mineral (FeS_2). Also known as "Fool's Gold."

pyroclastic—The general term for clastic rock material ejected during a volcanic eruption.

pyroxene—A group of dark, rock-forming, igneous silicate minerals common in mafic rocks.

quartz—The second most common rock-forming mineral (SiO_2). It is an extremely stable mineral and is more common than feldspars in sedimentary rocks.

quartz monzonite—An intrusive igneous rock with 5 to 20% quartz and 35 to 65% plagioclase and albite.

quartzite—A metamorphic rock formed by heat and compression of quartz sandstone.

radiogenic—Said of a product of a radioactive decay process.

radiometric—Pertaining to the measurement of geologic time. Often imprecisely used to refer to isotopic dating.

redbed—Sedimentary rocks of sandstone, siltstone, or shale that are red colored due to containing ferric oxide (hematite) coating individual grains.

rhyolite—A compositional term that refers to generally light colored, finely crystalline igneous rocks that are rich in quartz and potassium feldspar, similar in composition to granite.

rift—A long, narrow series of basins that are bounded on either side by normal faults.

sabkha—A supratidal sedimentary environment, which often forms in arid or semi-arid climates, characterized by saline evaporite deposits.

sandstone—A cemented sedimentary deposit consisting of sand-sized grains. Most sandstones consist largely of quartz (SiO_2).

scarp—A line of cliffs produced by faulting or erosion, or both.

scissor fault—A fault on which there is increasing offset or separation along the strike from an initial point of no offset, with reverse offset in the opposite directions.

sedimentary rock—A rock that results from the consolidation of loose sediment, such as clastic sediments (as in the case of sandstone) or chemically/organically precipitated sediments (as in the case of limestone).

shale—A finely laminated, hardened sediment composed of silt and clay particles.

schist—A strongly foliated crystalline rock, formed by metamorphism, that can be readily split into thin flakes or slabs because of the well-developed parallelism.

scoria—Irregular, bomb-sized pyroclastic material that is generally very vesicular. Also known as a cinder.

shoal—A submerged marine bank or bar consisting of or covered by sand or other unconsolidated material.

seismic—Movements related to shock waves generated in association with fault activity or induced artificially (as in seismic exploration).

silicic—Said of a silica-rich igneous rock or magma.

sill—A flat sheet of igneous rock intruding between, and parallel with, the existing rocks.

silt—Unconsolidated grains ranging in size from 0.0039 to 0.0625 mm.

siltstone—A hardened sedimentary deposit consisting of silt-size grains.

sinkhole—A closed depression in a karst or pseudokarst area, commonly with a circular or ellipsoidal pattern.

skarn—A type of silicate waste rock associated with iron-ore bearing sulfide deposits.

speleothems—Any mineral deposit that is formed in a cave.

stalactites—A speleothem that hangs from the ceiling or wall of a cave. It is deposited from evaporating drops of water and is usually composed of calcite.

stalagmites—A speleothem that grows upward from the floor of a cave by the action of dripping water. It is usually composed of calcite.

stock—An igneous intrusion that has a surface exposure of less than 40 mi^2 and thus is smaller than a batholith.

strata—Layers of rock, usually sedimentary but may include volcanic flows.

stratovolcano—A volcano that is constructed of alternating layers of lava and pyroclastic deposits, along with abundant dikes and sills.

subduction—The process in plate tectonics where a denser lithospheric plate is forced beneath another.

surge deposits—A pyroclastic flow deposit formed by a mix of gas and rock fragments.

syenite—A intrusive igneous rock with less than 5% quartz and 10 to 35% plagioclase and albite.

syncline—A fold of which the core contains the stratigraphically younger rocks; it is generally concave upward.

talus—A sloping heap of rock fragments at the foot of a cliff or steep slope.

terrace—Relatively flat bench on a hillside, generally remnants of former stream valleys or lake shorelines.

terrane—A general term for a body of rock or group of rocks of regional extent that are characterized by a geologic history different from that of surrounding terranes.

tidal flat—An extensive, nearly horizontal, tract of shoreline that is alternately covered and uncovered by tides and consisting of unconsolidated sediment (mostly mud and sand)

till—Unsorted and unstratified material, generally unconsolidated, deposited by a glacier without subsequent reworking by meltwater. Consists of a mixture of clay, silt, sand, gravel, and boulders.

transuranic—A material with a higher atomic number than uranium (92).

travertine—A variety of limestone formed from the deposition of layers of calcium carbonate from solution in both surface and ground waters, typically in caves or around the mouths of hot springs.

tufa—A variety of travertine that is commonly spongy or porous due to precipitation around a variety of plant structures (such as reeds, plant roots, and leaves) that later decompose.

tuff—Consolidated or cemented volcanic ash and lapilli.

turbidity current—A bottom-flowing current with suspended sediment, moving swiftly down an underwater slope and spreading horizontally on the floor of the body of water.

unconformity—A surface that separates older and younger rock units and represents a gap in the geologic record. It represents a time during which deposition was interrupted by a period of erosion or a time of no deposition.

varved—Couplets of sediment each representing one year of deposition.

vesicles—Cavites in a lava formed by the entrapment of gas bubbles during solidification.

vesicular—Refers to the texture of a lava characterized by an abundance of vesicles or cavities, formed by the expansion and entrapment of gas bubbles as molten lava approaches the surface.

volcaniclastic—A general term for volcanic materials that have been reworked by erosion and sedimentary processes.

welded tuff—A pyroclastic rock in which the volcanic deposits have been fused through heat and pressure during deposition.

Contributors

Bruce Allen is a geologist with the NMBGMR. He works on a variety of projects involving aspects of hydrology, chemistry, paleontology, stratigraphy, and paleoclimatology. He earned his doctorate from the University of New Mexico, studying late Quaternary lake beds and associated landforms in the Estancia basin of central New Mexico.

Paul Bauer is an emeritus principal geologist at the NMBGMR and has spent 38 years enjoying the sublime geology and landscapes of New Mexico. He has published geologic maps, books, and articles on the geology of north-central New Mexico, and has led field trips for professionals and the general public. Paul received an M.S. in geology (1983) from the University of New Mexico, and a Ph.D. in geology (1988) from New Mexico Tech.

Ron Broadhead has been a petroleum geologist for his entire career, studying petroleum source rocks, reservoirs, play and trend analyses, basin analysis and the genesis and occurrence of helium- and CO_2-rich natural gases. He has worked for the NMBGMR since 1981. He has a B.S. in geology from New Mexico Tech and an M.S. in geology from the University of Cincinnati, where he studied unconventional gas shales.

David Bustos is a graduate of New Mexico State University. He has worked at White Sands National Monument (now Park) since 2005 as the biologist and resource program manager. He is responsible for monitoring the cultural and natural resources at the park; this includes the plants, animals, dunes, water, fossils, and artifacts left from those that have passed through the park. He also works closely with researchers to learn more about the resources of the park and share the information gathered with the public.

Steven Cather is an emeritus Senior Field Geologist at the NMBGMR. During the past 40 years, Steven has specialized in the study of evolution of sedimentary basins from a perspective of stratigraphy and structural geology, and in regional tectonics. He has three degrees in geology: a B.S. degree from New Mexico Tech, and M.A. and Ph.D. degrees from the University of Texas at Austin.

Richard Chamberlin is an emeritus senior field geologist with the NMBGMR. He has mapped extensively in the volcanic terrains of central and southwestern New Mexico. Richard leads geological hikes for science teachers and the general public within the central Rio Grande rift, including the Sevilleta National Wildlife Refuge. He has two geology degrees from New Mexico Tech and a Ph.D. in geology from the Colorado School of Mines.

Colin Cikoski was a field geologist with the NMBGMR. His fieldwork experience spans several geologic settings in the state, including the Rio Grande rift, the Sierra Blanca volcanic field, and the southeastern Colorado Plateau. He also conducted geologic hazards studies and used GIS to develop three-dimensional hydrologic surfaces for southeast New Mexico.

George Crawford holds an M.A. in Anthropology from the University of Missouri and did much of his graduate work in Oxford, England. He is a consulting archaeologist currently based in the Midwest with a focus on the early peopling of North America. He has worked across most of the central and western portions of the U.S., primarily in the non-profit and university-based archaeological sector.

Nelia Dunbar is Director of the NMBGMR and the State Geologist for New Mexico. She has worked for the Bureau since 1992, focusing on geochemistry of volcanic rocks, particularly volcanic ashes and other explosive eruptions, mainly in New Mexico and Antarctica. Nelia completed a B.A. in geology at Mount Holyoke College (1983) and a Ph.D. in geochemistry at New Mexico Tech (1989).

Robert Eveleth is a 1969 graduate of New Mexico Institute of Mining and Technology. For over twenty years, he has researched and documented New Mexico's history relating to mining, the city of Socorro, and the School of Mines. He has been with the NMBGMR for over 35 years (now retired) and remains their Senior Mining Engineer, specializing in Mining Technology, Mining Law, and Mining History

Andy Jochems was a field geologist at the NMBGMR from 2013 to 2019. His professional and research interests include geologic mapping, arid land geomorphology, Quaternary faulting, and basin-fill stratigraphy of the southern Rio Grande rift. He holds an M.S. in geology from Utah State University and a B.S. in environmental science from the University of New Mexico.

Dave Love is the emeritus principal senior environmental geologist at the NMB-GMR. He started at the Bureau in 1980 where his work focused on environmental geology and geologic hazards. He enjoys geological outreach with teachers, students, and field trip groups. Dave holds a B.S. from Beloit College (anthropology and geology), and an M.S. and a Ph.D. in geology from the University of New Mexico.

Shari Kelley works for the NMBGMR as a senior geophysicist and field geologist. Shari has spent most of her professional career studying the uplift and erosion history and tectonic evolution of mountain ranges in the southwestern United States. She earned a B.S. in geological sciences at New Mexico State University in 1979 and a Ph.D. in geophysics at Southern Methodist University in 1984.

Dan Koning is a senior field geologist for the NMBGMR. His primary duties are geologic mapping and providing the geologic framework for groundwater studies. Dan's professional interests include studying the paleoclimatic and tectonic history of basins, landscape evolution, and fault hazards and evolution. He obtained a B.S. from the University of California, Riverside, and M.S. from the University of New Mexico.

Barry Kues is an emeritus professor in the Department of Earth and Planetary Sciences, University of New Mexico. He received his doctoral degree from Indiana University in 1974. Until retiring in 2011, he taught paleontology and researched the late Paleozoic and Cretaceous paleontology and stratigraphy of New Mexico and the U.S. Southwest. Of several books he's written, The Paleontology of New Mexico (UNM Press, 2008) is probably best known.

Lewis Land is a hydrogeologist and geophysical investigator with the National Cave and Karst Research Institute in Carlsbad. Lewis' current research focuses on regional investigations of karstic aquifers and karst geohazards (sinkholes) in southeastern New Mexico. He received his B.S. and M.S. degrees from the University of Oklahoma, and his Ph.D. at the University of North Carolina-Chapel Hill.

Spencer Lucas is Curator of Geology and Paleontology at the New Mexico Museum of Natural History and Science. He is a paleontologist and stratigrapher who has investigated New Mexico's Phanerozoic fossils and strata for more than 30 years. Spencer headed the scientific team that led to the creation of the Prehistoric Trackways National Monument in 2009. He has a M.S. and Ph.D. from Yale University.

Virgil Lueth is the Director of the Mineral Museum at the NMBGMR, with a background in geochemistry, mineralogy, and economic geology. He routinely interacts with professional geoscientists, mineral enthusiasts, and the general public. He received his M.S. and Ph.D. at The University of Texas at El Paso.

David McCraw worked in cartography and surficial geologic mapping with the NMBGMR STATEMAP program, specializing in the Lower Pecos Valley. He has a B.S. from Western Carolina University, a M.A. from the University of New Mexico, and took post-graduate classes at Louisiana State University while employed as a cartographer and mapper with the Louisiana Geological Survey.

Virginia McLemore is a senior economic geologist with the NMBGMR and has spent 36 years studying the geology, mineralogy, and environmental geology of mineral deposits and geology of the state parks in New Mexico. Ginger received a B.S. in geology and a B.S. in geophysics in 1977 and an M.S. in geology in 1980 from New Mexico Tech. She received a Ph.D. in 1993 from the University of Texas at El Paso.

William McIntosh is an emeritus senior volcanologist and geochronologist at the NMBGMR. His M.S. is from University of Colorado, Boulder and his Ph.D. is from New Mexico Tech, where he developed the stratigraphic framework of the ignimbrites in the Mogollon-Datil volcanic field.

Arthur Palmer is professor emeritus of hydrology, geochemistry, and geophysics at the State University of New York (Oneonta). He and his wife Peggy have spent 50 years studying the geology and origin of caves throughout the world. Art has received numerous awards for his cave studies. He is author of the books *Cave Geology and A Geological Guide to Mammoth Cave National Park*, and he and Peggy are editors of Caves and Karst of the USA.

James Ratté worked for the U.S. Geological Survey as a field geologist and retired in 1993. During his 40-year career with the USGS, nearly 30 years were spent mapping in southwestern New Mexico, primarily in the Mogollon-Datil volcanic field. He earned his B.S. at Michigan State in 1950 and his M.S. at Dartmouth in 1952.

Geoffrey Rawling is a senior field geologist at the NMBGMR. He has published numerous quadrangle-scale and regional geologic maps within the Sacramento Mountains of south-central New Mexico and completed several regional hydrology studies pertaining to the southern Sacramento Mountains and the High Plains of eastern New Mexico. He has an M.S. from SUNY Stony Brook, and a Ph.D. from New Mexico Tech.

Jason Ricketts is an assistant professor in the Department of Geological Sciences at The University of Texas at El Paso. His research focuses on structural geology, tectonics and brittle deformation in Earth's upper crust, and much of his current research is centered on the evolution of the Rio Grande rift. He earned an M.S. degree in 2010 from San Diego State University, and his Ph.D. degree in 2014 from the University of New Mexico.

Alex Rinehart is an assistant professor of hydrology at New Mexico Tech, where he studies the hydrogeology of basins and how rocks in the subsurface deform and fracture. Previously, he worked as a hydrogeologist at the NMBGMR (2014 to 2019), and worked on his Ph.D. in experimental rock mechanics at Sandia National Labs. He also has a M.S. in hydrology from New Mexico Tech.

Peter Scholle received a B.S. in geology from Yale in 1965 and a Ph.D. in geology from Princeton in 1970. His career as a sedimentologist includes stints with oil companies and the U.S. Geological Survey. He taught at the University of Texas at Dallas and was Albritton Professor of Geology at Southern Methodist University (14 years). From 1999 to 2011, he was the New Mexico State Geologist and Director of the NMBGMR.

William Seager retired in 1999 after 33 years of teaching and geologic research at New Mexico State University. His field work resulted in geologic maps and interpretive studies of most of the mountain ranges in south-central New Mexico. Bill completed his M.S. at University of New Mexico in 1961 and Ph.D. at University of Arizona in 1966. Painting desert landscapes has become one of his favorite pastimes.

Dana Ulmer-Scholle is an Associate Research Professor at New Mexico Tech, an Adjunct Geologist at NMBGMR, and head of Scholle Petrographic, LLC. She received a B.S. in geology from the University of Cincinnati in 1981 and an M.S. and Ph.D. in geology from Southern Methodist University in 1984 and 1992. As a carbonate sedimentologist and geochemist, she has worked on depleted uranium as well as carbon sequestration.

Matt Zimmerer is a field geologist at the NMBGMR (since 2015) and adjunct faculty at New Mexico Tech. His work involves studying the timescales of volcanic and magmatic processes in the southwest, including many parks and monuments in New Mexico. He received his B.S. in geology from the University of Kentucky (2005), and his M.S. (2005) and Ph.D. in geochemistry (2011) from New Mexico Tech

Acknowledgments

Thanks first and foremost to all of our authors, who provided the framework on which we built this volume, and all the staff that worked diligently to make this volume a reality.

Special thanks to Bruce Heise, Jason Kenworthy and the National Park Service for their support of this project from the beginning. Thanks also to David Certain (New Mexico State Parks) for facilitating the help of the state parks. The BLM and the National Park Service (White Sands) provided numerous photographs and artwork. Also, the New Mexico Museum of Natural History and Science kindly allowed us to use their illustrations.

Photo and Illustration Credits

Individual photographers, artists, and archives hold copyright to their works, which are used here with permission. All rights reserved. Unless otherwise noted, all maps and illustrations were produced by the staff of the New Mexico Bureau of Geology and Mineral Resources. When modified from existing maps, original creator is credited below.

Front cover	Wayne Suggs
iii	Peter A. Scholle
9	After Christopher Scotese
12	After Eugene Humphreys

PART 1

Photos by Peter A. Scholle: 18, 27, 28, 29, 30, 32, 34, 38, 40, 42, 43, 44, 45

24	Matthew J. Zimmerer
26, 41	Leo Gabaldon
37	Mark Mansell

PART 2

Photos by Peter A. Scholle: 46, 54, 60, 67, 68, 75, 76, 79, 80, 81, 84, 88, 89, 90
Photos by Matthew J. Zimmerer: 56, 57, 58, 59, 61, 63, 64, 65, 71, 73, 74, 78, 80

52	Leo Gabaldon
53, 84	New Mexico Bureau of Geology archives
68	James C. Ratté
72	William Radke, USFWS
73	U.S. Geological Survey
77	T.G. Barnes, USFWS
83	NASA/U.S. Geological Survey
85	Virginia T. McLemore
87	David J. McCraw

PART 3

Photos by Peter A. Scholle: 98, 104, 105, 108, 109, 111, 113, 114, 115, 117, 118, 120

92	Robert Wick, BLM
97	Leo Gabaldon
101 top, 103	David J. McCraw
101 bottom	Virginia T. McLemore
107	NASA/U.S. Geological Survey
117 bottom	Library of Congress, Prints and Photographs Division
119 top	Virginia T. McLemore
121	Library of Congress, Prints and Photographs Division
125, 126	Robert Wick, BLM
129, 131, 133	David J. McCraw
132 top	Robert Wick, BLM

PART 4

Photos by Peter A. Scholle: 138, 142, 143, 144, 149, 152, 153, 154, 158, 159, 166, 169, 172, 173, 177, 178, 180, 182, 193, 197, 198, 199, 206, 211, 215, 218, 220, 224, 225, 227, 228, 230, 231, 232, 233, 234, 238, 241, 244, 246, 248, 249, 250, 251, 252, 255, 260, 262, 263, 281, 282

134, 168	Matthew J. Zimmerer
137, 138, 140	Leo Gabaldon
141	NASA/U.S. Geological Survey
145, 147, 157	Leo Gabaldon
146, 148	Dave W. Love
149	Dana S. Ulmer-Scholle
150	Satellite image courtesy of Google Earth
153, 161	Richard M. Chamberlin
167 top	C.T. Brown photo courtesy of M.L. Brown Dillard; NMBGMR No.1505
167 bottom	Jeff Scovil, NMBGMR Museum No.11048
170	New Mexico Bureau of Geology archives
171	Jeff Scovil
174, 176	Colin Cikoski
183	U.S. Geological Survey
184	NASA/U.S. Geological Survey
187, 189, 191	Virginia T. McLemore
190	USDA Forest Service
194	Bureau of Land Management
202, 205, 207	Geoffrey C. Rawling
212, 213	Shari A. Kelley
216 top	Barry S. Kues
216	Illustration modified from New Mexico Office of Archaeological studies, Stephen C. Lentz
217	Dana S. Ulmer-Scholle
221, 224	NASA/U.S. Geological Survey
226	Leo Gabaldon
227	David Bustos
228	Dana Ulmer-Scholle
228 bottom	Karen Karr
236, 252, 256	Shari A. Kelley
237	Colin Cikoski
258	Robert Wick, BLM
262 top, 265	Matthew J. Zimmerer

264, 266, 270	Robert Wick, BLM
271, 272	Spencer G. Lucas
275, 277	Spencer G. Lucas
279	Photo courtesy of Akanawa, Wikimedia

PART 5

Photos by Peter A. Scholle: 284, 286, 290, 292, 296, 299, 301, 308, 309, 311, 313, 314, 319, 320, 322, 325, 328, 333, 335, 336 top, 342, 344

287, 288	Ronald Blakey
290	Dana S. Ulmer-Scholle
294	Virginia T. McLemore
297	Modified from Art Palmer & National Park Service
298	Modified from Hayes 1964, U.S. Geological Survey
299 bottom	adapted W.W. Tyrrell, Jr.
300	adapted from R. Wood, courtesy of SEPM
302–305	Art Palmer
307	adapted from C.A. Harman and C. Kerans
316, 317	Lewis Land
324, 326, 327	Lewis Land
329, 330, 338	NASA/U.S. Geological Survey
332	Lewis Land
336, 340	David J. McCraw
339	Dana S. Ulmer-Scholle

PART 6

Photos by Peter A. Scholle: 346, 353, 354, 356, 357, 358, 361, 362, 363, 365, 366, 368, 369, 372, 373, 375, 378, 383

347, 349	Leo Gabaldon
350, 351 top	Steven M. Cather
351 bottom	NASA/U.S. Geological Survey
352	U.S. Geological Survey
355, 357	Bruce D. Allen
361 top	Spencer G. Lucas
364, 367	NASA/U.S. Geological Survey
371	Ronald Blakey
379, 380	George Crawford
382	Blackwater Draw Museum
Back cover	Peter A. Scholle